Linguistische
Arbeiten 77

Herausgegeben von Herbert E. Brekle, Hans Jürgen Heringer,
Christian Rohrer, Heinz Vater und Otmar Werner

Bedeutung, Sprechakte und Texte

Akten des 13. Linguistischen Kolloquiums,
Gent 1978
Band 2

Herausgegeben von Willy Vandeweghe
und Marc Van de Velde

Max Niemeyer Verlag
Tübingen 1979

CIP-Kurztitelaufnahme der Deutschen Bibliothek

Linguistisches Kolloquium <13, 1978, Gand>:
Akten des 13. [Dreizehnten] Linguistischen Kolloquiums : Gent 1978 / hrsg. von Willy Vandeweghe u. Marc van de Velde. – Tübingen : Niemeyer.
(Linguistische Arbeiten ; ...)
Bd. 1 hrsg. von Marc van de Velde u. Willy Vandeweghe.

NE: Vandeweghe, Willy [Hrsg.] ; Velde, Marc van de [Hrsg.]

Bd. 2 → Bedeutung, Sprechakte und Texte

Bedeutung, Sprechakte und Texte / hrsg. von Willy Vandeweghe u. Marc van de Velde. – Tübingen : Niemeyer, 1979.
(Akten des 13. Linguistischen Kolloquiums ; Bd. 2) (Linguistische Arbeiten; 77)
ISBN 3-484-10343-4

NE: Vandeweghe, Willy [Hrsg.]

ISBN 3-484-10343-4 ISSN 0344-6727

INHALTSVERZEICHNIS

INHALTSVERZEICHNIS ZU BAND 1

VORWORT

Das 13. Linguistische Kolloquium fand vom 20. bis zum 23. September an der Staatlichen Universität Gent (R.U.G.) statt. Von den über 90 Referenten stellten die meisten uns auch ihren Text zur Veröffentlichung bereit. Die hohe Anzahl von Referaten, die nach der Tradition der Linguistischen Kolloquien keiner zensierenden Auswahl unterlagen, machte es notwendig, die Beiträge auf zwei Bände zu verteilen.

Der vorliegende Band "Bedeutung, Sprechakte und Texte" befasst sich hauptsächlich mit Studien zur Bedeutung und zu Sprechhandlungen. Daran schliessen sich einige Beiträge an, die die Herausarbeitung der Theorie von semantischen Netzen für die Künstliche Intelligenz anstreben. Der Umfang der Sektion "Pragmatik" - die auch Beiträge über Argumentationstheorie und Konversationsanalyse enthält - dokumentiert das Interesse, das diesem Forschungsgebiet heutzutage entgegengebracht wird. Die Beiträge zur Textlinguistik, deren Aufschwung durch die pragmatische Betrachtungsweise gefördert wurde, schliessen den zweiten Band ab.

Der erste Band "Sprachstruktur, Individuum und Gesellschaft" enthält die Beiträge, die sich mit der Sprachstruktur aus phonologischer, morphologischer und syntaktischer Sicht befassen. Er umfasst weiter die Vorträge über theoretische oder methodologische Probleme, und die Referate aus den Bereichen Fehleranalyse und Übersetzungswissenschaft. In den letzten zwei Sektionen werden die Beiträge Psycholinguistik ('Individuum') und Soziolinguistik ('Gesellschaft') zusammengefasst.

Nach der seit dem 10. Linguistischen Kolloquium (Tübingen 1975) bewährten Arbeitsweise schrieben die Verfasser ihre Beiträge selbst. Trotz einer detaillierten Schreibanleitung und Korrektur der meisten Unvollkommenheiten - durch die Autoren oder in Zusammenarbeit mit ihnen - liessen sich bestimmte formale Inkonsequenzen nicht ganz beseitigen. Diese erschienen uns aber weniger wichtig als die Einhaltung des Erscheinungstermins.

Zum Schluss möchten wir allen danken, die zum Gelingen des 13. Linguistischen Kolloquiums beigetragen haben, nicht zuletzt den zahlreichen Teilnehmern, die durch ihr reges Interesse und ihre aktive Beteiligung den besten Beweis für den Sinn und die Notwendigkeit eines solchen Forums für die jüngere (und nicht mehr so junge) Generation von europäischen Linguisten erbrachten. Unser Dank gilt auch dem Niemeyer Verlag und den Herausgebern der 'Linguistischen Arbeiten' für die Aufnahme der Akten in ihr Veröffentlichungsprogramm.

Gent, im Dezember 1978

Willy Vandeweghe
Marc Van de Velde

1. SEMANTIK UND KÜNSTLICHE INTELLIGENZ

TOPIC AND THE SCOPE OF NEGATION

Nico W.M. Apeldoorn

1. Topic

There is no generally accepted definition of the notion "topic". Traditionally,the topic is that part of the sentence about which the rest of the sentence says something. Topic is a pragmatic function within the predication proper, and is, according to DIK (1978), to be distinguished from "theme" and "tail". Dik starts from formula (1) :

(1) $(x_i)_{theme}$, Predication, $(x_j)_{tail}$

A constituent is theme if it "presents a domain or universe of discourse with respect to which it is relevant to pronounce the following 'Predication'." (DIK 1978: 130) In (2) that man is a typical theme :

(2) That man, we gave the book to him yesterday. (DIK 1978: 135).

A constituent is tail if it "presents, as an 'afterthought' to the Predication, information meant to clarify or modify (some constituent contained in) the Predication." (DIK 1978 :130).

In (3) your brother is a typical tail :

(3) He's a nice chap, your brother. (DIK 1978/ 153).

In contrast to theme and tail, the topic is an integral part of the predication. In his topic definition Dik accepts the traditional line of thinking :"A constituent with topic function refers to any entity assumed by the speaker to be 'known' or 'given' to the addressee,and about which the predication predicates something". (DIK 1978: 92).

BARRY (1975) elaborates the notion of topic, although she doesn't make a distinction between theme and topic. She shows that KEENAN's (1974) subject definition is a topic definition in the first place; topic then is the argument (of a functional expression) of which the referent can be determined independently of the meaning or the referent of the function symbol. (BARRY 1975: 2). This definition, however, is not sufficient, since in sentences like (4), (5) and (6) the specific indefinite een secretaresse must be topic, and its specificness can only be seen as a c o n s e q u e n c e of the topic function. The topic function imposes a specific interpretation of the indefinite constituent. Pragmatically, this means that the speaker requests the addressee to pretend as if the referent were known to him, while in fact this cannot be the case.

If the topic can create specificity, this is only possible if the topic can also be derived from the position in the sentence : if a non-stressed indefinite NP is not preceded by one or more definite NPs, then this NP has the topic function.[1.]

(4) Een secretaresse heeft me gisteren het boek overhandigd

(5) Gisteren heeft een secretaresse me het boek overhandigd

(6) A secretary handed me the book yesterday

2. Topic and truth value

The topic of a sentence refers to the reality on which the speaker and the addressee agree, and through this link with reality a sentence gets its truth value. KÄLLGREN (1976 :134) argues along these lines that every sentence must have "a 'pivot' of truth" (the topic), while the rest of the sentence is a statement about this true, or at least existing, part of the sentence.

A proposition is true if the referent of the argument is interpreted as a member of the set of objects that constitute the interpretation of the predicate (KEENAN 1972 :415), i.e. of the set of objects having the property referred to in the predicate. If the proposition is false, either the argument becomes meaningless with regard to the predicate, or the predicate becomes meaningless with regard to the argument. In principle both formulations have the same plausibility, if we base ourselves on Keenan's truth definition.

A sentence gets its truth value through its link with reality; this link is realized by the argument rather than by the predicate. Firstly, the interpretation of the predicate often isn't determinable independently from the interpretation of the argument, and never the other way round. This is the essence of KEENAN's (1974) "Functional Principle". Secondly, if the topic has no reference, the truth value of the sentence can't be established, while it càn be established if the predicate does not refer to something:

(7) Dat beest lijkt op een eenhoorn
that animal looks like a unicorn

(8) Die eenhoorn lijkt op zijn schoonvader
that unicorn looks like his father-in-law

In a world where no unicorn exist, sentence (7) may be true, while (8) doesn't have a truth value.

It is the topic that links the sentence up with reality. The sentence is false if the independently referring topic loses its reference through the predication. A sentence is true if the independently determined reference of the topic can remain the interpretation of the topic in the process of predication. Schematically :

(9)

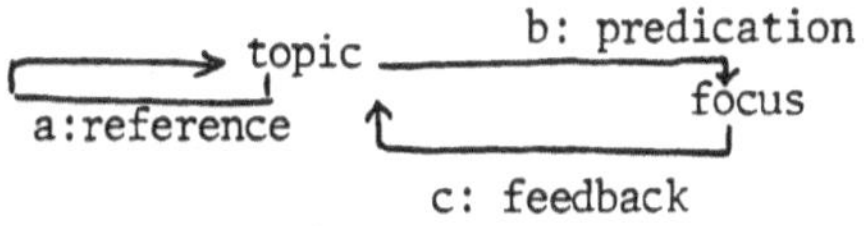

Truth is established by process c; b and c are in fact the same process but viewed form a different angle. A true sentence must therefore satisfy two conditions :

(10) the reference must be clearly defined (phase a in (9))

(11) the topic retains its reference if it is - via the whole predication - related to the focus [2].

In a false sentence condition (10) is met, but condition (11) is not : the falseness results from a failing feedback process; when related to the focus, the topic doesn't refer any more, the link with reality is destroyed, and so is the characteristic of a true sentence.

3. Negation

Our hypothesis is that negation affects (11), i.e. the scope of negation is the topic, more specifically its reference instance. In a sentence like (12) the negation means that while condition (10) is met, condition (11) is not. In (12) the topics hij and het boek cannot refer in view of their relation to the focus.

(12) Hij heeft het boek niet gelezen
he hasn't read the book

The main argument for our hypothesis is based on specific indefinite topic constituents, of which the specificity is suggested by its topic characteristics. In (4) and (5) een secretaresse meets condition (10) only half-way, which means that the truth value of these sentences is quite doubtful : they can only be considered as true insofar as the addressee, just like the speaker, wishes to pretend that the precise reference doesn't matter. Truth in this sentence depends on pragmatical considerations.

With the corresponding negative sentences (13), (15) and (17) there is no denying, as in the case of (12), of what was established earlier, but a denying of what was suggested as to be taken for granted. In these sentences the speaker wants the reference to be of no importance, while at the time making it clear that the reference is crucial in not being linkable to the focus. Sentences (13), (15) and (17) are pragmatically contradictory, and therefore unacceptable. [3].

(13) * Een secretaresse heeft me gisteren het boek niet overhandigd

(4) Een secretaresse heeft me gisteren het boek overhandigd

(13a) * A secretary didn't hand me the book yesterday

(14) Een glazenwasser is de ramen komen schoonmaken

(15) * Een glazenwasser is de ramen niet komen schoonmaken

(16) Een student is gisteren te laat gekomen

(17) * Een student is gisteren niet te laat gekomen

(18) * A student didn't come late yesterday

The unacceptability of (13), (15) and (17) can only explained if negation affects the topic.

Condition (10), however, is not as necessary a condition for negation as it is for truth or falseness. Negation is merely with (11), regardless of (10). This is clear from the acceptability of both (19) and (20).

(19) The present king of France is bald .

(20) The present king of France isn't bald.

Before tackling the problem of (20), let us consider what it means for a negative sentence to be true. Because (11) is the object of negation, the truth of a negative sentence is conditioned by an extra feedback circuit. A negative sentence is true if the following conditions are met :

(10)	the topic has a reference	(phase 1)
(21)	the topic has no reference when linked to focus	(phase 2 neg)
(22)	the topic has no reference any longer	(phase 3)
(23)	the no longer referring topic remains non-referring when linked to focus	(phase 4)

Condition (22) doesn't remove all truth values, because it is preceded by phase 1. A negative sentence is false if condition (23) is not met, while (10), (21) and (22) are.

Returning to sentence (20), this sentence doesn't meet (10), and consequently has no truth value. Whether (20) has also an interpretation on the basis of which this sentence can be seen as true, as perhaps in (24), is a question I shall not discuss here any further. Presumably, the negation here is the focus, and sentence (19) the topic; and if the sentence is the topic, then it is not necessarily its truth value.

(24) The present king of France ISN'T bald

4. Focus and the scope of negation

In &2 I have discussed two formulations of a true proposition in accordance with KEENAN's (1972 : 415) definition. The first assumed that the given argument is the most important factor, and this assumption was worked out in the preceding paragraphs. The second considers the role of the predicate to be central. In the case of sentences this would mean that it is the focus that constitutes the scope of negation. In the Prague School this theory has been developped particularly by Hajičová (1973, and in SGALL et al. 1973) :

> If the contextually bound part of the sentence is called the topic of the sentence and the contextually non-bound part its focus, then we can say that what is negated is the relation between the topic and the focus, what is identified by the topic, is presupposed (and as such is not touched by sentence negation, it is outside the scope of negation) ... (HAJICOVA 1973 : 91-2).

The same view has also been hinted upon by KÄLLGREN (1976 : 134) :

> In a way, it is wrong to talk about sentence negation, as the topicalized part is always outside the scope of that negation. Every sentence must have a 'pivot' of truth. The rest of the sentence is a negative statement about the true, or at least existing, part.

The main argument for considering focus as the scope of negation has been based on the fact that the topic can't possibly fall within it. In (25), where he answered is the presupposition and naturally the focus, it is not denied that he answered but that he did so in natural way. The presupposition is not affected by the negation, thus only the focus can be within the scope of negation. As the topic is at least a part of the presupposition, it falls outside the scope of negation.

(25) He didn't answer naturally (BUYSSCHAERT 1979)

Against this reasoning two arguments can be put forward :

1. WILSON (1975) showed that presuppositions are not necessarily preserved in negative sentences, as is the case when the negative particle is given stress :

(26) I DON'T regret that she missed the boat, because she didn't miss it.

In (26) the normal presupposition of regret, i.e. the factivity of the direct object, is not preserved, as is clear from the acceptability of this sentence. Wilson concludes from such cases that in a sentence like (25) he answered is not a presupposition, but just an entailment. Since entailment is not a part of the interpretation of the sentence, it can't be used as an argument in this discussion.

2. If the presupposition argument were true, then it is not possible to relate the unacceptability of (13), (15) and (17) to negation; since the

positive sentences are entirely acceptable, however, it appears to be necessary to relate the non-stressed specific indefinite topic to negation.

5. Topic and the "functional principle"

In the second paragraph we have seen that the topic may be important in order to identify the specific referent of an indefinite NP in the predicate; this view has been adopted from KEENAN's (1974) "functional principle". This principle holds that the referent of an argument expression must be determined independently from the meaning of the function symbol, while functions may vary with the choice of the argument. This principle can be reformulated more strongly : the topic determines the specificity of the indefinite arguments that follow, unless the lexical meaning of the predicate opposes itself to it and requires non-specificity, as in want-contexts. We assume here that Keenan's principle and our reformulation in fact boil down to the same thing.

The reversed process, creating non-specificness, occurs in JACKENDOFF's (1972) negation analysis. According to Jackendoff the negation element functions as a modal operator : all indefinite NPs that are preceded and commanded by the negative element have to be interpreted as non-specific. This would be an argument to consider the focus again as the scope of negation. However, according to our negation analysis, Jackendoff doesn't go far enough. Instead of attributing this modal operator function directly to the negation element, we can let it function indirectly, i.e. via the topic. The functional principle, reformulated above, predicts that it is the topic that creates specificness in indefinite NPs; our negation hypothesis claims that negation destroys the reference of the topic in view of the focus; this means that negation blocks at the same time the specifying influence of the topic, i. e. it blocks the functional principle. In (27) een boek is specific because of the topic Karel. In (28) een boek remains nonspecific, because the negation nooit blocks the specifying process before een boek can have undergone it. In (29), on the other hand, een boek stands before the negation element, and is therefore to be interpreted as specific, in accordance with the functional principle.

(27) Karel heeft een boek gelezen
Charles has read a book

(28) Karel heeft nog nooit een boek gelezen
Charles has never read a book

(29) Karel heeft een boek niet gelezen
there is a book Charles has not read

If we expand Jackendoff's modal operator 'negation' as indicated above, we can maintain our negation hypothesis.

6. Topic and the scope of negation [4.]

Summarizing, we have shown that the hypothesis that the focus is the scope of negation has serious shortcomings. Considering topic as the scope of negation we are able to explain the unacceptability of negative sentences with a specific indefinite topic, while at the same time we can account for the non-specificity of indefinite NPs that follow the negation element.

Notes

1. Sentences like (4) and (5) are to be clearly distinguished from the corresponding sentences with the numeral <u>één</u>.
2. I agree with BARRY (1975) that topic and focus aren't exclusive categories. Focus is determined by sentence stress, and is to be taken as the decisive function.
3. The specific indefinites are to be pronounced without any stress.
4. Additional evidence can be gathered from the interference of negation and questioning, if a question is also considered as a topic operator. For details and discussion, see my forthcoming dissertation (probably Leuven 1979).

Literature

BARRY, Roberta (1975) : "Topic in Chinese: an overlap of meaning, grammar and discourse function". GROSSMAN, R.E./San, L. James/VANCE,T.J.(eds): Papers from the parasession on functionalism. Chicago : Chigaco Linguistic Society. 1-9.

BUYSSCHAERT, Joost (1979) : "Remaining problems in English adverbial analysis". VAN DE VELDE, Marc / VANDEWEGHE,Willy (eds): Sprachstruktur, Individuum und Gesellschaft. Akten des 13. Linguistischen Kolloquiums, Gent 1978, Bd 1. Tübingen: Niemeyer.

DIK, Simon (1978) : Functional Grammar. Amsterdam: North Holland

HAJICOVA, Eva (1973) : "Negation and topic vs comment". Philologica Pragensia 16 : 81-93

JACKENDOFF, Ray S. (1972) : Semantic interpretation in generative grammar. Cambridge (Mass.) : MIT-Press.

KÄLLGREN, Gunnel (1976) : "On negation - a possible analysis". Studia Linguistica 30: 95-138.

KEENAN, Edward L. (1972) : "On semantically based grammar". Linguistic Inquiry 3: 413-462.

--- (1974) : "The functional principle : generalizing the notion 'subject of'". LA GALY, Michael W./FOX, Robert A./BRUCK, Anthony (eds) : Papers from the tenth regional meeting of the Chicago Linguistic Society. Chicago : CLS. 298-309.

SGALL, Petr/ HAJIČOVÁ, Eva / BENEŠOVÁ, Eva (1973) : Topic, focus and generative semantics. Kronberg Taunus : Scriptor Verlag Gmbh.

WILSON, Deirdre (1975) : Presuppositions and non-truth conditional semantics. London etc: Academic Press.

ON THE HIERARCHY OF OPERATORS
in a universal base component, especially with reference to Negation

Sonia Baghdikian

Having studied the realizations of negation that occur in 16c English it seemed of interest to make a comparative analysis of negation in several languages.

According to Slobin (1971:129), " . . . probably the same basic functions are performed by all languages--making and negating assertions, asking questions, giving commands, and so on. Certainly the basic form of human language is universal."

One could basically vindicate that negation is universal on the ground that it is an innate idea. Common universal notions can exist because of the uniformity of human nature. J. Mehler, psychologist, observes that some form of communication develops spontaneously between children in a group when they are left to themselves without being taught any language. (Lecture at the Université Libre de Bruxelles, 1977)

Chomsky thinks that the child has some innate form of grammar in advance, which accounts for his ability to learn a language since it is innate. Kiparsky claims that the child's language is closer to the DS in that it lacks the transformational rules the child has to learn in order to speak his particular language.

Languages offer, in fact, a great variety of expressions of negation and what is common between different languages is at a deeper level and is consequently best brought out by TG Grammar. As Chomsky says (1966: 34-35), "The DS that expresses the meaning is common to all languages, so it is claimed being a simple reflection of the forms of thought. The transformational rules that convert deep to surface structure may differ from language to language."

Chomsky says very little about Negation (1965), but in the same line of thought Seuren suggests the use of the category of Operators. The hypothesis formulated by Seuren (1969: 113-114) entails that "every sentence has at least two operators" and here the notion of a hierarchy of operators comes in. With the two operators just mentioned, he means more precisely, "a sentence qualifier such as Assertion, Question, Command, and a tense operator." Negation is also part of this first set of operators. Seuren further suggests that in English the operator of tense is subordinate to the operator of Negation or, in other words, that the operator of Negation commands the node for the Tense operator.

It may well be that his proposal of an order Neg-Tense is optional with respect to simplicity and efficiency for the English language. Since it was basically assumed that Negation is a universal category, it was deemed that it should be possible to apply Seuren's theory to all other languages. According to Hockett (1963 in Greenberg, p. 1) a language universal is "a feature or property shared by all languages or by all language." Chomsky claims that "any hypothesis as to the nature of linguistic universals must meet the empirical condition that it is not falsified by any natural language, any language acquired and used by humans in the normal way."

It was then attempted to use Seuren's theory including his hierarchical order of Negation and Tense, in the case of other languages. The method chosen to make this analysis is the "classical" method of using a corpus. Chomsky (in Lester 1970: 57-58) considers the possibility of there being "two distinct ways of approaching . . . the question of linguistic universals. One way is by an investigation of a wide range of languages. But there is also another and, for the time being, somewhat more promising way of studying the problem of universals. This is by deep investigation of a language, investigation directed towards establishing principles of organization of great abstractness in this language." He consequently analyses just one language.

Although reliance on a corpus is not always perfectly adequate as a method of investigation because of its inherent limitations, I nevertheless decided that such a method would offer a stronger basis. I thus chose to collect translations of a small number of English sentences into some 35 languages[1] (including 17 Indo-European languages, 6 African languages,

10 Asian languages, Hungarian, Finnish). Some sentences I had invented myself; others were taken from my readings, especially from Klima (1964).

(1) I like water.
(2) I don't like water.
(3) I liked water.
(4) I didn't like water.
(5) I am reading a book.
(6) Do you like water?
(7) Don't you like water?
(8) I'll go to the cinema tomorrow.
(9) I won't go to the cinema tomorrow.
(10) Brutus did not kill Caesar because he hated him (but because he wanted his place.)
(11) John did not kill his brother because he liked him.
(12) Brutus did not kill Caesar with a revolver.

Some sentences which seemed awkward to a number of informants for cultural reasons, for instance, the sentences about Brutus and Caesar, were supplemented with the whole historical context. Even the simple sentence I like water could give rise to certain problems. The verb like often had very peculiar syntactic properties, like that of being an impersonal verb as is the case in Hungarian, Spanish, and Greek, or being closer to the English to be fond of.

We thus tried to apply the order Neg-Tense to other languages. This hypothesis, it must be stressed, was primarily conceived of by Seuren (1969: 103) as applicable to the English language. Seuren does point out that his own conclusions and the model of a base component he puts forward is especially to be applied to the English language. "It should be kept in mind, at any rate, that there is, as yet, no convincing evidence of a deep structure underlying all languages." Seuren further states, "We are in the process of developing a deep structure hypothesis of English, and it would certainly be rash on our part if we assumed, on the basis of very scanty evidence indeed, that our ultimate model for English deep structure would be the most adequate for all other languages."

It was hoped it could also be a workable hypothesis for other languages so that it would ultimately be acceptable for a universal base component. The question then raised was to know whether this order could be acceptable from a logical point of view.

In fact, one can argue that just as "the negation applies to the truth of the following proposition, including the specific time in which it is said to occur or to be true," as Seuren does, in the same way the negated proposition only has validity for a given portion of time, hence the scope of the operator *Tense* would extend over the operator *Neg* and would precede it.

And that is exactly what is found in *Arabic*:

I like water	Aḥibbu 'l-ma'
I don't like water	*La* aḥibbu 'l-ma'
I liked water	Aḥbabtu 'l-ma'
I didn't like water	*Lam* aḥibbu 'l-ma' / *Ma* aḥbabtu 'l-ma'
I'll go tomorrow	Sa - adhhabu ila 'l-cinema ghaden
I won't go tomorrow	Sawfa *la* adhhabu ghadan

In Arabic the negation *la* is positioned before the verb. We have two possibilities for the past tense: a slightly different form of the negative *lam* which contains the mark for the past and an unchanged verbal form--or the verb having a marked past tense form (perfectum) with a different negation *ma*. This would suggest that the operator *Neg* is conditional upon Tense.

Similar kind of evidence is found in *Japanese*, where the normal negative form for the present tense is *nai* with a morphological variant *naka* but where a completely unrelated form is found with the future tense *sen*:

I like water	watashi wa mizu ga suki desu I water like
I don't like	watashi wa mizu ga suki dewa *nai*
I liked	watashi wa mizu ga suki deshita (ta = past)
I didn't like	watashi wa mizu ga suki dewa *nakatta* (tta = past)
I'll go	watashi wa asu eiga o mini yukimasu I tomorrow cinema watch will go
I won't go	watashi wa asu eiga o mini yukima*sen*

The hierarchy Neg-Tense would thus have to be reversed for Arabic and Japanese.

These observations concerning the possible order of the operators of Tense and Negation led to other queries concerning other possible restrictions on Neg, in the same way as Seuren had considered the possible significant ordering of the numeral and the Negation. Seuren shows, for instance, that the Negation element can be preceded by the numeral in English, yielding a sentence with a different meaning (1969: 106):

"John did not eat two apples" where Neg precedes two

which he contrasts with

"Two apples were not eaten by John" where two precedes Neg.

It may be of interest here to recall the use of a special negative μη in Greek which is used specifically in imperative sentences, where the choice of the negation comes after the selection for the mood (a sentence qualifier ASS, Imp).

Another important factor which can operate when the selection of the negative morpheme is made is the person. In Ikirundi (Bantu language) the general negative morpheme is -nti (which is combined with the verbal form) for all persons except the first person singular where the form of the negative element is completely different: sin.

I like water	ndakund$_a$	amazi	(n-ra-kund$_a$ amazi) I like water
I don't like water	sinkund$_a$	amazi	(si-n-kund$_a$)
You like water	urakund$_a$	amazi	(u-ra-kund$_a$)
You don't like water	ntukund$_a$	amazi	(nti-u-kund$_a$)
He likes water	arakund$_a$	amazi	(a-ra-kund$_a$)
He doesn't like water	ntakund$_a$	amazi	(nti-a-kund$_a$)

In Armenian the formation of negation is quite special. As a consequence the form of the negation is quite different, depending on the person. In other words the negation is sort of conjugated:

```
I like water          tchoure ge sirem   (tchour e ge sire m)
                                          water the   like I
I don't like water    tchoure tchem sirer (tch-em-sirer)
I liked water         tchoure sirei       (sire-i)
I didn't like water   tchoure tchei sirer
                           / tchoure tchem sirad (ad = past)
I'll go               vaghe cinema bidi yertam   (yerta-m)
I won't go            vaghe cinema bidi tchertam
                                          / tche bidi yertam)
```

In tchem sirer, sirer is a sort of neutral participial form, which corresponds in a way to the English formation I am not liking. With the past tense there are two possibilities in the negative:

```
either   neg + be (= ei)   (marked for past and person)
                                  + sir-(root) er(particle)
or       neg + be (= em)   (present + person) + sir-(verbal
                        root) ad (special affix for the past)
```

Here again, the form of the negation is determined by other elements in the sentence.

This kind of phenomenon can be brought into connection with the facts found in Tamil which support the hypothesis of a possible precedence of Tense over Neg:

I like	pidikkum
I don't like	pidikkadu
I liked	pidikadu
I didn't like	pidikkavillai
I will go	poven
I won't go	pogamatten

Tamil incorporates the negation into the verbal form, thus combining tense, negation and person so that the end result is a set of completely different forms. The choice of the verbal form depends simultaneously on several factors.

In those different languages (Arabic, Japanese, Greek, Ikirundi, Armenian, Tamil), the speaker must know in advance which tense or which mood or which person he is going to use before starting to speak in order to decide on the form of the operator of Negation. (Notice that in Greek the negation is uttered first.)

It may well be worthwhile to draw the parallel with another similar problem, viz. the normal process of a Dutch or German speaker who has to decide on the preposition of a prepositional object which is uttered before the verb in a sub-clause.[2]

Another possibility of restriction on Negation is the choice depending on the lexical verb: the Negation may in some languages be completely different depending upon whether the lexical verb is a purely action-verb or a verb that expresses a state, or a similar kind of distinction. This is the case in Chinese, Cantonese, and Malay.

In Malay there are two negative words tidak and bukan:

I like water	Saya suka air
I don't like water	Saya tidak suka air
I was reading a book	Saya sedang membacha buku
I was not reading a book	Saya bukan sedang membacha buku
I won't go	Saya tidak akan pergi
John did not kill C. because he hated him but . . .	B. membunuh C. bukan kerana in benchican dia
John did not kill his brother because . . .	John tidak bunuh abang nya kerana in sayangkan dia
B. did not kill C. with a revolver.	B. membunuh C. bukan dengan "revolver."

In Cantonese and Chinese the distinction is a more subtle one and is rather difficult to grasp: there are two possible negative forms ng and mo. Here one informant clearly did not realize that the meanings of the sentence relating to Brutus (10) and of that relating to John (11) were quite different. As a consequence, in the two sentences he used the same negation mo, although he used ng in most of the other negative sentences. The informant for the other Chinese dialect gives very similar translations, but he does make a difference between the two sentences:

B. did not kill C. because he hated him but . . .	B. ng hai wai chor tsang C. yee sat kui . . .
John did not kill his brother because he liked him.	John mo sat kui kei sai lo yan kill him wei . . . because

In conclusion, it does seem to be problematic to set up a universal hierarchy of Operators since the possible hierarchies encountered seem to be language-specific rather than universal.

It would look as if, in fact, it would be much better to develop a complex network in which one Operator can influence one or more other Operator(s), or where several operators can influence separately or jointly one Operator, for instance, the Operator of Negation we have been considering more specifically.

One would thus have to construct a much more intricate base component to account for all the possible variations and restrictions. Such a cross-linguistic investigation provides actual data which can serve as evidence to construct a principled descriptively adequate system. It made it possible to discover that the Neg-insertion rules operate under very complex conditions. Such a structure may best be represented as follows:

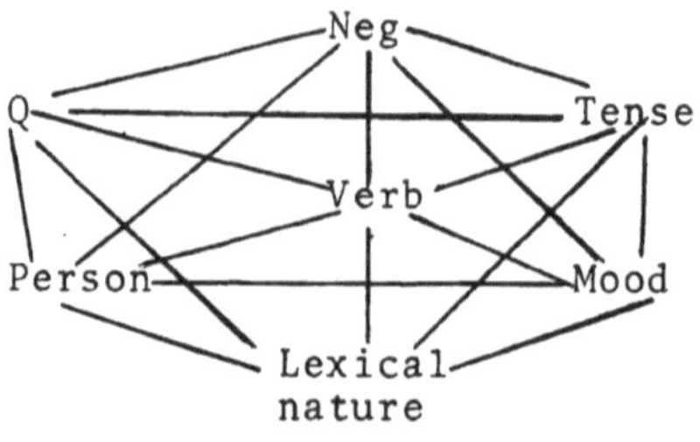

Structure of Operators

Footnotes

1 Informants for this research were many overseas students studying for a diploma in education at the University of Newcastle upon Tyne, U.K.

2 E.g. Dutch: Ik weet dat hij van haar houdt.
German: Ich weiss, dass er es für möglich hält

Bibliography

BACH, E. and HARMS, R.T. (eds.) (1968): Universals in linguistic theory. New York: Holt, Rinehart and Winston.

CHOMSKY, NOAM (1965): Aspects of the theory of syntax. Cambridge, Mass.: M.I.T. Press.

--- (1966): Cartesian linguistics. New York and London: Harper and Row.

GREENBERG, J.H. (ed.) (1963): Universals of language. Cambridge, Mass.: M.I.T. Press.

KALLGREN, G. (1976): "On negation--a possible analysis." Studia Linguistica. Lund: Gleerup.

KIPARSKY, P. (1968): "Linguistic universals and linguistic change." Bach/Harms (eds.): 171-202.

KLIMA, E.S. (1964): "Negation in English." J.A. Fodor and J.J. Katz (eds.): The philosophy of language. Englewood Cliffs, N.J.: Prentice Hall.

LAK OFF, G (1968): "Instrumental adverbs and the concept of deep structure." Foundations of Language, 4-29.

LENNEBERG, E.H. (ed.) (1964): New directions in the study of language. Cambridge, Mass.: M.I.T. Press.

LESTER, M. (1970): Readings in applied transformational grammar. New York: Holt, Rinehard and Winston.

McCAWLEY, J.D. (1968): "The role of semantics in a grammar." Bach/Harms (eds.): 125-171.

McNEILL, D. (1971): "Are there specifically linguistic universals?" D.D. Steinberg and L.A. Jakobovits, (eds.): Semantics: an interdisciplinary reader in philosophy, linguistics and psychology. Cambridge: University Press.

McNEILL, D. and McNEILL, N.B. (1972): "What does a child mean when he says 'no'. P.A. Adams (ed.): Language in thinking. Harmondsworth, Middlesex, England: Penguin.

PARTEE, B. HALL (1970): "Negation, conjunction and quantifiers: syntax vs. semantics." Foundations of Language, 153-165.

RIVERO, M.L. (1970): "A SS constraint on negation in Spanish." Language 46.

SEUREN, P.A.M. (1969): Operators and nucleus. Cambridge: University Press.

--- (1972): "Autonomous versus semantic syntax." Foundations of Language 8, 237-265.

--- (ed.) (1974): Semantic syntax. London: Oxford University Press.

SLOBIN, D.E. (1974): Psycholinguistics. Glenview, Illinois: Scott Foresman.

STAAL, J.F. (1967): Word order in Sanskrit and universal grammar. Dordrecht, Holland: D. Reidel.

MEANING IN ADVERTISING: A CASE FOR FUNCTIONAL SEMANTICS AND FOR A PRINCIPLE OF FUNCTIONAL COOPERATION

Louis Goossens

0.1. The starting-point for this paper is the question whether a procedure can be set up to interpret printed consumer advertisements in a consistent and coherent manner.

With printed consumer advertisements we mean the type that is found in magazines of all sorts. They occupy a well-outlined space (though varying from ad to ad) and combine as a rule text and illustration. They are therefore no texts in the strict sense of the word, since they incorporate a non-linguistic element as an essential constitutive component, but they have at least this in common with texts that they are clearly recognizable meaning units incorporating(as rule) textual elements.

0.1. It is claimed here that in the interpretation of such ads, we are aware of their function, of the purpose that is behind them. Only if we are, can we impose coherence on what otherwise may appear as largely incoherent and therefore difficult to interpret. In other words interpreting advertisements is a case for functional semantics.

More generally, we claim the existence of a principle of functional cooperation, which takes different forms according to the register of language we are operating in. Grice's cooperative principle is to be viewed as an instantiation of this more general principle which relates to conversational interaction. This paper provides an illustration which is concerned with advertising.

0.3. As regards advertising the principle of functional cooperation boils down to an awareness that ads are acts of recommendation to the reader to consume the product advertised(henceforth X) (to 'consume' it on a larger scale, etc.). This is worked out in 1. Most of the time this act of recommendation is not made explicit. On the other hand, the ad builds up meaning around X which provides both an identification of X as well as a justification ('reasons') for the recommendation(see 2). Taking this into account a procedure will be set up to reformulate the meaning in an ad as a series of (positive) properties of X, giving due consideration to salience and degree of explicitness(see 3).

0.4. By way of conclusion an indication will be added as to how the principle of functional cooperation can be used to approach other registers besides advertising.

1. From a functional point of view advertisements can be regarded as a c t s on part of the advertiser vis à vis the readers of the advertisements, comparable to and analysable as(pretended) acts of recommendation. In other words, a convenient way to sum up advertisements is some formula like 'We recommend you to consume (drink, smoke, travel by, use, buy, etc.)(more of) X' or 'Why not consume X' or 'Consume X'.

This act of recommendation can be explicated by providing a parallel to Searle's analysis of the speech act ADVISE(SEARLE 1969: 67). Searle's rules are quoted between brackets: in view of the distinction, hinted at by Searle, between ADVISE and RECOMMEND, I have slightly adapted the essential rule.

(Propositional content: 'Future act A of H') In advertising the future act is the consumption of X by the reader of the advertisement.

(Preparatory rule 1: 'S has some reason to believe A will benefit H') The advertiser has reasons(or purports to have them) to think that consumption of X will benefit the prospective consumer.

(Preparatory rule 2: 'It is not obvious to both S and H that H will do A in the normal course of events') Without agreeing that in the SA the non-obviousness also holds for H, I offer the following parallel: advertising only makes sense because the consuming public needs special incentives to get interested in the advertiser's product, or to consume it on a larger scale, or to keep consuming it in spite of present or future alternatives to X.

(Sincerity rule: 'S believes A will benefit H') We may have to speak of pretended rather than real sincerity, but as such it is an essential ingredient of the advertising game.

(Essential rule:'Counts as an undertaking to the effect that H, in his own interest, ought to do A') An advertisement counts as an undertaking to the effect that the potential consumer, in his own interest, ought to consume X (ought to consume it on a larger scale etc.).

One important point should be added to this analysis. To be successful in his recommendation, the advertiser will, as a rule, have to go out of his way to bring home to the potential consumer that consuming X is in his interest. In other words, he will have to give reasons why 'A will benefit H'. This is why in most ads of the type with which we are concerned here an *act of trying to convince* merges with the more basic act of recommendation.

2. Occasionally ads contain an explicit expression of recommendation or at least an explicit directive expression (e.g."Drink Coca Cola"; "Go Belgian" - in an ad for SABENA, the Belgian airline company). If they do not (which is the rule), we nevertheless approach them as recommendations because (again as a rule) we are functionally cooperative.

Looking for what always *is* present we find that in all ads you get at least the name of X.[1] As a rule, some indication as to what X is will be added, textually or pictorially, or in both these ways. Sometimes, the advertiser takes it for granted that the addressee will be able to supply an identification on his own; this is often the case in neon advertising, but extremely rare in the type of advertising that we are concerned with here.

Whatever meaning is established by the advertisement in addition to this, is to be regarded, in the functional perspective adopted, as so many reasons for recommending X, to convince the potential consumer that the recommendation is well-founded.

These reasons necessarily hang together with X. It should, therefore, be possible to (re)formulate this meaning in terms of (positive) properties predicated over X.[2] The functional links, as we have already indicated, are: properties of X - reasons justifying recommendation - act of recommendation on part of advertiser vis à vis potential consumer with respect of consumption of X.

3. Accepting the position that advertisements are recommendations to 'consume' some X with which we have to link up the meaning created in the ad either as identification or as positive properties of X justifying the recommendation, we can establish a procedure to decode advertisements.

This procedure amounts to drawing up a matrix in which the ad is interpreted taking X, the product advertised, as a starting-point. The c o l u m n s in this matrix are headed by a specification of the subparts of the ad. On the one hand the illustration (possibly distinguishing main from secondary illustration), on the other textual subdivisions (headline, body-copy - if necessary with a number of subparts -, signature line, standing details).[3] For the salience column which is added to this see 3.3. The r o w s list the successive properties of X which result from a careful scanning and interpretation of the different components making up the advertisement, together with an indication in which part of the ad they are to be found. A letter E is used in the appropriate column when the feature is explicitly in the text, a letter I when it is implied (as always when it comes from an illustration, but also quite often when it results from an interpretation of the text).

D i a g r a m

Advertisement for X	Illustration (one or more)	Headline	Body-copy (possibly subsections)	Signature line	Standing details	SALIENCE
Identifying feature 1						
.						
.						
. Identifying feature n						
Recommendation feature 1						
.						
.						
. Recommendation feature n						

Drawing up such a matrix presents us with a number of problems. The first of these is typical of the type of advertisements with which we are concerned here (see 3.1). The others have their parallels in the interpretation of other meaningful wholes (3.2, 3.3, and 3.4).

3.1. A minor difficulty is connected with the division of the ad into its different parts. Obviously one or more (but not all) parts may be missing. In cases where the text is reduced to a minimum it is sometimes not clear whether it is to be assigned to headline, body-copy, or signature line. Actually this complication does not really affect the interpretation: it is merely a matter of locating the (functional) meaning in the advertisement as a whole.

3.2. A more serious problem is the phrasing of the 'properties' on the horizontal lines, both for those that derive from the text and those based on the illustration.
a) Explicit predications over X can be taken over as such. "Every Poggenpohl kitchen is individually designed and built" directly yields the property "Individually designed and built". The same holds for disjunctive predications like "PAN AM. The world's most experienced airline". (In this particular instance we get an identifying feature "airline" and a property feature "the world's most experienced".)

We can also derive properties from attributive adjuncts, as in "The most coveted - and copied - kitchen system in Europe", which provides us with "the most coveted in Europe" and "the most copied in Europe". Actually, we may also summarize these two features under a common denominator, such as "No. 1 in Europe" (at the risk of diminishing the precision while expressing ourselves with greater conciseness).
b) Let's turn to implicit textual features now. We might as well say implicated, to the extent that they are intended meanings derived from literal ones.

As a first instance of an implied feature, I offer another Poggenpohl example, which may be regarded to contain both an explicit and an implicit predication: "Poggenpohl kitchens - for people who refuse to let a little thing like money come between them and the very best". The feature based on the literal meaning is something like "For people who want the very best even if it costs more"; the implied one could be formulated as "(A little) more expensive, but the very best".

Another example. A great many ads for cigarettes contain the phrase "King size flavour" as part of the text. Implied features: "Having large size" and "Having some special flavour"(i.e. the special flavour that goes with King size; the fact that it does not make sense to associate flavour with size is of no importance for our purposes).

c) Turning to the illustration we shall only find implied features. There is one instance, though, which comes close to explicit predication in directness: the picture of X with the brand name striking the eye(the Minolta camera shiningly dominating the page, the latest Volvo filling three quarters of the advertisement, the bottle of Martini flanked by two glasses in the bottom right hand corner), in each case providing us with an an identification feature for X. In a similar way an illustration showing us certain new gadgets of a car, e.g. is directly interpretable in terms of properties which can be predicated over the X advertised.

More often, such a direct link with X cannot be established. Most of the time, though, we have little difficulty in finding the implied feature. When a couple is represented smoking their filter cigarette with happy, contented faces, we may interpret something like "giving pleasure". The elderly gentleman relaxing in his arm-chair with a glass of Black & White whisky suggests a feature "relaxing"(the actual wording of the feature is, of course, arbitrary if we have no textual clues to go by).

d) Often features can only be established accurately(as a reconstruction of the advertiser's intention)[4] after a consideration of the advertisement as a whole, taking into account how text and picture interact, or even after a comparison with a few other ads belonging to the same campaign. For an illustration see GOOSSENS (1974: 177-8).

3.3. All the meaning we thus assemble as a series of predications over X is not equally prominent. A step to be included therefore is a judgement of the salience of the features that we establish. In the procedure proposed here this can be captured by providing an additional column in which each of the features is marked with respect to a three-point scale: S(alient),MS(medium salient), NS (non-salient). Such a saliency marking is essential when it comes to formulating the functional meaning of an ad.

The question is now what criteria we have to determine saliency. There is no difficulty for simple advertisements such as the one which promotes the use of brick by filling a wole page with a picture of a brick wall and no other textual elements than the slogan(headline?) BRICK IS BEAUTIFUL. Obviously, the feature "Beautiful" will have to be marked as S. As the complexity of an ad increases we have to rely on such indications as are available to determine relative salience. Recurrence of a feature in several parts of the advertisement is usually a clear sign of saliency. Otherwise a feature summing up the main illustration or one that is clearly present in the headline stands a greater chance of being salient than what can be derived from the body-copy only or from the standing details. The standing deatails for that matter typically contain non-salient information. Again, a consideration of several ads forming one campaign will help us to record more adequately what the salient feature(s) of an individual ad is(are). It may therefore be required to set up campaign matrices and to determine saliency for a campaign as a whole.

3.4. A last parameter that may be introduced to deepen our insight into the way in which ads function is the distinction denotative(objective) vs. connotative(subjective). It can easily be incorporated into our matrices by grouping the features in two blocks.

The problem that arises here is that the distinction is sometimes difficult to make. In a number of advertisements for Johnnie Walker e.g., "World's No. 1" comes out as a recommendation feature. We are probably aware, for one thing, that Johnnie Walker is a widely selling whisky, and that it can boast on a long tradition, for another(as the ads make clear, J.W. has existed since 1820). Yet there is no way for an average reader to check the objectivity of the feature. Generally speaking, though, it is fairly easy to sort out the features into those that are grounded in reality and those that are simply imposed on X in the universe of meaning created by the ad.

Note also that thereis, at least up to a point, a functional distinction between denotative and connotative features. The denotative features include all identifying ones(such as X is "cigarettes" or an "airline" etc.) and some of the recommendation features. Connotative features, on the other hand, are necessarily recommendation features: they are always offered as reasons for (increased) consumption and, when the ad is approached cooperatively, interpreted as such.

4. By way of conclusion, I should like to place the foregoing in a somewhat broader perspective.

With respect to advertising the functional approach sketched here will have to be enriched to account for a number of characteristics which we have not, or only partially, integrated. I merely give some indication as to what this enrichment would have to be like. It can be formulated with reference to the Gricean maxims of Quantity, Quality, Relevance and Manner. What we have been concerned with brings out the general relevance of an ad. It has little or nothing to say about the ways in which advertisements live up to QN, QL or M. Actually, these maxims are only very partially followed in advertising. The amount of information(QN) varies from very little to an abundance of salient, medium salient and non-salient features of X, depending on what in the opinion of the advertiser stands most chances to impress the potential consumer. In other words, there is no generalizable specificity of purpose against which we can measure what would be the ideal quantity. Again, QL does not appear to determine what an advertiser proclaims about his X. The very presence of connotative features indicates that QL is not really followed(it would require a rather strict legislation to eliminate the often very imaginative approaches that predominate in advertising).As far as M is concerned, we find that in a lot of cases prolixity and non-standard ways of using language are afeature of advertising. It would seem, in other words, that Grice's maxims are largely abandoned in favour of something else. This something else can be captured under such headings as 'Attention value'(the ad must draw attention to itself),Readability(it must sustain the interest it has attracted), 'Memorability' (it must be remembered, or at any rate recognized as familiar), 'Selling power'(it must prompt to the right kind of action)(cp. LEECH 1966: 27 and ff.). How and to what extent these special advertiser's 'maxims' counteract QN, QL, and M and help to determine what is regarded as Relevant, is a subject of considerable interest, but outside the scope of this paper.

Let me also briefly indicate how a principle of functional cooperation plays a part in other registers. In fact polite conversation is already an instance of such a register. Grice's maxims can therefore be looked upon as a particularization of a more general principle. It has been pointed out by others that the cooperative principle is subject to modification according specific shifts in purpose that may occur in conversational interaction (KASHER 1976 and DASCAL 1977, although the main emphasis in both papers is on something else), we do not go further into that question here. When interpreting legal language, we are functionally cooperative to the extent that we take into account that total disambiguation within a rather esoteric tradition is one of the aims of the people that write legal texts. Hence we do not start looking for implicatures whenever the Gricean Maxims of Quantity or Manner appear to be violated. Similarly, our attitude to much theological language is different according to whether we areaware of the fact that it tries to explain the depth of human attitudes, feelings and aspirations in terms of a specific belief system, or not. At present, I cannot go beyond merely pointing this out as another interesting area for investigation in the functional perspective.

Summing up. The acceptance of a principle of functional cooperation was seen to be of help when we want to arrive at coherent interpretations of consumer ads. A procedure was set up to capture the meaning in an ad according to this principle. Finally, both with respect to advertising and with respect to other registers the principle of functional cooperation was shown to call for further investigation.

Notes

1 Perhaps, after a long and successful campaign, an advertiser might choose to imply the name of X, but then the implication should be extremely clear(e.g. What does life go better with?)

2 Obviously these properties need not always be dependent on 'real world' properties; they reflect the universe of meaning created by the ad.

3 Cp. LEECH 1966.

4 The procedure can, of course, be used to explicate one's interpretation without bothering about what the advertiser intends. The principle of functional cooperation, however, tells us to work towards what the advertiser is trying to communicate.

References

DASCAL, Mario(1977):"Conversational Relevance". Journal of Pragmatics 1: 309-28

GOOSSENS, Louis(1974): "Reference in advertising: a functional approach". Handelingen van de Zuidnederlandse maatschappij voor taal- en letterkunde en geschiedenis XXVIII, 171-79.

GRICE H. Paul(1967): "Logic and conversation". COLE, Peter and MORGAN Jerry L.(eds.)(1975): Speech Acts. Syntax and Semantics, Vol. III. New York, Academic Press: 41-58.

KASHER, Asa(1976): "Conversational Maxims and Rationality" in: KASHER(ed.): Language in Focus. Dordrecht, Reidel: 197-216.

LEECH, Geoffrey(1966): English in Advertising. A Linguistic Study of Advertising in Great Britain. London, Longmans.

SEARLE, John R.(1969): Speech Acts. An Essay in the Philosophy of Language. Cambridge University Press.

EINE MODALLOGISCHE REPRÄSENTATIONSSPRACHE ZUR DARSTELLUNG VON WISSEN [+]

Christopher Habel / Arno Schmidt

1. Zur Stellung semantischer Repräsentationssprachen in Systemen der künstlichen Intelligenz

Als gemeinsames Schema für die semantische Interpretation innerhalb der linguistischen Semantik und für die maschinelle Verarbeitung von Sätzen natürlicher Sprachen ergibt sich:

$$\text{NL} \xrightarrow{\ t\ } \text{SRL} \xrightarrow{\ i\ } \wp(\text{PW})$$

Abb. 1: Schema der semantischen Sprachverarbeitung

(Diese Schematisierung wird in HABEL/SCHMIDT/SCHWEPPE 1978 ausführlich erläutert.) Folgt man der Terminologie PETÖFIS (1977), so stellt die Übersetzung t, die Sätze der natürlichen Sprache NL in Ausdrücke einer semantischen Repräsentationssprache überführt, den Bereich der primären Verarbeitung dar, während die Interpretation i der Ausdrücke der SRL über der Potenzmenge möglicher Welten (PW = Menge der möglichen Welten) - vgl. Habel/ Schmidt/Schweppe 1978 - die sekundäre Verarbeitung ausmacht.

Die Schematisierung der Abbildung 1 macht deutlich, daß die semantische Repräsentationssprache SRL sinnvollerweise als Schlüsselpunkt zwischen primärer und sekundärer Verarbeitung bezeichnet werden kann, und daß ihr somit eine besondere Bedeutung innerhalb linguistischer - und auch "computer-linguistischer" - Untersuchungen zukommt. Anforderungen an die semantische Repräsentationssprache ergeben sich aufgrund der zentralen und verbindenden Stellung sowohl aus den Bedürfnissen der primären als auch der sekundären Verarbeitung. Da sich die linguistische Semantik überweigend mit der Übersetzung t, d.h. mit dem Vorgang des Repräsentierens eines natürlich-sprachlichen Satzes innerhalb einer semantischen Repräsentationssprache[1] beschäftigt, sind die Bedürfnisse, die sich aus dem Bereich der sekundären Verarbeitung ergeben, bei der Auswahl der SRL zumeist nicht berücksichtigt worden.

Im weiteren sollen insbesondere die Anforderungen, die sich innerhalb einer Frage-Antwort-Umgebung ergeben, und die für die Eigenschaften und die Interpretation der SRL relevant sind, untersucht und beschrieben werden. Ausgegangen wird hierbei von folgenden Voraussetzungen bzw. Forderungen:

- Informationen über die Welt, d.h. Wissen über die Welt, kann in SRL repräsentiert werden[2]. Insofern können Aussagesätze als Eingaben für ein Frage-Antwort-System verwendet werden.
- Frage-Sätze können unter Verwendung von Ausdrücken der SRL innerhalb eines Frage-Antwort-Systems beantwortet werden.
- Ausdrücke der SRL müssen somit nicht nur in einer "Wissens-Basis" abspeicherbar sein, sie sollten darüberhinaus den Beantwortungs-Prozeß unterstützen und isnbesondere "intelligente"[3] und befriedigende Antworten ermöglichen.

Die im folgenden vorgestellte semantische Repräsentationssprache MSRL wird innerhalb des experimentellen Frage-Antwort-Systems BACON[4] verwendet. BACON ist entsprechend dem Schema der Abb. 1 aufgebaut, bestehend aus einer Übersetzungskomponente BEAST und einer Auswertungskomponente SIREN:

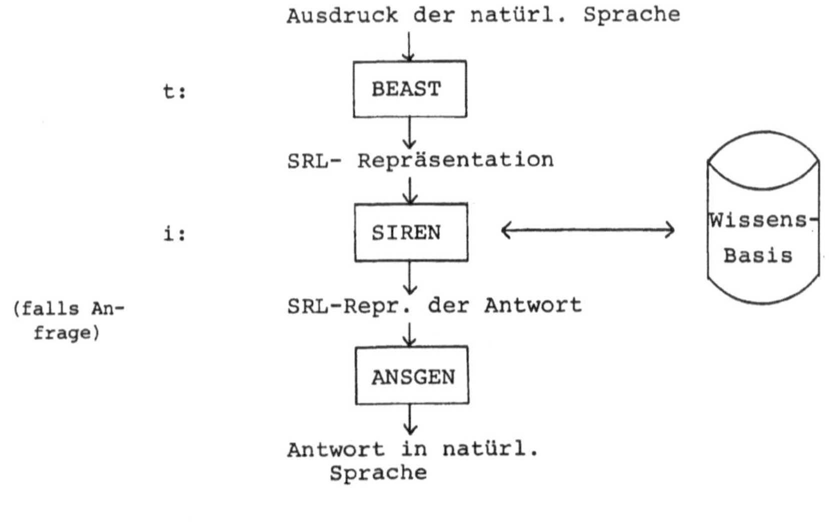

Abb. 2: Systemarchitektur des Frage-Antwort-Systems BACON

2. MSRL, eine modallogische Repräsentationssprache

Die innerhalb des BACON-Systems verwendete Repräsentationssprache kann als "logische Repräsentationssprache" bezeichnet werden; da sie auch über Modal-Operatoren verfügt, sogar als modallogische Repräsentationssprache (MSRL). Die Beispiele dieses Abschnitts werden jedoch zeigen, daß die Operatoren dieser Sprache nicht nur auf Modal-Operatoren beschränkt bleiben. Ein einfaches Beispiel[5] soll MSRL charakterisieren:

Individiuenkonstante[6]: H = Hans , P = Peter , M = Maria,
HH = Hamburg , C = Cambridge

Prädikate: t = fahren , l = Vortrag halten
(jeweils zwei Argumente: 1. Arg. - Person
2. Arg. - Ort)

Operatoren: B = glauben , K = wissen
(jeweils zwei Argumente: 1. Arg. - Person
2. Arg. - Formel)
zusätzlich Junktoren und Quantoren der Prädikaten-Logik

i Maria hält in Cambridge einen Vortrag.
l (M, C)

ii Hans oder Peter fahren nach Hamburg.
t (H , HH) $\vee$ t (P , HH)

iii Wenn Peter nach Hamburg fährt, hält er dort einen Vortrag.
t (P , HH) $\Longrightarrow$ l (P , HH)

iv Maria glaubt, daß Hans nach Hamburg fährt.
B (M , t (H , HH))

v Jemand weiß, daß Hans nicht nach Hamburg fährt.
ex x : K (x , not t (H , HH))

Abb. 3: Beispiel: "Eine Erzählung"

Es handelt sich bei MSRL - offensichtlicherweise - um eine symbolische Sprache mit Operatoren, wie sie etwa zur Beschreibung formaler Theorien Verwendung findet; die hier vorgelegte Sprache stützt sich in ihrer Syntax auf die Definitionen von KALISH/ MONTAGUE (1964); eine ausführliche Beschreibung wird in HABEL/

ROLLINGER/SCHMIDT (1978c) gegeben. Eine einheitliche Behandlung der Operatoren, etwa durch Präfix-Notation, wird durchgeführt, um den Übergang zur maschinen-orientierten Darstellung in der Wissens-Basis zu veranschaulichen:

```
    i'    l ( M , C )
   ii'    or ( t(H,HH) , t(P,HH) )
  iii'    impl ( t(P,HH) , l(P,HH) )
   iv'    B ( M , t(H,HH) )
    v'    ex ( x , K ( x , not ( t(H,HH) ) ) )
```

Abb. 4: Beispiel in Präfix- Notation

An den Beispielsätzen in Präfix- Notation - wie sie in Abb. 4 aufgeführt sind - läßt sich der wesentliche Unterschied in der Interpretation logischer Ausdrücke innerhalb der beiden "Kontexte" Frage-Antwort-System bzw. Logik beschreiben: Betrachtet man etwa ii', so ist eine übliche Formulierung - etwa innerhalb von Lehrbüchern zur elementaren Logik - "ii' ist genau dann wahr, wenn wenigstens einer der Ausdrücke t(H,HH) bzw. t(P,HH) wahr ist", d.h. der Wahrheitswert von ii' ist aus den Wahrheitswerten der eingebetteten Formeln zu ermitteln. In der Frage-Antwort-Umgebung eines Informationssystems wird jedoch von der natürlichsprachlichen Entsprechung von ii' (siehe Abb. 3) ausgegangen, der entsprechende Satz der natürlichen Sprache wird als Aussage aufgefaßt und somit wird ihm der Wahrheitswert "wahr" durch das System zugewiesen. Die Wahrheitswerte der eingebetteten Formeln hingegen sind hier nicht (!) bekannt; darüberhinaus sind diese Wahrheitswerte nicht aus dem Wahrheitswert von ii' ableitbar.

Ein Problem "intelligenter" Frage-Antwort-Systeme liegt z.B. darin, auf die Entscheidungs-Frage "Fährt Peter nach Hamburg?" eine befriedigende Antwort zu finden. Geht man hier von der Information aus, die durch den Ausdruck ii' repräsentiert ist, so kann man die Antworten "Ja" bzw. "Nein" nicht als befriedigend ansehen, da beide Antworten falsch sein können; die Antwort "Keine hinreichende Information vorhanden" ist ebenso schlecht, da das Informationsbedürfnis des Fragenden nicht gedeckt wird, obwohl - wenn auch unvollständige - Information[7] verfügbar ist. Die vermutlich beste Antwort[8] (ausschließlich bezüglich ii') ist:

"Die Frage ist nicht beantwortbar, als relevante Information ist jedoch bekannt: Hans oder Peter fahren nach Hamburg". Unter Verwendung weiterer Information - etwa des durch i-v repräsentierten Wissens-und unter Verwendung von Inferenz-Verfahren, kann die oben gestellte Beispielsfrage sogar vollständig beantwortet werden, vgl. HABEL/ROLLINGER/SCHMIDT (1978a) bzw. ROLLINGER/GAHN (1978).

Die bedeutungstragenden Elemente der MSRL sind in zwei Kategorien zu unterscheiden, in Terme und Formeln[9], wobei die Formeln die wahrheitswertfähigen Konstituenten sind. (Im Rahmen der vorliegenden Arbeit werden nur Individuen-Konstanten als Beispiele für Terme verwendet; der Aufbau einer allgemeineren"Term-Semantik"[1o] ist gegenwärtig in Entwicklung.) Die oben am Beispiel beschriebene Problematik der "Wahrheitswerte eingebetteter Formeln" führt zu einer weiteren Unterscheidung innerhalb der Kategorie der Formeln, in die der wahrheitswertbelegten und die der nichtwahrheitswertbelegten. Diese Unterscheidung, die auf der semantischen Ebene durchgeführt wird, wird auf der syntaktischen Ebene durch die Einteilung in nicht-eingebettete und eingebettete Formeln widergespiegelt; hierbei ist jedoch darauf zu achten, daß auch eingebettete wahrheitswertbelegte Formeln existieren, so etwa beim Auftreten einer Konjunktion in einem wahren Aussagesatz: "Hans und Peter reisen nach Hamburg".[11] Im weiteren werden nicht-eingebettete, und somit wahrheitswertbelegte Formeln als "Tatsachen", eingebettete Formeln, d.h. solche, bei denen nicht von einer Wahrheitswertbelegung ausgegangen werden kann, als "Sachverhalte" bezeichnet.[12] Das in Abb. 3 bzw. 4 repräsentierte Wissen läßt sich unter Verwendung einer Unterteilung in Tatsachen und Sachverhalte folgendermaßen darstellen.

	Tatsachen		Schaverhalte
T_1:	l (M , C)	S_1:	t (H , HH)
T_2:	or (S_1 , S_2)	S_2:	t (P , HH)
T_3:	impl (S_2 , S_3)	S_3:	l (P , HH)
T_4:	B (M , S_1)	S_4:	K (x , S_5)
T_5:	ex (x , S_4)	S_5:	not (S_1)

Abb. 5: Tatsachen - Sachverhalte

Die Beziehungen zwischen Tatsachen und den Sachverhalten, die diese "konstituieren", sind innerhalb von Abb. 5 durch Verweise auf Sachverhalte - etwa "S_1" gegeben -; substituiert man die durch S_1 bzw. S_2 gekennzeichneten Ausdrücke in T_2, so erhält man ii' (aus Abb. 4).

Darüberhinaus bestehen auch Verweise innerhalb der Sachverhalte, so etwa von S_4 auf S_5 und von S_5 auf S_1. Da Sachverhalte über keine Wahrheitswerte verfügen - es handelt sich um die nicht-wahrheitswertbelegten Formeln - stellen Inkonsistenzien innerhalb der Sachverhalte, etwa zwischen S_1 und S_5 kein Problem dar. (Konsistenz innerhalb der Tatsachen ist dagegen zu fordern).

Die in Abb. 5 nur implizit vorliegenden Verweisstrukturen lassen sich auch graphisch (und zwar als gerichteter Graph) explizieren:

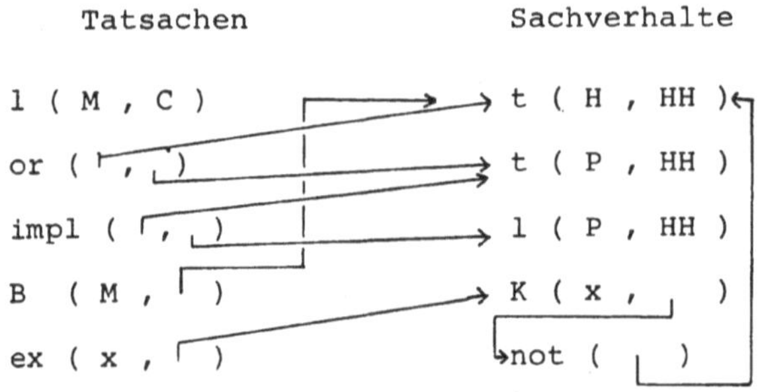

Abb. 6 : Darstellung von Wissen in einer geteilten Datenbank [13]

3. Strategien zur Beantwortung von Fragen und zur Dialogführung

Die semantische Repräsentationssprache MSRL wurde im Abschnitt 2 in zwei Ausprägungen dargestellt; die in Abb. 3 erläuterte ist offensichtlicherweise an der logisch-linguistischen Semantik orientiert; die in Abb. 6 an einer für Antwort-Algorithmen geeig neten Daten-Strukturierung. Aufgrund der Äquivalenz der beiden Darstellungsformen (äquivalent per Übersetzung) wird die Eignung von MSRL für die sekundäre Sprachverarbeitung (vgl. Abschnitt 1) an Hand der maschinen-orientierten Darstellung aus Abb. 6 untersucht, und zwar inweiweit die Repräsentationssprache bzw. Repräsentationsform die Beantwortung von Fragen und die Dialogführung zwischen Mensch und Maschine innerhalb eines Frage-Antwort-Prozesses unterstützt.[14]

Ohne die Syntax der Anfrage-Sprache[15] zu explizieren, können einige wesentliche Antwort-Strategien erläutert werden: Handelt es sich um Entscheidungsfragen, etwa um die Frage "Reist Peter nach Hamburg?" (Repräsentation: ? t (P,HH), so wird die entsprechende logische Formel in der Menge der Tatsachen gesucht (das Ergebnis ist in diesem Fall negativ). Ein nächster möglicher Schritt beim Versuch der Beantwortung (vgl. ausführlicher: HABEL/ROLLINGER/SCHMIDT 1978a,Rollinger/Gahn 1978) besteht in der Suche über der Menge der Sachverhalte (diese verläuft in diesem Fall positiv). Falls auch diese Suche negativ verläuft, muß entweder der Beantwortungsprozeß abgebrochen werden oder es müssen-unter Umständen äußerst-mächtige (d.h. aufwendige) Inferenz-Mechanismen in Aktion gesetzt werden. Falls jedoch (wie im Beispiel) die Suche über den Sachverhalten positiv verläuft, ist aufgrund der bestehenden Verweisstruktur eine Beantwortung durch "Hans oder Peter reisen nach Hamburg".) möglich, womit dem Benutzer des Systems (d.h. dem Fragenden) eine wenigstens einigermaßen befriedigende Antwort zur Verfügung gestellt wird. Durch Verwendung inferenzieller Methoden, die etwa den Schluß-Verfahren der Logik entsprechen können[16], kann dem Fragenden - in diesem Fall - sogar die vollständige Antwort gegeben werden (vgl. Habel/Rollinger/Schmidt 1978a bzw. Rollinger/Gahn 1978). Auf die Entscheidungsfrage "Hält Peter in Hamburg einen Vortrag?" würden die entsprechenden Verfahren zur Antwort "Falls er nach Hamburg fährt, wird er dort einen Vortrag halten." führen können; darüberhinaus besteht auch eine weitere "intelligentere" Lösungs-Strategie: Da für die Beantwortung der Frage ein Wissen über den Sachverhalt S_2, d.h. darüber, ob Peter nach Hamburg reist, benötigt wird, kann beim Fragenden (dem Benutzer also) zurückgefragt werden, ob dieses, dem System fehlende Wissen, bei ihm existiert; das System kann also zurückfragen: "Fährt Peter nach Hamburg?". Durch die Zusammenarbeit zwischen Mensch und Maschine kann in einem derartigen Verfahren das Informationsbedürfnis befriedigt werden.

Durch die Rückfrage der Maschine und eine eventuell vom Menschen erfolgende Beantwortung besteht darüberhinaus die Möglichkeit, daß im Rahmen eines Beantwortungs-Prozesses das System Information hinzugewinnt.

Ebenso wie in den vorangegangenen Beispielen wird die Beantwortung der Fragen "Was glaubt Maria?" (Repr.: B(M, ?)) und "Wer glaubt, daß Hans nach Hamburg reist?" (Rep.: B(?,t(H,HH))) durch die Verweisstrukturen der Darstellung, die durch die Einteilung in Tatsachen und Sachverhalte gegeben ist (Abb. 5 bzw. 6), gesteuert. Insofern ist die hier beschriebene semantische Repräsentationssprache in der Interpretation der Teilung in Tatsachen und Sachverhalte geeignet, die sekundäre Sprachverarbeitung zu unterstützen. Hiermit sind die Forderungen des Abschnitts 1., daß die semantische Repräsentationssprache für die Verwendung innerhalb gewisser Kommunikationsprozesse (hier zwischen Mensch und Maschine) geeignet sein muß, erfüllt.

Anmerkungen

+ Diese Arbeit entstand im Rahmen des vom BMFT geförderten Projekts "Automatische Erstellung semantischer Netze".

1 Das in Habel/Schmidt/Schweppe (1978) vorgestellte BEAST-System (Berlin Automatic Semantic-oriented Translation) stellt eine Algorithmisierung der Abbildung t dar: Sätze einer Teilmenge des Deutschen werden in die semantische Repräsentationssprache überführt.

2 D.h., daß die semantische Repräsentationssprache als Repräsentationssprache für Wissen im Sinne des Teilgebietes "Representation of Knowledge" der "Künstlichen Intelligenz" zu fungieren in der Lage ist.

3 Wenn hier von Intelligenz gesprochen wird, so geschieht dies in einer informellen und intuitiven Verwendungsweise, insbesondere ohne den Anspruch erheben zu wollen, daß das vorliegende System menschliche Intelligenz simuliere.

4 MSRL = Modal Semantic Representation Language
BACON = Berlin Automatic Construction of Semantic Networks
SIREN = Semantic Representation- and Net-Evaluator
ANSGEN = Answer Generation

5 Die innerhalb des Beispiels vorgeschlagenen Übersetzungen erheben nicht den Anspruch der Adäquatheit; so ist vielmehr von einer differenzierteren Behandlung der Junktoren auszugehen,

(verschiedene Übersetzungsmöglichkeiten für das natürlichsprachliche "und", etc.). Die hier intuitiv vorgenommene Übersetzung entspricht im wesentlichen dem gegenwärtigen Stand des Übersetzungssystems BEAST (vgl. Habel/Rollinger/Schmidt 1978b).

6 Der Individuenbereich ist explizit sortiert, d.h. es liegt eine Sortenlogik vor.

7 Hier werden die Begriffe "Information" und "Wissen" als weitgehend synonym angesetzt; eine Erläuterung verschiedener Typen von Wissen wird in Rollinger/Gahn 1978 vorgenommen.

8 Man bedenke, daß auch ein menschlicher Antworter - bei gleichem Informationsstand - nicht besser reagieren kann, als es hier für das System angestrebt wird.

9 Man vergleiche etwa Kalish/Montague 1964.

10 D.h. einer maschinen-orientierten Term- und Referenzsemantik.

11 Vgl. Anmerkung 5., falls das "und" als "Gemeinsamkeits-Operator" interpretiert wird, handelt es sich z.B. um ein Termbildendes "und".

12 Die Unterscheidung Tatsache - Sachverhalt ist in Habel/Rollinger/Schmidt (1978a) eingeführt worden, und zwar durch die Dichotomie fact - factual.

13 Die in Abb. 6 dargestellte Struktur ist das semantische Netz der "Erzählung" aus Abb. 2.

14 Die Stadien eines derartigen Beantwortungsprozesses sind ausführlich dargestellt in Habel/Rollinger/Schmidt 1978a bzw. Rollinger/Gahn 1978.

15 Eine adäquate Anfrage-Sprache müßte in etwa über die von Belnap/Steel 1976 beschriebenen Eigenschaften verfügen.

16 Derartige inferentielle Methoden sollen keine Simulationen mentaler Prozesse darstellen; vgl. Anm. 3.

Literatur

BELNAP, N.D. / STEEL, T.B. (1976): The Logic of Questions and Answers. New Haven

HABEL, Ch. / ROLLINGER, C.-R. / SCHMIDT, A. (1978a): Firsthand and Secondhand Knowledge. Semantic Network Report 1/78. TU Berlin.

--- (1978b) : Machine Translation of Natural Language into a Modal Representation Language . CIS Bericht 13/78 , TU Berlin. (Erscheint in den Proceedings der COLING 78, Bergen)

--- (1978c) : Eine semantische Repräsentationssprache zur Darstellung modaler Konzepte . Semantic Network Report 6/78 . TU Berlin.

HABEL, Ch. / SCHMIDT, A. / SCHWEPPE, H. (1978): "Zur automatischen Paraphrasierung natürlich-sprachlicher Sätze" in: Conte/Ramat/Ramat (Hrsg.): Wortstellung und Bedeutung. Tübingen. 239-49.

KALISH, D. / MONTAGUE, R. (1964): Logic:Techniques of Formal Reasoning. New York.

PETÖFI, J. (1977): Textrepräsentation und Lexikon als semantische Netzwerke. in: Heger, K. / Petöfi, J. (Hrsg.): Kasustheorie, Klassifikation, semantische Interpretation. Hamburg.

ROLLINGER, C.-R. / GAHN, A. (1979): "Zur Erschließung impliziten Wissens aus semantischen Netzen." In diesem Band : VANDEWEGHE, Willy/VAN DE VELDE, Marc (Hrsg.)(1979) :Bedeutung, Sprechakte Und Texte. Akten des 13. Ling. Koll., Gent 1978,Bd. 2. Tübingen : Niemeyer.

ZUR DARSTELLUNG EXPLIZITEN UND IMPLIZITEN WISSENS IN SEMANTISCHEN NETZEN [+]

Claus-Rainer Rollinger/Annette Gahn

1. Die Darstellung von Wissen in SIREN (Semantic Representation and Net Evaluator) [1]

Die Architektur des SIREN-Systems[2] basiert auf einer 4-Teilung der Datenbank. Die für die Abspeicherung des semantischen Netzes wesentlichen Teile sind hierbei die"geteilte Datenbank" (Aufteilung in Tatsachen und Sachverhalte) und die Wissensbasis der Inferenzregeln. Das durch die Methodenbank bereitgestellte Wissen ist in diesem Zusammenhang nicht relevant, da es sich hier um Methoden bzw. Strategien für Updating und Retrieving handelt.

Die Besonderheit der "geteilten Datenbank" liegt darin, zwischen Tatsachen und Sachverhalten zu unterscheiden. Dieses Begriffspaar steht in enger Beziehung zu dem Paar explizites Wissen - implizites Wissen. Unter Tatsachen werden solche Sätze der MSRL[3] verstanden, die sich aus der Übersetzung eines kompletten natürlichsprachlichen Satzes ergeben. Sachverhalte sind die Teile dieser Sätze, die wohlgeformte Formeln der MSRL ausmachen und in den kompletten Satz eingebettet sind. Tatsachen sind Formeln und als solche wahrheitswertbelegt, während Sachverhalte nicht direkt mit einem Wahrheitswert belegt sind. Somit repräsentiert durch Tatsachen abgespeichertes Wissen immer implizites Wissen und als Sachverhalt aufgenommenes Wissen immer implizites Wissen.

Betrachtet man diese Aufteilung unter dem Gesichtspunkt der Semantik möglicher Welten, dann sind die eingebetteten Formeln Bestandteil der möglichen Welt, die durch den entsprechenden Operator eröffnet wird. Jeder Sachverhalt ist somit nur bezüglich der durch die zugehörige Tatsache konstituierten möglichen Welt wahrheitswertbelegt. Da wir innerhalb der Sachverhalte keine Unterscheidung der möglichen Welten vornehmen und keine Wahrheitswerte zuweisen, ist es unproblematisch, widersprüchliche Sachverhalte zuzulassen.

In der Inferenzen-Bank ist inferentielles Wissen abgespeichert. Unter inferentiellem Wissen ist regelhaftes Wissen zu verstehen, mit dem implizites Wissen zu explizitem Wissen erhoben wird. Es sind sowohl Deduktionsregeln der Logik, als auch außerlogische Schlußregeln beteiligt. Die Anwendung der Regeln wird durch das in der Methodenbank gespeicherte Wissen gesteuert. Es ist geplant, auch das inferentielle Wissen automatisch durch natürlichsprachliche Kommunikation upzudaten und zu retrieven.

2. Explizites und implizites Wissen

Unter explizitem Wissen ist im Rahmen dieses Systems nun das Tatsachen-Wissen sowie das inferentielle Wissen zu verstehen. Implizites Wissen ist dagegen all das Wissen, das aus dem tatsächlich vorhandenen Tatsachen-Wissen und den damit verknüpften Sachverhalten unter Verwendung der explizit vorhandenen Inferenzregeln abgeleitet werden kann. Es besteht somit eine enge Beziehung zwischen implizitem Wissen und Tatsachenwissen (ohne Tatsachen gibt es kein implizites Wissen) und zwischen dem inferentiellen und dem impliziten Wissen (ohne Inferenzregeln ist kein implizites Wissen ableitbar).

Beispiele für einige ausgewählte Deduktionsregeln der Logik sind die folgenden, wobei S_1 und S_2 wohlgeformte und wahrheitswertbelegte Ausdrücke der MSRL sind.

(1) $S_1 \; , \; S_1 \Longrightarrow S_2 \vdash\!\!\!\!-\!\!\!-\; S_2$

(2) $\neg S_1 \; , \; S_1 \vee S_2 \vdash\!\!\!\!-\!\!\!-\; S_2$

(3) $S_1 \wedge S_2 \vdash\!\!\!\!-\!\!\!-\; S_1 \; , \; S_2$

Eine außerlogische Regel (im Sinne von nicht prädikatenlogisch) ist beispielsweise die der epistemischen Logik, daß all das, was gewußt wird auch wahr (also Tatsache) ist. Wenn ein Individuum einen Sachverhalt weiß, dann folgt daraus, daß es jemanden gibt, für den gilt, daß er den Sachverhalt weiß und daraus folgt, daß der Sachverhalt zu einer Tatsache erhoben werden kann. (vgl. Regel 4):

K (Ind.Konst. , Sachverhalt)

↓

ex x: K(x , Sachverhalt)

↓

Sachverhalt wird Tatsache

Inferenzschema für Wissen (Regel 4).

An dieser Stelle werden zwei Fragen relevant:

- Was passiert mit dem abgeleiteten Wissen? und
- Wann wird das implizite Wissen abgeleitet?

Beide Fragen stehen in engem Zusammenhang zueinander. Die erste ist nicht ohne die zweite zu beantworten. Die Motivation zum Inferieren ist trivialerweise klar; man erschließt das vorhandene Wissen, um möglichst viele Anfragen so gut wie möglich beantworten zu können. Das inferierte Wissen wird also dem Benutzer zur Verfügung gestellt; die Frage meint aber etwas anderes, und zwar ob einmal zu einer Tatsache erhobenes Wissen auch als Tatsache in die Menge der Tatsachen aufgenommen wird, so daß man bei einer zweiten Nachfrage nicht erneut dieses Wissen ableiten muß, oder ob man es gleich wieder "vergessen" soll. Bei einer flüchtigen Betrachtungsweise mag man dazu neigen zu sagen, daß es doch absurd wäre, abgeleitetes Wissen wieder zu vergessen, zum einen da man froh sein sollte, es jetzt doch explizit vorliegen zu haben und zum anderen müßte man ja den ganzen aufwendigen Inferenzprozeß bei einer erneuten Anfrage ein weiteres Mal durchführen. Dies ist zwar ein wesentlicher und auch richtiger Gesichtspunkt, nur treten enorme Schwierigkeiten auf, wenn eine der Prämissen, die am Inferenzprozeß beteiligt war, nachträglich geändert wird. Dann nämlich ist das ursprünglich abgeleitete Wissen nicht mehr ableitbar und es steht in der Datenbank eventuell etwas explizit falsches, d.h. widersprüchliches.

An dem folgenden Beispiel soll diese Problematik verdeutlicht werden. Gehen wir davon aus, daß das System über das Wissen verfügt:

(1) Fritz ist ein Mann.

(2) Fritz ist erwachsen.

(3) Fritz ist unverheiratet.

und über die Inferenzregel (Regel 5):

> Wenn es ein Individuum gibt, das unverheiratet, erwachsen und ein Mann ist, dann ist dieses Individuum ein Junggeselle.

Stellt man dem System nun die Frage, ob Fritz ein Junggeselle sei, dann sollte es in der Lage sein, aus seinem Wissen heraus diese Frage zu beantworten. Wird nun aber diese "Erkenntnis" des Systems, daß nämlich Fritz tatsächlich ein Junggeselle ist, in das Tatsachen-Wissen mit aufgenommen, dann verliert das System die direkte Information, aus welchem Wissen das neue Wissen abgeleitet worden ist; indirekt ist sie natürlich durch die Inferenzregel vorhanden. Sollte Fritz nun heiraten und teilt man dies dem System auch mit, dann enthält die Datenbank folgendes widersprüchliches Wissen:

(1) Fritz ist ein Mann.
(2) Fritz ist erwachsen.
(3) Fritz ist verheiratet.
(4) Fritz ist Junggeselle.

Die Frage, ob Fritz Junggeselle sei, wird weiterhin mit ja beantwortet, da man zuerst aus der Menge des explizit vorhandenen Wissens heraus die Frage zu beantworten sucht. Will man diesen Effekt vermeiden, müßten alle anderen Tatsachen dahingehend überprüft werden, ob sie im Widerspruch stehen zu der geänderten bzw. neu aufgenommenen Tatsache. Da dies, falls man eine große Menge von Wissen gespeichert hat, einen enormen Aufwand bedeutet, handelt es sich um keine praktikable Lösung; wir belassen die Bewahrung der Widerspruchsfreiheit in der Verantwortung des Benutzers. So kommen wir vorerst zu dem Schluß, abgeleitetes Wissen nicht in die Datenbasis mitaufzunehmen. Bei dem Versuch, die zweite Frage zu beantworten, werden wir allerdings zu einem etwas anderen Ergebnis gelangen.

3. Readtime- und Questiontime-Inferenzen

Es gibt grundsätzlich zwei Zeitpunkte, zu denen es möglich ist, Inferenzen zu ziehen. Zum einen, der auch schon oben intuitiv vorausgesetzte Zeitpunkt, zu dem eine Frage gestellt wird, im weiteren als zur "Questiontime" genannt. Die andere Möglichkeit besteht bei der Neuaufnahme von Information, zur sogenannten

"Readtime". Die oben beschriebene Problematik entspricht im wesentlichen der, die bei Readtime-Inferenzen auftreten. Der Inferenzprozeß wird nur etwas weiter vorverlagert, man wartet nicht, bis eine Anfrage kommt, leitet das Wissen ab und trägt es dann in die Datenbasis ein, sondern zieht gleich bei der Eingabe neuen Wissens alle möglichen Inferenzen und nimmt die Ergebnisse in die Datenbasis auf. Der Vorteil einer solchen Verfahrensweise liegt darin, daß das System sehr schnell Antworten erzeugen kann, sich bezüglich des Antwortverhaltens benutzerfreundlich zeigt. Entsprechend verlangsamt sich aber das Updaten, da eben zum einen sämtliche (und möglicherweise nicht benötigte) Inferenzen gezogen werden und zum anderen, soll die Datenbasis zusätzlich widerspruchsfrei sein, ein enormer Aufwand getrieben werden muß, um für jede neu aufgenommene oder geänderte Tatsache zu überprüfen, ob dadurch innerhalb des Wissens Widersprüche entstehen. Überläßt man dem Benutzer die Verantwortung für die Widerspruchsfreiheit, wird es für ihn - läßt man Readtime-Inferenzen zu - sehr schwer, wie oben versucht wurde darzustellen, die Datenbasis von Widersprüchen frei zu halten, schwerer auf jeden Fall, als wenn nur Inferenzen zur Questiontime zugelassen wären.

Man kann im Bereich der Readtime-Inferenzen eine Verbesserung erreichen, schaltet man hinter die die Inferenzen ziehende Systemkomponente eine Art Filter, der diejenigen Prädikate festhält und somit nicht an die Datenbasis weiterleitet, die durch den Inferenzprozeß vollständig aufgelöst worden sind. Was unter vollständigem Auflösen eines Prädikats gemeint ist, soll an dem bereits erwähnten Beispiel des Junggesellen erläutert werden. Gibt man in das BEAST-System (Habel/Schmidt/Schweppe 1978) den natürlich-sprachlichen Satz "Fritz ist ein Junggeselle" ein, so wird als mögliche Repräsentation in der MSRL der Ausdruck

Junggeselle(Fritz).

erzeugt. Ein inferentielles System, das die Umkehrung von Regel 5 kennt, ist in der Lage, aus diesem Satz der MSRL abzuleiten:

NOT(VERHEIRATET(Fritz))
ERWACHSEN(Fritz)
MANN(Fritz)

Es ist hierbei nicht relevant, ob diese Ausdrücke konjunktiv verknüpft sind oder nicht, wesentlich ist vielmehr, daß sie zusammen dieselbe Bedeutung beinhalten, wie der Satz Junggeselle (Fritz). Dies bedeutet, daß die Proposition Junggeselle(Fritz) durch die drei abgeleiteten Sätze ersetzt werden kann. Somit kann sie auch als vollständig aufgelöst betrachtet werden. Ist dies nun der Fall, besteht auch kein Grund mehr, die Proposition selbst in die Datenbasis aufzunehmen. Stellt man in dem Filter eine solche Struktur fest, wird die Weitergabe des Prädikats verhindert.

Hiermit hat man erreicht, daß der Grad der Widerspruchsfreiheit des Wissens genau dem entspricht, als hätte man direkt die Aussage Junggeselle(Fritz) in die Datenbasis eingetragen, zusätzlich aber alle aus dem Prädikat zu ziehenden Inferenzen bereits gezogen. Wesentlich ist in diesem Zusammenhang natürlich, daß die entsprechenden Anfragen in ähnlicher Form aufgelöst werden, d.h. also, daß bei der Anfrage "Ist Fritz ein Junggeselle?" an die Datenbasis die Frage gestellt wird:

NOT(VERHEIRATET)Fritz)) AND ERWACHSEN(Fritz) AND MANN(Fritz).

Der Vorteil besteht darin, daß bei Readtime-Inferenzen mit Filter man von den möglichen Fragen nach den vier Aussagen 75% direkt beantworten kann, während bei 25% ein Inferenzprozeß durchgeführt werden muß, nämlich bei Fragen nach Junggeselle(Fritz). Ohne Readtime-Inferenzen sieht das Verhältnis genau umgekehrt aus.

Besonders zu eignen scheinen sich demnach für Readtime-Inferenzen derartige Konstruktionen der natürlichen Sprache, die sich durch eine Proposition ausdrücken lassen, die sogenannten Meaning Postulates z.B. Es wird somit sinnvoll sein, zumindest für diese Readtime-Inferenzen durchzuführen, eine Ausdehnung auf Definitionen und ähnliches ist dabei möglich.

Questiontime-Inferenzen sind Inferenzen, die - wie oben erwähnt - erst bei der Beantwortung einer Frage gezogen werden. Dies natürlich auch nur dann, wenn das nachgefragte Wissen nicht explizit in der Datenbasis vorgefunden wird. Questiontime-Inferenzen sind somit immer in Beziehung zu einer Frage zu sehen, d.h., daß der Inferenzprozeß zielgerichtet zur Beantwortung der Frage durchgeführt wird. Readtime-Inferenzen stellen dementgegen eine bessere "Aufbereitung" der Information dar. Wie ein solches zielgerichtetes Inferieren durchgeführt werden kann, soll das folgende Beispiel[4] erläutern:

(1) Maria hält in Cambridge einen Vortrag.
    l(M,C)

(2) Hans oder Peter fahren nach Hamburg.
    t(H,HH) $\vee$ t(P,HH)

(3) Wenn Peter nach Hamburg fährt, hält er dort einen Vortrag.
    t(P,HH) $\Rightarrow$ l(P,HH)

(4) Maria glaubt, daß Hans nach Hamburg fährt.
    B(M,t(H,HH))

(5) Jemand weiß, daß Hans nicht nach Hamburg fährt.
    ex x: K(x,NOT(t(H,HH)))

Hinzu kommen die oben aufgeführten Regeln (1) bis (4) als Inferenzregeln. Zur Verdeutlichung wird die Unterteilung in Tatsachen und Sachverhalte vorgenommen (geteilte Datenbasis).

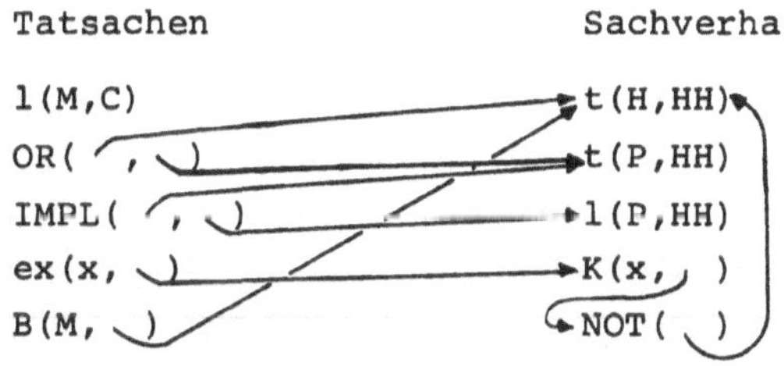

Stellt man dem System die Frage:

"Hält Peter in Hamburg einen Vortrag?" ; l(P,HH)?

so wird durch einen Algorithmus versucht, die Antwort aus dem Wissen abzuleiten.

Als erstes versucht man, die Frage aus der Menge der Tatsachen heraus zu beantworten; dies bedeutet, daß man nachschaut, ob der Satz "Peter hält in Hamburg einen Vortrag." in der Menge der Tatsachen explizit aufgeführt ist. Da dies nicht der Fall ist, muß nun zur Beantwortung ein inferentieller Prozeß einsetzen.

Dieser beginnt damit, in der Menge der Sachverhalte nach der Formel l(P,HH) zu suchen. Ist die Suche erfolgreich, werden die Formeln aufgebaut, in die l(P,HH) eingebettet ist und die den

Status einer Tatsache im System haben. Ist die Suche unergiebig, muß der Beantwortungsprozeß abgebrochen werden mit der Auskunft des Systems, über keine Information bezüglich l(P,HH) zu verfügen.

Konnte man 'Tatsachen-Formeln' gewinnen, in denen l(P,HH) eingebettet ist, in unserem Beispiel t(P,HH)⇒l(P,HH), versucht man als nächstes, eine Inferenzregel zu finden, die aus der Tatsachen-Formel l(P,HH) als Tatsache abzuleiten vermag. Dies wäre durch die Deduktionsregel 1 möglich, allerdings nur dann, wenn die Prämisse t(P,HH) wahr - in unserer Terminologie eine Tatsache - ist. Folglich untersucht man mit dem nächsten Schritt, ob t(P,HH) dem System als Tatsache bekannt ist.

Der für l(P,HH) beschriebene Vorgang wiederholt sich, mit dem Unterschied jedoch, daß t(P,HH) aus der Datenbasis minus der Formel t(P,HH)⇒l(P,HH) (als eine Prämisse der Deduktionsregel) heraus beantwortet werden soll. Das Ergebnis der Suche ist in diesem Fall die Tatsachen-Formel t(H,HH) ∨ t(P,HH). t(P,HH) könnte aus dieser Formel als Tatsache mit der Deduktionsregel 2 abgeleitet werden, wäre NOT(t(H,HH)) eine Tatsache. Der Prozeß wiederholt sich ein zweites Mal und liefert die Tatsachen-Formel ex(x,K(x,NOT(t(H,HH)))). Unter Anwendung der Regel 4 ergibt sich NOT(t(H,HH)) als wahrheitswertbelegt und somit als Tatsache, und man ist nun in der Lage, aus NOT(t(H,HH)) und t(H,HH) ∨ t(P,HH) t(P,HH) als Tatsache abzuleiten. Mittels t(P,HH) und t(P,HH)⇒ l(P,HH) kann l(P,HH) inferiert werden. Man hat so gezeigt, daß die Frage, ob Peter in Hamburg einen Vortrag hält, aus dem Wissen des Systems heraus mit "Ja" beantwortet werden kann.

Unser Ansatz besteht darin, sich nicht ideologisch auf nur Readtime- oder nur Questiontime-Inferenzen festzulegen, sondern zu beiden Zeitpunkten Teile der möglichen Schlüsse zu ziehen, wobei das Auflösen von Meaning Postulates zur Readtime und zielgerichtetes Inferieren zur Questiontime stattfinden sollte.

4. Vollständiges und unvollständiges Wissen

Wir begreifen hier den Vollständigkeitsbegriff nur im Wechselspiel von Frage und Wissen des Systems. Ist eine Frage aus dem Wissen heraus nicht ableitbar, ist das Wissen bezüglich der Frage unvollständig. Bei dieser Betrachtungsweise ist explizites

Wissen gleichzusetzen mit der Menge der Antworten, die auf eine Menge von Fragen gegeben werden können, ohne einen Inferenzprozeß durchzuführen. Sollen die Definitionen konsistent gehalten werden, beinhaltet dies, daß auch das in der Inferenzenbank gespeicherte Wissen abgefragt werden kann. Das implizite Wissen steht in Korrespondenz zu der Menge der Fragen, die nur unter Zuhilfenahme von Inferenzen beantwortet werden können. Das unvollständige Wissen des Systems ist somit das Gesamtwissen abzüglich sowohl des expliziten als auch des impliziten Wissens entsprechend der obigen Definition.

In dem Sinne, daß wir ein natürlichsprachliches Informationssystem als Hilfsmittel des menschlichen Benutzers sehen, sollte ein solches System auch "bemüht" sein, eine Anfrage so gut wie möglich zu beantworten. Wird einem Menschen eine Frage gestellt, wird er, falls er nicht in der Lage ist, die Frage vollständig zu beantworten, sich - gegebenenfalls - bemühen, dem Fragenden die Teile seines Wissens zur Verfügung zu stellen, von denen er meint, daß sie ihm bei der Beantwortung der Frage behilflich sind. Übertragen wir dies auf ein System - das quasi zur Hilfe verpflichtet ist - sollte es fähig sein, dem Benutzer das bezüglich einer Frage unvollständige Wissen vorzulegen.
Man kann noch einen Schritt weiter gehen bei der Mensch-Mensch-Kommunikation und sich die Verhaltensweise überlegen, daß, wenn einer der Kommunikationspartner eine Frage aufwirft, die von dem anderen nicht vollständig beantwortet werden kann, sich beide gemeinsam dadurch an die Beantwortung der Frage machen, daß sie ihr gemeinsames Wissen bezüglich der Frage untersuchen und so bei dem Beantwortungsprozeß weiterzukommen versuchen. Wollen wir diese Fähigkeit auf ein intelligentes System übertragen, muß es die zur Beantwortung der Frage relevanten Informationen aus seinem Wissen extrahieren können und vor allem dem Benutzer eine so gezielte Frage stellen können, daß es in der Lage sein sollte, wenn der Benutzer die Frage zu beantworten weiß, mit dem Inferenzprozeß fortzufahren.

Man könnte mit einem derartigen Verfahren außerdem die zielgerichteten Inferenzen zur Questiontime unterstützen, indem man nach einer bestimmten Anzahl von Inferenzschritten einfach annimmt, auf unvollständiges Wissen gestoßen zu sein und versucht, das Wissen des Benutzers in den Inferenzprozeß miteinzubeziehen. Dies hätte vor allem den Vorteil, die Antwortzeiten zu verkürzen, indem man mit dem Benutzer einen echten Dialog führt und das Wissen des Systems durch den Dialog zu erweitern.

Anmerkungen

\+ Diese Arbeit entstand in dem vom BMFT geförderten Projekt "Automatische Erstellung semantischer Netze", in dem der erste der Autoren angestellt ist.

1 Dieser Abschnitt baut auf dem Beitrag HABEL/SCHMIDT (1979) auf.

2 Siehe dazu HABEL/ROLLINGER/SCHMIDT (1978a)

3 Siehe dazu HABEL/ROLLINGER/SCHMIDT (1978b)

4 HABEL/SCHMIDT (1979)

Literatur

HABEL,Ch./ROLLINGER,C.-R./SCHMIDT,A. (1978a) : "Firsthand and secondhand knowledge - processing of modal concepts." SNR 1/78 Berlin.

HABEL,Ch./ROLLINGER,C.-R./SCHMIDT,A. (1978b) : "Eine semantische repräsentationssprache zur Darstellung modaler Konzepte." SNR 6/78 Berlin.

HABEL,Chr./SCHMIDT,A. (1979) : "Eine modallogische Repräsentationssprache zur Darstellung von Wissen." In diesem Band : VANDEWEGHE,Willy/VAN DE VELDE,Marc (Hrsg.) (1979) : Bedeutung, Sprechakte und Texte. Akten des 13. Ling. Koll.,Gent 1978, Bd.2. Tübingen : Niemeyer.

HABEL,Ch./SCHMIDT,A./SCHWEPPE,H. (1978) : "Zur automatischen Paraphrasierung natürlichsprachlicher Sätze." CONTE,M.E. / RAMAT, A. / RAMAT,P.(Hrsg.) (1978) : Wortstellung und Bedeutung.Akten des 12. Ling. Koll., Pavia 1977, Bd. 1.Tübingen : Niemeyer.

THE SEMANTICS OF THE DUTCH UNIVERSAL QUANTIFIERS <u>ELK</u> AND <u>IEDER</u>

Camiel Hamans

This paper is about universal quantifiers, expecially Dutch universal quantifiers. This implies nearly all examples are from Dutch and my conclusions are also only important - if they are - for the Dutch quantifying system. But, if my conclusions are correct, semantic theory has to take account of the scope of domains. That is why this investigation has been made.
I have to apologize for making a strange use of the notions unacceptable and even ungrammatical. The intuitions about the sentences I discuss are not very clear among Dutch native speakers. Therefore an asterisk before a sentence does not always mean that sentence is ungrammatical in Dutch. Sometimes it only means I think this sentence is unacceptable in this reading. A question mark before a sentence means I prefer not to use this sentence in this reading, although this sentence could be used in such a reading.
In Dutch there are, as in English, at least three words for the universal quantifier. These words are <u>alle</u>, which means <u>all</u>, and <u>elk</u> and <u>ieder</u>. <u>Elk</u> can be translated by <u>each</u>, <u>ieder</u> by <u>every</u>. As Vendler (1967) and Dik (1974) pointed out, there is a difference, both in Dutch and English, between <u>all(alle)</u> on the one hand, and <u>each(elk)</u> and <u>every(ieder)</u> on the other hand.
<u>Alle(all)</u> is used as an universal quantifier quantifying over groups as a totality:

(1) Alle levende wezens zijn sterfelijk
All animate beings are mortal

(2) Alle (al de) eieren zijn vers
All (the) eggs are new-laid

(3) Hij geeft alle (al de) meisjes een kans
He gives all ((of) the) girls a chance

(4) Alle (al de) honden blaften tegelijk (tegelijkertijd)
All (the) dogs barked together
Alle Hunde bellten gleichzeitig.

(5) Alle (al de) collega's haten elkaar
All (the) colleagues hate each other

(6) Al die lijnen lopen parallel
All those lines run parallel.

In the first three examples the words elk/each or ieder/every may be used. The examples (4), (5) and (6) become ungrammatical for every Dutchman in changing alle/all for elk/each or ieder/every. The same applies to English:

(7) Elk levend wezen is sterfelijk
Ieder levend wezen is sterfelijk
Each animate being is mortal
Every animate being is mortal

(8) Elk ei is vers (elk van de eieren is vers)
Ieder ei is vers (ieder van de eieren is vers)
Each egg is new-laid (each of the eggs is new-laid)
Every egg is new-laid (every of the eggs is new-laid)

(9) Hij geeft elk meisje een kans (elk van de meisjes)
Hij geeft ieder meisje een kans (ieder van de meisjes)
He gives each girl a chance (each of the girls)
He gives every girl a chance (every(one) of the girls)

(10)*Elke hond blafte tegelijk
*Iedere hond blafte tegelijk
*Each dog barked together (*each of the dogs)
*Every dog barked together (*every of the dogs)
*Jeder Hund bellte gleichzeitig

(11)*Elke collega haat elkaar (*elk van de collega's)
*Iedere collega haat elkaar (*ieder van de collega's)
*Each colleague hates each other (*each of the colleagues)
*Every colleague hates each other (*every (one))

(12)*Elk van die lijnen loopt parallel
*Ieder van die lijnen loopt parallel
*Each of those lines runs parallel
*Every of those lines runs parallel

In the last three examples it becomes clear what the difference in meaning between alle/all and elk/each-ieder/every is and why these examples are ungrammatical. In the examples (4), (5), (6) and (10), (11) and (12) the predicates require as subjects more than one individual, namely a group. Therefore alle/all can be combined with these predicates, whereas elk/each and ieder/every cannot.

Alle/all stresses the totality of the group, whereas elk/each and ieder/every do the individuality of each of the members of a set that is implied. Vendler (1967) calls all collective and that is the reason why a sentence like (4) is grammatical and (10) not. In (4) there is a collectivity of dogs barking; of course this implies that each individual dog barks. In using each-every we ascribe qualities or attributes to the members of the group one by one. Therefore we cannot use a predicate like bark together in (10). Vendler (1967) calls each-every distributive universal quantifiers. This difference becomes more clear in examples like:

(13) Hij haat alle muziek
He hates all music.

(14)*? Hij haat elke muziek
*He hates each music.

(15) Hij haat elke soort muziek
He hates each kind of music

(16) Hij verslindt alle groente
He devours all green stuff

(17) *Hij verslindt elke groente
*He devours each green stuff

(18) Hij verslindt elke soort groente
He devours each kind of green stuff

In these examples alle/all has been used with mass nouns. Alle/all combined with mass nouns can only be interpreted as referring to the whole species, to the whole universal set.

Elk/each and also ieder/every cannot be used with mass nouns. Therefore (14) and (17) are ungrammatical and that is why (15) and (18), where countable expressions have been used, are grammatical. Nevertheless any native speaker of Dutch may object that (17) gets a grammatical reading, if groente is interpreted as a count noun. Of course this objection is correct. However, sentence (17) under this reading may not be contrasted with sentence (16) but with

(16a) Hij verslindt alle groentes
He devours all greens/vegetables

However, not all mass nouns have such countable counterparts, compare music.

That _elk/each_ and _ieder/every_ distribute an attribute over the members of a set one by one can be seen in:

(19) Elk van die pionnen is te vervangen door elk van de anderen

Each of those pawns can be replaced by each one of the others

(20) Ieder van de spelers van Ajax is inzetbaar voor ieder van de anderen

Everyone of the players of Ajax van be exchanged for everyone of the others

Thusfar, there are no problems. We met three universal quantifiers. One of them -_alle/all_- must be separated from the two others. However, Hogg (1977) claims there are only two words for the universal quantifier in English. Hogg believes _all_ and _each_ to be universal quantifiers and the remaining quantifier -_every_- an existential. His argument looks like that of John Anderson (1973), although Anderson does not speak about _every_ as an existential quantifier.

I am not convinced by Hogg and Anderson, as I have shown elsewhere (Hamans 1977). I maintain that these three words can be used only as universal quantifiers. However, the universal quantifier may be used for different domains, and that is what causes Hogg's misunderstanding:

In a sentence like:

(21) All children like cream

all refers to all children of all times, everywhere.

This interpretation Hogg calls generic. He claims that _all_ is always generic. Therefore, he have to call a sentence like (22) ungrammatical:

(22) All boys have kissed the girls

In my opinion this sentence is grammatical, at least in Dutch.

(22a) Alle jongens hebben de meisjes gekust.

The interpretation of this sentence is exactly the same as the interpretation of (23)

(23) Al de jongens hebben de meisjes gekust

All the boys have kissed the girls

All of the boys have kissed the girls

Al/All in (23) refers to all the boys of a specified set, a finite set. The same applies to elk/each and ieder/every:

(24) Ieder meisje droomt van jongens
Elk meisje droomt van jongens
Every girl dreams about boys
Each girl dreams about boys

(25) Ieder van de meisjes droomt van jongens
Elk van de meisjes droomt van jongens
Everyone of the girls dreams about boys
Each of the girls dreams about boys

(26) Deze meisjes dromen ieder van jongens
Deze meisjes dromen elk van jongens
These girls dream about boys each

We may conclude that the three words for the universal quantifier - at least in Dutch - may be used for two kinds of sets: infinite and finite. The interpretation of the universal quantifier in finite sets is non - generic. However, the quantifier remains universal, because it quantifies about all members of the set. Van Langendonck (1971) calls such an interpretation non-dispersive, which means the quantification is only about entities of the same kind at one and the same space and time. In infinite sets the universal quantifier will be interpreted as generic, which means that the quantifier refers to entities of the same kind at all moments and at all places. Simplifying Van Langendonck's feature we may call this interpretation dispersive.

In a non-generic reading the universal quantifier can be accompagnied by a definite article, a pronoun or partitive construction, as we saw in the sentences (23) through (26). However, it is not obligatory. An universal quantifier without for instance a definite article may be interpreted as a non-generic, according to its context:

(27) All men are mortal

(28) All Americans vote for Carter (after Camp David)

All (27) is generic, all (28) is not.

We know elk/each - ieder/every are universal quantifiers, neither of them is an existential. Furthermore we conclude these words attribute qualities to the members of a set one by one.

There is a slight difference between Dutch _elk_ - _ieder_. However, this difference is so small that it is sometimes difficult to perceive. There is only one expression in which the difference is clear and unquestionable.

(29) Elk van beiden heeft dit gedaan
Elk van beide jongens heeft dit gedaan
Both of them did this
Both of the boys did this

(30) *Ieder van beiden heeft dit gedaan
*Ieder van beide jongens heeft dit gedaan

Comparably Vendler (1967) tells us that in English one prefers _each of the two_ to _every of the two_. The latter he calls pompous. In Dutch (30) is positively impossible.

From the following sentences I conclude that _ieder_ prefers a large or even infinite set. This use of _ieder_ agrees with the existence of the substantival pronoun _iedereen_ (= _everyone_), where a corresponding pronoun deriving from _elk_ does not exist in modern standard Dutch. _Elk_ suggests a small set, which corresponds with the existence of the pronoun _elkaar_ (= _each other_). In Dutch there is no word, meaning each other, which corresponds to _ieder_. Also in English the expression _every_ _other_ - under the reading intended - does not occur.

(31) Zij draagt aan elk oor een ring
She wears a ring at each ear
She wears a ring at both ears

(32) ?Zij draagt aan ieder oor een ring
?She wears a ring at every ear

The Dutch sentence (32) implies she has a lot of ears.

(33) Zij is blind aan elk oog
She is blind in each eye
She is blind at both eyes

(34) ?Zij is blind aan ieder oog
?She is blind in every eye

The Dutch sentence (34) implies she has at least more than two eyes.

(35) Zij had een zweer aan iedere tepel
She had an abscess at every teat

(36) Zij had een zweer aan elke tepel
She had an abscess at each nipple.

The Dutch sentences (35) en (36) are both acceptable, but there is a difference in interpretation. Sentence (36) refers to a woman with only two nipples, or, if you want, an animal like a cow with her four teats; (35) suggests we are speaking about an animal, for instance a sow, a female pig, with a lot of teats, or about a many nippled goddess of fertility.

In this sentences elk is used for small groups, mostly containing two members. However, elk may also be used for sets consisting of more than two members.

(37) (Jan heeft drie huizen.) Elk (daarvan) wordt goed onderhouden
(John has three houses.) Each (of them) is in good repair.

(38) (Jan heeft drie huizen) Ieder (daarvan) wordt goed onderhouden
(John has three houses) Everyone (of them) is in good repair.

Not all examples are so clear:

(39) Elke poes houdt ervan geaaid te worden
Each cat likes to be chucked

(40) Iedere kat houdt ervan geaaid te worden
Every cat likes to be chucked.

Elke poes/each cat in (39) does not refer to only two, three or four cats. This noun phrase can refer to all cats like iedere poes/every cat in (40) can. However, there is a slight difference between these two sentences. Usually they are interchangeable, but, in my opinion (40) has a slight preference, if you want to say any cat likes to be chucked. If you want to say the cats in this street or this house like to be chucked, you should use preferably (37), I suppose.

Sometimes elk does not suggest a small group:

(41) Jan heeft 1000 huizen. Elk daarvan is goed onderhouden
?Ieder daarvan is goed onderhouden
John owes a thousand houses. Each of them is in good repair
?Everyone of them is in good repair.

Elk, as against ieder, suggests an already introduced or already known group, in these examples. Ieder does not:

(42) Ieder slot is geolied
?Elk slot is geolied
Every lock has been oiled
?Each lock has been oiled

Sometimes you can not say whether _elk_ refers to a small or to an already introduced group:

(43) ?Elk van de vrienden van de Nederlandse Opera is na de brand om een bijdrage gevraagd
Ieder...
?Each of the friends of the Dutch Opera has been asked for a gift after the fire
Every...

(44) Elk van mijn vrienden heeft naar vermogen bijgedragen
?Ieder...
Each of my friends contributed to the best of their abilities
?Every...

It is not so clear why in (43) _ieder_ is preferable and why in (44) _elk_. It could be that the number of my friends is smaller than that of the Dutch Opera; it also could be the set of my friends is known whereas the set of friends of the Dutch Opera is not.

Honselaar (1978) explains the difference in meaning of my Dutch examples by referring to a deictic element _de_ (= _the_). _Elk_ contains this element, he claims, whereas _ieder_ does not.

In accordance with Ebeling (1977) he defines the meaning of '_de/the_': "something that, given context and situation, is most likely to be meant by the speaker", I cannot understand exactly what is meant by this definition, but I infer from his explanation of the meaning of (31) through (34) that he means the '_de/the_'-reading refers to the referent of _she_. Usually a she is a woman, a woman has only two ears and two eyes. Therefore we have to prefer (31) and (33), he claims.

I do not agree with Honselaar, because of (35) and (36). Both of these sentences are acceptable but they have a different reference. This only depends on the use of _elk_ or _ieder_.

My explanation says _elk_ refers to a small group, and _ieder_ to a bigger one. Therefore a pig with her lot of teats cannot be referred to by elk. Honselaar says elk suggests the most likely hypothesis; but what is most likely a woman, a cow or a pig? Honselaar has to explain why she in (35), as against she in (36), refers to an animal and not to a woman. There are more differences between elk and ieder. Ieder has a preference for human beings, see Dik (1974) and my already mentioned paper (Hamans 1977). Also it has been shown in my former paper that a difference Vendler (1967: 78) states between each and every can be found between elk and ieder. Vendler explains this difference as follows: "every stresses completeness or, rather, exhaustiness (...); each, on the other hand, directs one's attention to the individuals they appear, in some succession or other, one by one". However, both of these differences are unimportant for the problem under consideration.

On the contrary, what has to be stated, is that the differences between elk and ieder I established are very, very small. If there is any stylistic reason for not using elk for small groups, we can easily use ieder instead of elk. Only in the case of elk van beide (both of them/each of the two) it is impossible to interchange elk and ieder.

Do these reflections make sense, if we can interchange ieder and elk almost always? These question is not so problematic. If my conclusions are correct we have to account for the scope of a group in the description of this Dutch universal quantifiers. May be we have to account for scope in English also:

(45) He was sitting with a child on each side of him

is not exactly the same as

(46) He was sitting with a child on every side of him

Greenberg (1966) claims there is a hierarchy for numerals; lower numerals being unmarked in relation to higher. He gives examples from a few languages that there is a difference in gender distinctions, declination etc. between low and higher numerals - high ones being for instance three or five. Most probably it is the same problem as for the Dutch universal quantifiers elk and ieder.

In claiming an semantic universal - scope of domains - I have to explain why a lot of Dutch dialects only use one of the two words elk and ieder. For example in the dialects of Limburg there is only one word eder, a local variant of ieder. This expression can be used for all instances of universal quantifying corresponding to Dutch elk and ieder. However, if one wants to say something like (31), one uses a word like the standard Dutch allebei. There is a difference between (31) and a corresponding Limburgian sentence with allebei. The Limburgian sentence have exactly the same meaning as a standard Dutch sentence with allebei.

(31) Zij draagt aan elk oor een ring
She wears a ring at each ear.

(47) Zij draagt een ring aan allebei haar oren
(de)
She wears a ring at both ears.

In (31) elk stresses the point that she wears a ring at each of her two ears, seen one by one (compare Vendler's quotation (1967: 78)). In (47) we speak about only two ears; however, the rings have been seen together. Nevertheless it is easily possible to translate (31) or (49) in the dialects of Limburg using a phrase corresponding to the standard Dutch stuk voor stuk or apart (= one by one).

(de)
(48) Ze draagt aan allebei haar oren een ring, stuk voor stuk
She wears a ring at both of her ears one by one

(49) De koe heeft een zweer aan elke tepel
The cow has an abscess at each teat

(de)
(50) De koe heeft een zweer aan alle (vier) haar tepels apart
The cow has an abscess at all (of the) (four) teats one by one.

De koe heeft een zweer aan iedere tepel apart
The cow has an abscess at every teat one by one.

In these dialects one uses the same semantic distinction, although not expressed in the same words. However, the system is similar. The same applies to the Dutch dialects of the Middle Ages. In those dialects ieder did not occur. At the end of the Middle Ages ieder gradually toke the place of a couple of words mainly formed by a pronoun or adverb suffixed by (ge)lijc: haergelijc, manlijc, mallic, iegelijc, etc.. These forms, partly derived from the same original form as elk (= een(ge)lijk), fulfilled together the same functions as elk and ieder in modern standard Dutch (see Hamans, forthcoming). The absence of ieder in Middledutch does not offer any counterargument against this analysis. Therefore I claim semantic theory has to take account of the scope of domains.

Bibliography

ANDERSON, J.M. (1973):"Universal quantifiers". Lingua 31: 125-176.

DIK, S.C. (1974): Universal Quantifiers in Dutch. Publication no. 6, Institute of General Linguistics, University of Amsterdam.

EBELING, C.L. (1977): Syllabus structurele syntaxis. Unpublished manuscript University of Amsterdam.

HAMANS, Camiel (1977): "De semantiek van elk en ieder". Paper read at TIN 1978. To be published in ZONNEVELD (ed.)

HOGG, R.M. (1977): English quantifier systems.Amsterdam: North-Holland.

HONSELAAR, Wim (1978): "De betekenis van elk en ieder". To be published in Spektator 8.

LANGENDONCK, W. van (1971): "Over zgn. generische determinatoren". Studia Neerlandica 2: 351-382.

VENDLER, Zeno (1967): "Each and every, any and all". Linguistics in Philosophy. Ithaca: Cornell U.P.

ZONNEVELD, Wim (ed.) (in prep.): Linguistics in the Netherlands 1977-9. Dordrecht: Foris.

ENTWEDER-ODER ODER ODER, ODER? ODER NICHT?

Zu einigen Gebrauchsweisen einer deutschen Konjunktion

Manfred Kohrt

In einführenden Darstellungen, die logische Grundbegriffe erläutern, findet man dort, wo es um die Verknüpfung von Aussagen mittels *oder* geht, gewöhnlich zweierlei: Da ist zunächst eine Wahrheitswerttabelle, die die Verwendung des logischen Zeichens 'v' illustriert, d.h. die angibt, wann ein komplexer Ausdruck, der aus zwei durch 'v' verbundenen Teilaussagen gebildet ist, wahr ist und wann nicht. Sodann wird ausdrücklich darauf verwiesen, daß ein Ausdruck 'p v q' auch in dem Fall wahr ist, wenn sowohl *p* als auch *q* wahr ist, und dieser Hinweis bildet zumeist den Anstoß dazu, dem Gebrauch dieses logischen Verknüpfungszeichens den Gebrauch des Wortes *oder* in der Umgangssprache gegenüberzustellen. (Als dritter Teilpunkt folgt oftmals noch der Hinweis, daß das Zeichen 'v' durchaus entbehrlich ist, wenn nur Negation und Konjunktion vorhanden sind, und daß 'p v q' ohne weiteres durch '$\neg(\neg p \,.\, \neg q)$' ersetzt werden kann.)

Dem sogenannten adjungierenden *oder*, von dem hier bislang die Rede war, ist nun ein sogenanntes disjungierendes *oder* entgegenzustellen, wenn wir die gängigen Analysen zugrundelegen. Beim disjungierenden bzw. 'ausschließenden' *oder* dürfen die durch diesen Ausdruck verbundenen Konjunkte nicht den gleichen Wahrheitswert haben, während dies beim adjungierenden bzw. 'nicht-ausschließenden' *oder* durchaus möglich ist. Um eben diesen Gegensatz zwischen Adjunktion und Disjunktion geht es dort, wo logische Notierungen mit Sätzen der natürlichen Sprache kontrastiert werden. Wenigstens drei Annahmen scheinen in diesem Zusammenhang allgemein verbreitet zu sein. Zunächst einmal - und diese Annahme dürfte wohl unbestritten bleiben - geht man davon aus, daß in der Umgangssprache sowohl ein ausschließender als auch ein nicht-ausschließender Gebrauch des Wortes *oder* zu beobachten ist. Die zweite dieser Annahmen betrifft die Gewichtung dieser beiden Gebrauchsweisen und läßt sich anhand eines Zitats aus KAMLAH/LORENZEN (1967:154) belegen:

"In der deutschen Umgangssprache kommt das 'oder' im Sinne von 'vel', das eine Adjunktion darstellt, weit seltener vor als das ausschließende 'oder', das die Disjunktion ausdrückt."[1] Nicht das Wort oder allein, sondern die Verbindung entweder-oder betrifft die dritte Annahme: sie besagt, daß dieses entweder-oder grundsätzlich ausschließenden Charakter hat[2] und daß es somit dem exklusiven Gebrauch des alleinstehenden oder zur Seite gestellt werden muß.

Es gibt also gewisse Grundvoraussetzungen bei der Analyse von oder, die weitestgehend akzeptiert zu sein scheinen; dennoch bleibt einigermaßen unklar, wie diese grundsätzlichen Annahmen in einer konkreten Darstellung umzusetzen sind. Nehmen wir eine recht elementare Frage als Beispiel: Gibt es ein einziges ambiges oder, das doppeldeutig ist zwischen inklusiver und exklusiver Lesart, oder gibt es ein exklusives oder (nennen wir es: oder$_1$), das im inklusiven oder (nennen wir es: oder$_2$) ein homophones Gegenstück besitzt? Die vorherrschende Meinung, die sich an einer extensionalen Bestimmung der Bedeutung von oder orientiert, geht augenscheinlich dahin, daß zwei verschiedene oder anzusetzen sind und daß zwischen ihnen und entweder-oder der in (1) grob skizzierte Zusammenhang besteht.

(1)

Eine solche Auffassung spiegelt sich in den Logikhandbüchern, die für das exklusive und das inklusive oder verschiedene Symbole ansetzen, und sie findet sich (wenngleich auch selten ausformuliert) in vielen linguistischen Analysen: so trennt z.B. Lang (1977:178) - um nur ein jüngstes Beispiel zu erwähnen - "das inklusive 'oder' [...] von dem exklusiven 'oder', das wir durch den Zusatz entweder (-oder) verdeutlichen [...]." Man geht sicher nicht fehl, wenn man (1) als Repräsentation einer gewissen Standardanalyse von oder-Verknüpfungen im Deutschen auffaßt.

Sehen wir uns nun einige Sätze an, die bestimmte Standardtypen von Beispielen in verbreiteten Handbüchern illustrieren. Da der überwiegenden Meinung zufolge das nicht-ausschließende oder seltener vorkommt als das ausschließende, wird zumeist versucht, vor allem nicht-ausschließende Verwendungsweisen von oder in der Umgangssprache zu belegen.

(2) a. Ich will mir einen Mercedes oder einen Opel kaufen.
b. Herr Müller ißt zum Nachtisch Käse oder Obst.

Der Satz (2-a) stammt von GUTKNECHT/PANTHER (1973:13); die Autoren schreiben dazu, daß sich die Verwendung des inklusiven oder an diesem Beispielsatz "gut illustrieren" ließe. Wie diese Illustration allerdings genauer aussehen soll, dazu wird weiter nichts gesagt, und ich persönlich muß sagen, daß ich das oder beim Hören von (2-a) sicherlich als disjungierend interpetieren würde. (Vermutlich liegt das daran, daß mein eigener Bekanntenkreis keineswegs so beschaffen ist, daß sich diese Leute bisweilen mehrere Wagen der gehobenen Mittelklasse gleichzeitig kaufen würden - schlimmer noch: ich würde ihnen eine solche Handlungsweise noch nicht einmal zutrauen.) Vom Satz (2-b), der aus KAMLAH/LORENZEN (1967:154f.) stammt, wird immerhin gesagt, daß er ambig sei: "das hier verwendete 'oder' [kann] disjungierend, es kann aber auch adjungierend gemeint sein [...]." Ich selbst würde beim Hören dieses Satzes die exklusive Interpretation sicher vorziehen, wobei meine Auffassung zweifellos (auch) geprägt ist durch gewisse Erfahrungen mit dem Menu-Angebot bestimmter (vor allem französischer) Restaurants.

Als geradezu 'klassisches' Beispiel kann das 'Getränkeangebot' gelten, das immer wieder als eindeutiger Beleg für ein exklusives oder angeführt wird. So schreibt z.B. WALL (1974:34): "'Oder' bedeutet manchmal 'entweder das eine oder das andere, aber nicht beides'; das wäre die übliche Lesart von 'Du kannst Kaffee haben oder du kannst Tee haben'." Hinzuzusetzen wäre also noch, um die intendierte Bedeutung explizit zu machen: 'aber nicht beides'. Nun gehört es zu den Speisegewohnheiten zumindest in den europäischen Ländern, daß man Tee und Kaffee nicht nebeneinander und schon gar nicht gemixt trinkt: die Ausschließlichkeit ist also durch das (grob gesagt) kulturelle Umfeld bedingt und nicht durch einen besonderen Teil der Bedeutung von oder. Das wird vielleicht noch deutlicher, wenn man Sätze wie (3) und (4) gegenüberstellt.

(3) Möchtest Du Tee oder Kaffee?

(4) Nimmst Du Milch oder Zucker?

Tee und Kaffee gleichzeitig zu trinken, widerspricht den herrschenden Speisegewohnheiten, und auch wenn man z.B. die absonderliche Gewohnheit hat, Tee und Kaffee halbe-halbe zu trinken, so widerspräche es zudem den Geboten der Höflichkeit, dem Gastgeber zuzumuten, zweierlei Getränkezubereitung auf sich zu nehmen.[3] Anders steht es hingegen mit (4): Zumindest beim Kaffee ist das Nehmen von Milch <u>und</u> Zucker (genauso wie das Nehmen von <u>nur</u> Milch oder <u>nur</u> Zucker oder das "Schwarz-Trinken") alltägliche Erfahrung, und auch beim Tee werden ostfriesische Trinksitten zumindest in Betracht gezogen (denn sonst wäre die Äußerung von (4) sinnlos). Einer inklusiven Interpretation steht auch kein Höflichkeitsgebot entgegen, da man weiß, daß das Nehmen von Milch <u>und</u> Zucker dem Gastgeber keine weiteren Mühen auferlegt und ihn sicher auch nicht an den Rand des Bankrotts bringen wird.

Ob eine Verknüpfung mit <u>oder</u> exklusiv oder inklusiv interpretiert wird, hängt also von einer ganzen Reihe von Faktoren ab, die man als im weitesten Sinne 'pragmatisch' bezeichnen kann. Es erscheint wenig plausibel, wenn man z.B. den Unterschied zwischen (3) und (4) von den verschiedenen lexikalischen Bedeutungen zweier <u>oder</u> her begründen wollte oder wenn man ihn durch eine mögliche Ambiguität von <u>oder</u> zu erklären suchte.[4] Sinnvoller wäre es, von nur einem einzigen <u>oder</u> auszugehen, das durchaus nicht doppeldeutig ist. Dabei wäre allerdings noch zu klären, wie dieses <u>oder</u> genau beschaffen ist. Es könnte z.B. ein grundsätzlich inklusives <u>oder</u> sein[5], das durch Implikaturen im Sinne von GRICE (1975) uminterpretiert werden kann[6]; es könnte aber auch ein <u>oder</u> sein, das hinsichtlich der Distinktion 'inklusiv-exklusiv' unspezifiziert (vage) ist.

Im folgenden wollen wir uns allein auf die Beziehung des Wortes <u>oder</u> zu der Verbindung <u>entweder</u>-<u>oder</u> konzentrieren. Den obigen Überlegungen zufolge wäre der Analyserahmen (1), der den bisherigen Betrachtungen zugrunde lag, durch etwas wie (5) zu ersetzen.

(5) entweder-oder ------------------ oder

Unter der eingangs berufenen Annahme, daß entweder-oder grundsätzlich exklusiv wirkt, und der Einsicht, daß oder sowohl exklusiv als auch inklusiv interpretiert werden kann, scheint für das Problem der Beziehung zwischen entweder-oder und oder eine einfache Lösung möglich, auf die - zumindest meinem Wissen nach - in der Literatur bislang noch nicht hingewiesen worden ist. Diese Lösung bedient sich eines klassischen Begriffs aus der Prager Schule, nämlich des Begriffs der 'Markiertheit', und entweder-oder wird dabei als markierte Variante zum unmarkierten oder aufgefaßt.

Nehmen wir die 'klassische' Bestimmung des Begriffs der Markiertheit durch JAKOBSON (1932:74) zum Ausgangspunkt: "[...] falls die Kategorie I. das Vorhandensein von A ankündigt, so kündigt die Kategorie II. das Vorhandensein von A nicht an, d.h. sie besagt nicht, ob A anwesend ist oder nicht. Die allgemeine Bedeutung der Kategorie II. im Vergleich zu der Kategorie I. beschränkt sich auf den Mangel der 'A-Signalisierung'. Falls in einem gewissen Kontext die Kategorie II. das Nichtvorhandensein von A ankündigt, so ist es bloss eine der Anwendungen der gegebenen Kategorie: die Bedeutung wird hier durch die Situation bedingt [...]." JAKOBSON (1932: 74f.) bringt dazu folgendes Beispiel: "Das russische Wort têlka (Färse) kündigt das weibliche Geschlecht des Tieres an, wogegen die allgemeine Bedeutung des Wortes telênok (Kalb) keine Ankündigung des Geschlechtes des gemeinten Tieres enthält. Wenn ich telênok sage, bestimme ich nicht, ob es sich um ein Männchen oder um ein Weibchen handelt, aber fragt man mich "êto têlka?" und ich antworte "nêt, telênok", so wird hier das männliche Geschlecht angekündigt - das Wort ist in verengter Bedeutung angewandt." Dieses Beispiel läßt sich sicher nicht direkt auf die Gegenüberstellung in (5) projezieren, aber der gemeinsame Grundzug dürfte deutlich sein. Die Verbindung entweder-oder kündigt an, daß es sich um eine exklusive Verbindung handelt, während für oder allein zunächst nur gilt, daß es diese Exklusivität n i c h t signalisiert. (Daß oder aufgrund bestimmter situationeller Faktoren ebenfalls eine exklusive Interpretation erhalten kann, steht auf einem anderen Blatt.) Die Gegenüberstellung von entweder-oder und oder in (5) ist also nicht als krasser Schwarz-Weiß-Gegensatz zu verstehen; faßt man entweder-oder als markierte, oder hingegen als unmarkierte Form auf, so lassen sich die Spielräume bei der Interpretation von oder und auch die Rigidität bei der Interpretation von entweder-oder ohne weiteres erklären.

Trotz der unmittelbaren Plausibilität einer solchen Analyse bleiben zumindest noch zwei offene Probleme, die ihre Richtigkeit in Frage stellen. Das erste Problem besteht darin, daß diese Analyse keine Beziehung zwischen entweder-oder und oder in dem Sinne herstellt, daß sie erklären würde, warum die phonologische (bzw. morphologische) Form oder sowohl in der markierten als auch in der unmarkierten Version erscheint. Das zweite der Probleme wird in dem Moment offenbar, wo sich die allgemein behauptete Ausschließlichkeit bei der Interpretation von entweder-oder als keineswegs ausschließlich erweist - mit anderen Worten: m.E. gibt es durchaus Verwendungsweisen von entweder-oder, wo eine inklusive Interpretation angebracht erscheint.

Weiter oben hatten wir uns einige Verwendungsweisen von oder angesehen, die in der Literatur gemeinhin als exklusiv interpretiert werden. Paradebeispiel für eine i n k l u s i v e Interpretation von oder in umgangssprachlicher Verwendung ist in der logischen Literatur der 'Bedingungsfall', bei dem p und/oder q erfüllt sein müssen, damit ein bestimmtes menschliches Wesen etwas bekommt, als etwas angesehen wird usw.[7] Der folgende Beleg für diesen Beispieltyp ist WALL (1974:34) entnommen.

(6) Die Bewerber für dieses Stipendium müssen Waisen sein oder sie müssen ein körperliches Gebrechen haben.

Zweifellos ist es so, daß auch ein Bewerber, der z.B. beinamputiert ist und zudem noch seine Eltern verloren hat, für die Erteilung eines solchen Stipendiums (zumindest) in Frage kommt. Nehmen wir nun an, ich habe einen Aushang gelesen, der u.a. den Satz (6) enthält, und treffe wenig später meinen Freund A., der mir erklärt, daß er sich um ein solches Stipendium bewerben will. Daraufhin äußere ich ihm gegenüber den Satz (7):

(7) Um dieses Stipendium zu bekommen, muß man entweder Waise sein oder ein körperliches Gebrechen haben.

M.E. verpflichtet die ernsthafte Äußerung eines solchen Satzes noch keineswegs zu der Sichtweise, daß alleinstehende Körperbehinderte von der Erteilung eines solchen Stipendiums ausgeschlossen wären. (7) bringt vielmehr zum Ausdruck, daß es sich bei den in diesem Satz angeführten Bedingungen um zwei getrennte Voraussetzungen handelt und daß nicht damit gerechnet wird, daß A. auch nur eine einzige dieser Vorbedingungen erfüllt. Es gibt demnach auch Kontexte, bei denen die Verwendung von entweder-oder nicht unbedingt eine exklusive Interpretation mit sich bringt.

Die zweite Verwendungsweise von entweder-oder, auf die hier eingegangen werden soll, betrifft den Gebrauch dieser Verbindung bei Drohungen. In Arbeiten, die sich mit dem (sprachlichen) Handlungstyp des Drohens beschäftigen[8], werden immer wieder Sätze wie in (8) und in (9) als mögliche (und gleichzeitig typische) Äußerungsformen herausgestellt, mit denen dieses Handlungsmuster realisiert werden kann.

(8) a. Du benimmst Dich jetzt anständig, oder ich hau Dir eine runter.

b. Entweder Du benimmst Dich jetzt anständig, oder ich hau Dir eine runter.

(9) Wenn Du Dich jetzt nicht anständig benimmst, hau ich Dir eine runter.

In den Arbeiten zum Drohen stehen sich Äußerungsformen wie in (8) und solche wie in (9) gewöhnlich unverbunden gegenüber, und sie scheinen außer der Tatsache, daß sie jeweils Realisierungen ein und desselben Musters sind, keinerlei Berührungspunkte zu haben. Unbeachtet bleibt dabei die Tatsache, daß die Sätze in (8) und der Satz in (9) durch eine Tautologie verbunden sind, die hier als '$a \vee b \equiv \neg a \rightarrow b$' wiedergegeben werden kann. Auf der Grundlage dieser Tautologie ist es z.B. ohne weiteres möglich, Sätze wie in (8) in solche wie in (9) zu überführen. Dabei ist allerdings zu beachten, daß '$a \vee b \equiv \neg a \rightarrow b$' nur bei einer inklusiven Interpretation des ersten Verknüpfungszeichens tautologisch ist, während es sich bei exklusiver Interpretation keinesfalls um einen Ausdruck handelt, der unter allen möglichen Belegungen wahr wäre. Nur dann also, wenn die oder-Verknüpfungen in (8) inklusiv interpretiert werden, sind (8) und (9) über die angeführte Äquivalenzrelation miteinander in Beziehung zu setzen - und das heißt nichts anderes, als daß auch die entweder-oder-Verbindung in (8-b) nichtausschließend zu interpretieren ist.[9]

Eine dritte Klasse von Fällen, bei denen eine entweder-oder-Verbindung nicht exklusiv zu interpretieren ist, kann durch einen Satz wie (10) illustriert werden:[10]

(10) Ich nehme als Nachtisch entweder Käse oder Obst oder auch beides zusammen, wenn Birne mit Roquefort auf der Karte steht.

Für die herkömmliche Analyse, bei der entweder-oder grundsätzlich als Zeichen für Ausschließlichkeit gedeutet wird, stellen solche Sätze ein Problem dar, da die vermeintliche Ausschließlichkeit sich durch den Rest des Satzes als trügerisch erweist.

Wie die angeführten Beispiele demonstrieren, sind entweder-oder-Verbindungen nicht in jedem Fall exklusiv aufzufassen. Dennoch ist nicht zu leugnen, daß solche Verknüpfungen in einer weitaus überwiegenden Zahl von Fällen als ausschließend verstanden werden. Wie ist diesem Faktum Rechnung zu tragen? Eine Erklärung ist m.E. dann möglich, wenn man entweder-oder zu einer anderen Verwendungsweise von oder in Beziehung setzt, die ebenfalls eine exklusive Interpretation nahelegt; bei dieser Verwendungsweise handelt es sich um das oder mit kontrastiver Betonung.

(11) Ich will mir einen Mercedes óder einen Opel kaufen.

Dieser Satz ist nur auf eine Art und Weise zu verstehen - als Absichtserklärung nämlich, sich nur einen Mercedes oder nur einen Opel zu kaufen, keinesfalls aber beides.

Emphatische Hervorhebungen wie z.B. in (11) haben grundsätzlich zwei Aspekte, einen syntagmatischen und einen paradigmatischen. Auf der syntagmatischen Ebene wird das betreffende Element durch Ausdrucksvariation vom umgebenden Kontext abgehoben, auf der paradigmatischen von all den Elementen, die alternativ an seiner

Stelle stehen könnten. Es gibt also in jedem Fall eine Menge von paradigmatischen Kontrastelementen, die zu dem hervorgehobenen Element im Gegensatz stehen. Für oder nun scheint es im Deutschen nur ein einziges Element zu geben, das in etwa die gleiche Verteilung hat wie dieses Wort: nämlich die Konjunktion und. Wird oder kontrastiv hervorgehoben, so wäre das zu umschreiben als "Ich sage ganz bewußt oder und nicht und"; es wird also ausdrücklich der Gegensatz zu dem Fall herausgestellt, wo beide Konjunkte gleichzeitig gelten. Ähnlich wie eine solche kontrastive Betonung wirkt m.E. die zusätzliche Einführung von entweder in eine oder-Verknüpfung; die Verwendung von entweder-oder anstelle eines einfachen oder hebt dieses oder besonders hervor[11], und diese Hervorhebung ist durchaus vergleichbar mit dem, was eine kontrastive Betonung dieses Wortes bewirken würde. Die Verwendung von entweder-oder legt demnach über den Kontrastierungsmechanismus ebenfalls eine exklusive Interpretation der betreffenden Verbindung nahe.

Die Vorteile einer Analyse, wie sie hier kurz angedeutet wurde, liegen m.E. auf der Hand. Die bislang übliche Einstufung von entweder-oder als zwar syntaktisch diskontinuierlichem, semantisch aber eine Einheit bildendem Ausdruck läßt völlig unerklärt, warum dieser Ausdruck als entweder-oder (und nicht z.B. als *entweder-ider) erscheint. Wird entweder hingegen als Zusatz zu einem unmarkierten oder verstanden, dann ergibt sich die Übereinstimmung in der Form sozusagen von selbst. Für die herkömmliche Analyse, bei der entweder-oder grundsätzlich eine exklusive Charakterisierung erhielt, sind entweder-oder-Verbindungen mit inklusiver Interpretation ein Rätsel. Wenn jedoch diese gewöhnliche exklusive Interpretation über einen Kontrastierungsmechanismus erklärt wird, ist es theoretisch ohne weiteres möglich, daß ein solcher Mechanismus durch andere Interpretationsstrategien sozusagen überspielt wird.

Anmerkungen

1 Hier gibt es allerdings auch gegenteilige Ansichten; vgl. z.B. QUINE (1974:29).

2 So sprechen z.B. KAMLAH/LORENZEN (1967:154) schlichtweg vom

"ausschließenden 'entweder-oder'"; REICHENBACH (1947:23) nennt das "exclusive 'or', which [...] language expresses by 'either ...or'".

3 Es ist übrigens interessant zu bemerken, daß Gebote der Höflichlichkeit es grundsätzlich ausschließen, (3) als ausschließliches Angebot zu formulieren.

(i) a. Möchtest Du entweder Tee oder Kaffee?
b. Du kannst Tee oder Kaffee haben, aber nicht beides.

Beide Sätze in (i) sind sicher sprachlich wohlgeformt, widersprechen aber den normalen Gepflogenheiten der Gastlichkeit. Auch dies weist darauf hin, daß die exklusive Interpretation von (3) auf sehr vermittelten Schlußinstanzen beruht und nicht einfach auf eine lexikalische Bedeutung von oder gegründet werden kann.

4 "A difference in understanding is a necessary, but not a sufficient, condition for ambiuity." (ZWICKY/SADOCK 1973:3).

5 Diese Meinung vertritt z.B. PELLETIER (1977).

6 Vgl. zu einem solchen Ansatz etwa GAZDAR (1977).

7 GEIS/ZWICKY (1971:564) notieren für ähnliche Kontexte seltsamerweise eine inklusive Interpretation als vorherrschend; zu einer kritischen Auseinandersetzung damit s. BOER/LYCAN (1973:493ff.).

8 Vgl. z.B. WUNDERLICH (1976:280) und APELTAUER (1977).

9 Gegen eine solche Argumentation könnte man einwenden, daß '$\neg a \rightarrow b$' eine inkorrekte Formalisierung von (9) darstellt und daß '$\neg a \equiv b$' stattdessen richtig wäre. Die Form '$(\neg a \equiv b) \equiv (a \vee b)$' ist aber genau dann tautologisch, wenn 'v' exklusiv interpretiert wird. - Man beachte aber, daß die Uminterpretation von '$a \rightarrow b$' in '$a \equiv b$' eine Strategie ist, wie sie GEIS/ZWICKY (1971) als "conditional perfection" beschrieben haben, und es scheint durchaus sinnvoll, dann eine analoge "exclusive perfection" zu postulieren, aufgrund derer ursprünglich inklusive Verbindungen ausschließend uminterpretiert werden. Diese Überlegung weist in eine Richtung, die im folgenden eingeschlagen werden wird.

10 Auf Beispiele dieser Form hat mich Ewald Lang aufmerksam gemacht.

11 Vergleichbare Fälle in anderen Kategorien sind nicht schwer zu finden; man vergleiche etwa den Gebrauch von heute und am heutigen Tage in mündlicher Sprachverwendung. Die Auszeichnung der längeren Variante gegenüber der kürzeren, 'normalen' wäre wohl im Zusammenhang mit der Gesprächsmaxime des "Be brief" bei GRICE (1975) zu sehen.

Literatur

APELTAUER, Ernst (1977): "Drohen". SPRENGEL, Konrad / BALD, Wolf-Dietrich / VIETHEN, Heinz Werner (eds.) (1977): Semantik und Pragmatik. Akten des 11. Linguistischen Kolloquiums Aachen 1976. Bd.2. Tübingen: Niemeyer, 187-198.

BOER, Steven E. / LYCAN, William G. (1973): "Invited inferences and other unwelcome guests". Papers in Linguistics 6: 483-506.

GAZDAR, Gerald (1977): Implicature, presupposition, and logical form. Bloomington, Ind.: Indiana University Linguistics Club (vervielf.).

GEIS, Michael L. / ZWICKY, Arnold M. (1971): "On invited inferences". Linguistic Inquiry 2: 561-566.

GRICE, H.P. (1975): "Logic and conversation". COLE, Peter / MORGAN, Jerry L. (eds.) (1975): Syntax and semantics 3: Speech acts. New York: Academic Press, 41-58.

GUTKNECHT, Christoph / PANTHER, Klaus-Uwe (1973): Generative Linguistik. Ergebnisse moderner Sprachforschung. Stuttgart: Kohlhammer.

JAKOBSON, Roman (1932): "Zur Struktur des russischen Verbums". Charisteria Guilelmo Mathesio quinquagenario a discipulis et circuli linguistici Pragensis sodalibus oblata. Prag, 74-84.

KAMLAH, Wilhelm / LORENZEN, Paul (1967): Logische Propädeutik. Vorschule des vernünftigen Redens. Mannheim: Bibliographisches Institut.

LANG, Ewald (1977): Semantik der koordinativen Verknüpfung. Berlin: Akademie Verlag (studia grammatica, 14).

PELLETIER, Francis Jeffry (1977): "Or". Theoretical Linguistics 4: 61-74.

QUINE, Willard van Orman (1974): Grundzüge der Logik. Frankfurt: Suhrkamp.

REICHENBACH, Hans (1947): Elements of Symbolic Logic. New York: Macmillan.

WALL, Robert (1974): Einführung in die Logik und Mathematik für Linguisten. Bd.1: Logik und Mengenlehre. Kronberg: Scriptor.

WUNDERLICH, Dieter (1976): "Behauptungen, konditionale Sprechakte und praktische Schlüsse". WUNDERLICH, Dieter (1976): Studien zur Sprechakttheorie. Frankfurt: Suhrkamp, 251-292.

ZWICKY, Arnold M. / SADOCK, Jerrold (1973): "Ambiguity tests and how to fail them". Ohio Working Papers in Linguistics 16: 1-34.

PRESUPPOSITION DESTRUCTION IN NEGATIVE SENTENCES

Frank Platteau

1. This paper deals with a phenomenon which is well known in linguistics, the cancellation of presuppositions. It will be argued that one needs a pragmatic approach in order to deal with this problem, i.e. one has to consider the context in which sentences are uttered. Furthermore, it will be shown that a similar cancellation also occurs with other notions, such as implicatures, and even entailment.

2. One might wonder what is common to the following notions : generalized conversational implicatures, particularized conversational implicatures, conventional implicatures, semantic presupposition, pragmatic presupposition, entailment etc. Though the exact difference between these terms may not be clear at all, it is probably easier to state what they have in common : whenever a speaker utters an assertion (or any other speech act, but for the rest of this paper we shall only deal with assertions) then he should accept as true , beside the propositional content of the assertion, anything he implies, or implicates etc. This is perhaps not completely true: the speaker need not be aware of all the entailments of his assertion, because it might be an infinite list, but rather he should not be aware of there being any situation in which his assertion would be true and a particular entailment false (cp. VAN DER AUWERA (1978)). Any person who utters (1) or (3) will engage himself in at least accepting (2) or (4) as true, if only for the sake of conversational convenience.

(1) The murderer of Bonzo is insane.

(2) Bonzo has been murdered.

(3) The inhabitants of Tralfamadore are four-dimensional.

(4) Tralfamadore is inhabited.

Any of those terms, which have to be accepted as true, I shall henceforth call <u>consumptions.</u>

3. Suppose now, that one creates a set, say C^u, whose elements are all the implicatures (both conversational and conventional), entailments, presuppositions etc. that can be attached to a certain utterance U. We call the

elements of this set consumptions. Starting from this set we can create a set which contains the consumptions that can be attached to the negation of the same utterance U. This set is called C^{not-u}. In many cases the intercision between C^u and C^{not-u} is not empty: at least the subset of presuppositions belongs to both C^u and C^{not-u}. We also need the complement of a set of consumptions : not-C^u is defined as the set of the negations of the consumptions that can be attached to a certain assertion U with propositional content p. Finally, we need the following abbreviation : the union of two sets is defined as $C^{u \text{ and } v}$.

4. Let us now provide an example in order to clear this out:
   (5) Socrates managed to get out of prison.
   (6) Socrates did not manage to get out of prison.
   (7) Socrates did not manage to get out of prison, because he refused to escape.

The list of consumptions of (5) is :

a: There is someone called Socrates.

b: Socrates was in prison at a certain moment.

c: Socrates got out of prison.

d: Socrates managed to get out of prison.

e: Socrates tried to get out of prison.

C^p = (a,b,c,d,e).

For (6) we do not need any other consumptions, we only have to negate some:
C^{not-p} = (a,b,not-c,not-d,e).

This means that we can infer from (any normal interpretation of) (6) that Socrates existed, that he was in prison, that he tried to escape from prison, but that he did not escape, and so did not manage to escape.

Note that the sets of consumptions also contain the propositional content of the assertion, simply because any proposition entails itself. We can now also define not-C^p :it is (not-a,not-b,not-c,not-d,not-e). It is granted that these sets are not complete. Presumably there are a number of other consumptions that also could be mentioned (for one thing, the subset of entailments might be infinite), but this does not invalidate our argument; we only take into consideration the most important consumptions The interesting case is, of course, sentence (7). One cannot simply compute the set of consumptions of (7) by making the union between C^p and C^r, where C^r is (a,b,not-c,not-d,not-e,f) and f is : Socrates refused to escape from prison.

$C^{\text{not-p and r}}$ = (a,b,not-c,not-d,e,not-e,f).

This causes the paradox that Socrates both tried and did not try to escape out of prison. The rest of this paper will be an attempt to solve this paradox.

5. First of all, it is not too difficult to construct a formula which selects the required set of consumptions. For this we need the notion of complement of a consumption set, which I introduced above.

The formula is the following

$$C^{\text{not-p and r}} = C^r \cup ((C^p \cap C^{\text{not-p}}) / \text{not-}C^r)$$

This means the following : the set of all consumptions that can be attached to a complex utterance not-p and r (which we shall henceforth call a <u>denial</u> , is equal to the sum of all the consumptions of r and all the consumptions that belong to both p and not-p, minus all the negated consumptions of r. Or ,in other words: all the consumptions of the second part of the denial + all the presuppositions of the first part that are not explicitly negated by the second part.

$C^{\text{not-p and r}}$ = (a,b,not-c,not-d,not-e,f) $\cup$ (((a,b,c,d,e) $\cap$ (a,b,not-c, not-d,e)) / (not-a,not-b,c,d,e,not-f))

= (a,b,not-c,not-d,not-e,f) $\cup$ ((a,b,e) / (not-a,not-b,c,d,e, not-f))

= (a,b,not-c,not-d,not-e,f) $\cup$ (a,b)

= (a,b,not-c,not-d,not-e,f)

So, in this example the presupposition that Socrates tried to escape from prison is cancelled, though it is a presupposition of (6).

The question is, now: how can all this be explained? If one defines presupposition semantically, as a kind of entailment (p presupposes q iff p entails q and not-p entails q), then the only way to account for the cancellation is to accept that negation is ambiguous. But in ATLAS(1977) it is shown very clearly that negative sentences are not ambiguous, but rather semantically general, or non-specific, or vague. This claim is based upon the ambiguity-generality tests as they were conceived by ZWICKY/SADOCK (1975). I shall not repeat all the arguments here and simply accept that negation is non-ambiguous.

6. We could call (7) a denial, i.e. the combination of a negative sentence and the explanation why the positive counterpart is not true. Other examples of denials are:

(8) Robert Kennedy was not murdered in 1963, John Kennedy was.

(9) Muhammed Ali is not great, he is the greatest.

(10) Love does not rule the world, but money does.

(11) Holland isn't going to win the World Cup, Argentina or Holland is, since these two teams will be playin the final.

These and similar sentences have the logical form(not-p and r), where not-p is the negated sentence and r is the explanation of why asserting p is wrong. In each of the sentences a number of consumptions are cancelled. Note, though, that it are not only presuppositions that are cancelled. Take, for example, (11): (a similar sentence can be found in WILSON (1975: 149)). The problem here, as Wilson notes, is that this sentence logically implies that Argentina is going to win the World Cup since (11) has the logical form (not-p and (p or q)), which implies q. This, however leads us to a serious problem, since it is certainly not the intention of the speaker to inform the hearer of the fact that he believes that Argentina is going to the World Cup. The assertor of (11)merely says that it is not at all certain that Argentina will win, it might as well be Holland.
A denial normally arises only if someone,e.g. the hearer has previously suggested or indicated that it might be true that p (that Holland will win the World Cup), because otherwise the speaker would be uttering a very queer sentence. With this in mind we can set up the following precondition for denials:

(D) In order to deny a sentence the speaker should presume that it is possible or likely that the hearer believes that p (where p is the positive counterpart of the denied sentence).

7. Let me now focus on the second part of the denial, i.e. the explanation of why asserting p is wrong. There may be many reasons to deny a sentence, but I think that they can be classified into the following three subgroups:

1. The relationship between the predicate and the arguments simply does not hold (we could call this the level of proposition calculus).
2. One of the arguments does not exist(level of first order predicate logic, involving quantification over entities).

3. The predicate does not exist (= level of second order predicate logic, involving quantification over predicates).

(12), (13), (14) would be examples of 1., 2. and 3. respectively:

(12) My auntie Mildred does not have a moustache, what kind of family do you think I got?

(13) My auntie Mildred does not have a moustache, there is no one in our family of that name, but we do have an aunt Mollie, and she has got some beautiful whiskers!

(14) That duck did not queck. 'Queck' does not exist, 'quack' is the right word.

It appears that the most normal case of negation is 1.,whereas 2. and 3. are less frequent, though certainly not impossible. It are the consumptions of level 1. that most commonly are negated, because it is the most natural reaction first to assume that the relationship between predicate and arguments does not hold (if we look at predicates as being sets of individual entities, then in the case of 1. the arguments simply would not be elements of that particular predicate set, but would be elements of some other sets whereas in the cases 2. and 3. the respective argument or predicate would not be member of any set except some peculiar sets such as the set of all non-existent entities that have been referred to in particular situation etc.). In (13) the supposition is denied that there is an aunt Mildred, and one might wonder why in fact the speaker misleads the hearer, by using the expression 'my auntie Mildred'in the first place and thus presupposes the existence of this aunt.We can only explain this if we accept that the expression 'my auntie Mildred' is a kind of echo of the expression the hearer has used; because (13) clearly can only be uttered properly in a context in which the hearer has, in some way or other, suggested that 'your auntie Mildred has a moustache', or at least in a context in which the propositional content of his utterance contains this sentence.

This would be the only reason for the speaker to mention his so-called aunt Mildred. So the speaker does not really presuppose the existence of auntie Mildred, but rather that the expression'auntie Mildred' has been used previously, he uses the expression because this is the only way in which he can refer to it. In order to deny the existence of an aunt Mildred the speaker first uses the term, he takes it up again, he re-sumes the words the hearer has used (or implied) and then says that its referent does not exist.

So I would propose that in the case of a denial the speaker <u>resumes</u> the affirmative counterpart of the denied sentence, then negates it and (optionally) adds why he cannot believe the hearer's assertion. Thus, when resuming the speaker only accepts those consumptions that are not explicitly destroyed.

8. Let us apply this same scheme to another example:

(15) The present king of France is not bald, because France is a republic.

In order for (15) to be a denial, it should be presupposed that the hearer previously suggested in some way or other that the present king of France might be bald (the hearer may have asserted, implied, implicated questioned it etc.). In any case, the utterer of (15) has been able to infer that the hearer might believe that (16) is true.

(16) The present king of France is bald.

The speaker believes that (16) is false. The reason for this falsehood may be, as I showed above, threefold, and the speaker picks out reason 2.: there is something wrong with the identity of one of the arguments: the reference of the definite description 'the present king of France' is the empty set. France does not have a king at the moment, but rather a president. Therefore the speaker negates the previous statement of the hearer and adds his reasons why he believes that this sentence can never be true, given the world as it is today. The speaker does not simply believe or accept as true all the consumptions of (17):

(17) The present king of France is not bald.

He is denying a statement, which means that he only accepts as true those consumptions that are not explicitly negated in the second part of (15). This is, again, done by means of the formula $C^{not\text{-}p}$ and $r = C^r \cup ((C^p \cap C^{not\text{-}p}) / (not\text{-}C^r))$. We, again, agree that a,b,c,d,e... are variables for consumptions:

a: There is one and only one country called France.

b: There is one and only one king of France.

c: The king of France is bald.

d: The king of France has hair on his head.

e: France is a republic.

The following sets can be established:

C^p = (a,b,c,not-d...)

$C^{not\text{-}p}$ = (a,b,not-c,d...)

C^r= a,not-b,not-c,not-d,e...)

not-C^r = not-a, b,c,d,not-e...)

Then: $C^r \cup ((C^p \cap C^{not\text{-}p}) / (not\text{-}C^r)$

= (a,not-b,not-c,not-d,e) $\cup$ (((a,b,c,not-d) $\cap$ (a,b,not-c,d))/ (not-a,b,c,d,not-e))

=(a,not-b,not-c,not-d,e) $\cup$ ((a,b)/(not-a,b,c,d,not-e))

=(a,not-b,not-c,not-d,e) $\cup$ (a)

=(a,not-b,not-c,not-d,e)

which is exactly the set of consumptions we need: there is one country called France, it does not have a king (and consequently this king can neither be bald nor have hair) and France is a republic.

9. Up to now we have been dealing with the point of view of the speaker. If we choose the side of the hearer, then we come to a difficulty: how can the hearer know that the speaker is not consuming anything that is implied in some way or other in a negative sentence, but is resuming it? There are no explicit syntactic features that make it possible to distinguish between consumption and resumption.
Suppose now that a speaker utters (15) to a hearer H. It is clear that there comes a moment when H has decoded the first part of the sentence (the negative sentence) and still has not received the rest of the message. At that point the hearer believes that the speaker presupposes that there is a king of France and that he implies that this king has a wealth of hair, since for any entity in our world it is true that it is either in the set of bald entities or in the set of non-bald entities. So the set of consumptions of the hearer's utterance comprises the presupposition of the existence of a French monarch, and the entailment that the monarch has hair (at least according to the hearer). But after this first moment the speaker asserts that there is no king of France, and consequently that he cannot have hair either, and this urges the hearer to conclude that the speaker is contradicting himself, and that therefore the speaker has violated the conversational maxims. Unless, of course, something else is at stake. What could it be? Since the hearer knows, as well as the speaker, all the language rules the speaker knows, he is able to infer that the speaker in fact resumed the (potential) utterance of the hearer, negated it and added his comment.

10. The following scheme represents the point of view of the hearer:

(1) The speaker asserts that the king of France is not bald.

(2) The hearer assumes that the speaker believes that the king of France has hair.

(3) The speaker asserts that France is a republic.

(4) The hearer assumes that the speaker believes that France does not have a king.

(5) The hearer assumes that the speaker believes that this king of France cannot have hair.

(6) A contradiction arises.

(7) The hearer becomes aware (in an attempt to do away with the paradox) that the precondition for a denial has been fulfilled (i.e. that the speaker presupposes that the hearer might believe that the king of France is bald) and that, indeed, the speaker has uttered a denial.

(8) The hearer concludes that the speaker resumed the hearer's utterance, and that therefore the speaker did not necessarily consume that the king of France has hair.

(9) The hearer computes the right set of consumptions.

(10) The hearer realizes that the speaker does not believe either that the king of France is bald, nor that he has hair.

One should, of course, always bear in mind that the hearer first perceives the first part of (15), and has already internalized and interpreted it before he is fully conscious of the meaning of the second part of (15).

11. It is interesting to mention that a number of psycholinguists have also studied denials,e.g.JUST/CARPENTER(1976); they were able to show in experiments that in cases of denials people first check the truth of what we called the resumed sentence, which explains the fact that it takes about 0.1 secons longer to judge whether denials are true or false. In the experiments it was also proven that the interpretation of false denials takes less time than the interpretaion of true denials, again because it appeared that people first check the resumed sentence; if this appears to be true, then one immediately knows that the denial must be false. In the case of true denials people had to check the truth of the negative sentence as well. For further information I refer to CLARK/CLARK (1977:110).

12. Does all this mean that negation is ambiguous after all? Since there seem to be two cases: denials and non-denials. For example, (18) would be a negative sentence which would not be a denial:

(18) I'm not going to school today, I've got a stomach ache.

One can easily imagine a context in which nobody has previously suggested that the speaker is going to school, so the precondition for a denial does not seem to be fulfilled. But it would not be difficult to extend the definition of denials to any negative sentence, because why else would (18) ever be uttered , if it were not normal that the speaker would go to school, i.e. that the hearer would suppose him to go. If the speaker never went to school,(18) would be at least a very odd utterance.

So, the precondition for a denial would be a precondition for any negative sentence. Then where lies the difference between denials and normal negative sentences? I think that the solution should be found in the three reasons for negating a sentence that I suggested earlier in this paper.

A normal negation would then be a negation of level 1., the level which I called propositional calculus (because it is simply the negation of the proposition without bothering about the predicates or the arguments, their existence is not doubted). In the case of real denials, the reasons for negating the sentence are reasons of level 2 (the level of first order predicate calculus, because the existence of certain arguments is doubted) or of level 3. (the level of second order calculus because the existence (if it means anything at all for a predicate to exist) of the predicate is doubted. The formula I suggested for the computation of the right set of consumptions can be applied to normal negative sentences as well, but here there really is no problem, because the difference between the set of presuppositions of the negative sentence and the complement of the set of consumptions of the explanation why the positive counterpart of the negative sentence is wrong, is simply equal to the same set of presuppositions,i.e. no consumptions are destroyed. This immediately proves my point: negation is not ambiguous, denials are simply a special kind of negation, the set of denials is a subset of the set of negative sentences.

Note also that a similar precondition for affirmative sentences does not seem to exist: it is not so that when a person utters an affirmative sentence, he presupposes that the hearer probably believes that the negation of the affirmative sentence is true. So, if someone says (19),

(19) Muhammed Ali is the greatest.

it does not necessarily follow that someone has suggested that Ali is not the greatest. The same could be applied to other speech acts as well. Consider, e.g., (20) and (21):

(20) Don't touch that piano.

(21) Weren't you going to town tonight?

One appropriateness condition for negative imperatives could be, for instance, that the speaker presupposes that probably the hearer is going to perform the action which the speaker prohibits him to perform. An imperative such as (22) would not have this presupposition.

(22) Touch that piano!

In this case the speaker does not necessarily have the intention not not to touch the piano (he may have no intention at all as far as the piano is concerned). Similarly, the appropriateness condition for negative questions might be that the speaker presupposes that the hearer will give a positive answer to the questioned.

Literature

ATLAS, Jay David (1977): "Negation, Ambiguity and Presupposition". Linguistics and Philosophy 3:321-336.

CLARK, Herbert H./ CLARK, Eve V. (1976): Psychology and Language. New York: Harcourt.

JUST, M.A./ CARPENTER, P.A.(1976): "Eye Fixations and Cognitive Processes". Cognitive Psychology 8:441-480.

KIMBALL, John P. (ed.) (1975): Syntax and Semantics 4. London etc.: Academic Press.

ZWICKY, Arnold M./ SADOCK, Jerrold M.:"Ambiguity Tests and How to Fail them". KIMBALL (ed.) : 1-36.

KOMPARATIV UND NORMALVERTEILUNGSKURVE

Norbert Reiter

Gegenstand dieses Vortrages sind die sogen. "Positive" von Kontrastivadjektiven (auch: Polaradjektiven) oppositiv zu deren "echten Komparativen.

Es geht um die Bestimmung des NORMALWERTes auf einer Skala. Auf den NORMALWERT beziehen sich die sogen. Positive von KONTRASTIV-PAAREN, wie z.B. dtsch.: klein/gross; schmal/breit usw.
Sätze wie: Erna ist klein / Bruno ist gross werden gewöhnlich gedeutet als:

(1) Erna ist kleiner als normal
(2) Bruno ist grösser als normal

Zur Darstellung des Positivs wird hier eine Umschreibung mit dem "echten" Komparativ verwendet. POSITIV und "echter" KOMPARATIV sind nicht gleichwertig.

Um das zu erkennen, müssen wir uns mit dem NORMALWERT auseinandersetzen. Der NORMALWERT - so heisst es - beruhe auf Konvention, sei eine im weitesten umweltbedingte Grösse. Dagegen ist nichts einzuwenden. Jedoch:

Der NORMALWERT ist ein S k a l e n w e r t und als solcher eine natürliche Zahl. Er verlangt eine rechnerische Bestimmung. Umweltfaktoren sind nicht gleich den Werten einer Skala. Vielmehr sind diese durch jene bedingt.

Zwischen der rechnerischen Bestimmung des NORMALWERTes und den ihn beeinflussenden Umweltfaktoren ist streng zu unterscheiden. Jedoch sollte darauf geachtet werden, dass die rechnerische Bestimmung zu den Umweltfaktoren funktional in Beziehung steht.

BARTSCH-VENNEMANN haben versucht, den NORMALWERT rechnerisch zu bestimmen. Ihrer Meinung nach ist der NORMALWERT ein DURCHSCHNITT im s t a t i s t i s c h e n Sinne, also ein Mittel. Sie setzen ein a r i t h m e t i s c h e s Mittel an.

Das arithmetische Mittel aber, noch sonst irgendein anderes, hat zu den Umweltfaktoren keinen Bezug. Das ist nur einer seiner Nachteile. Demgegenüber meine ich:

Der NORMALWERT ist tatsächlich ein Durchschnitt, jedoch nicht im mathematischen, sondern im volkstümlichen Sinne. DURCHSCHNITT im volkstümlichen Sinne aber meint: H ä u f i g k e i t . Berücksichtigen wir die grösste Häufigkeit, so kommen wir zu einem zweidimensionalen System. Dieses besteht aus:

1. der Skala, auf die sich das jeweilige Kontrastivpaar bezieht, ich nenne sie die <u>spezielle Skala</u>, z.B. die der Geschwindigkeit, vertreten durch <u>langsam</u> / <u>schnell</u>

2. der Häufigkeitsskala.

Diese beiden Skalen können als Koordinatensystem dargestellt werden. Dabei ist die <u>spezielle Skala</u> gleich der <u>x-Achse</u>, die <u>Häufigkeitsskala</u> gleich der <u>y-Achse</u>.

Die Werte der x-Achse können mit denen der y-Achse korreliert werden. Mit diesem Vorgang wird gesagt, welcher Wert der speziellen Skala wie oft vorkommt, genauer: wie oft er repräsentiert ist.

Die Korrelation zwischen den Werten der einen und denen der anderen Skala, bzw. den Werten der x-Achse mit denen der y-Achse, führt zu einer Kurve, die - von einem einzigen Fall abgesehen - die äusseren Formen einer Glocke annimmt und deren ideale Ausformung die <u>Gauss'sche Glockenkurve</u> ist.

Derjenige Wert der speziellen Skala, der am häufigsten repräsentiert wird, dem also der grösste Wert auf der Häufigkeitsskala zukommt, ist der NORMALWERT der speziellen Skala.

Betrachten wir die Kurve, so stellen wir fest, dass sie bei diesem Wert ihre Richtung ändert. Wird diese Kurve auf die spezielle Skala projiziert, so wird aus der vorher ungerichteten nunmehr eine gerichtete Skala. Auf ihr gibt es jetzt einen Wert, von dem aus man in die eine oder die andere Richtung zählen kann. Dieser Wert gehört selbst k e i n e r der beiden Richtungen an. Er hat die Eigenschaft NULL. Anders gesagt:

Derjenige als natürliche Zahl zu verstehende Wert einer speziellen Skala, der seiner häufigsten Repräsentation wegen der NORMALWERT ist, fungiert auf der speziellen Skala als NULL. Die ihn flankierenden Nachbarwerte gelten dann als +1 bzw. -1. Die Anlage ist der auf dem Thermometer vergleichbar.

Über die Häufigkeitsskala wirken sich die Umweltfaktoren aus; denn welcher Wert der speziellen Skala wie oft repräsentiert ist, hängt im weitesten von diesen Faktoren ab. Damit zugleich hängt auch der gesamte Kurvenverlauf von diesen Faktoren ab. Eine Kurve kann steiler oder flacher sein. Es ist anzunehmen, dass der Kurvenverlauf seinen pragmatischen Niederschlag findet. Hierbei könnte folgende Regel gelten:

Je steiler die Kurve, desto ausgeprägter die Vorstellungen der Kommunikanten vom NORMALWERT und desto sicherer, vielleicht auch häufiger die Verwendung eines Kontrastivpaares - versteht sich: im POSITIV.

Je flacher die Kurve, desto grösser die Unsicherheit der Kommunikanten bei der Lokalisierung des NORMALWERTes und desto grösser auch die Unsicherheit im Gebrauch des Kontrastivpaares.

Ob diese Deduktion in der Sprachpraxis ihre Bestätigung findet, müsste untersucht werden. Immerhin kann gezeigt werden, dass die Lokalisierung des NORMALWERTES bei zunehmender Kurvenabflachung beeinträchtigt wird.

Würde nämlich die Kurve auf 180° zu einer Geraden abgeflacht, so gäbe es keinen NORMALWERT mehr. Das wäre der Fall, wenn alle Werte der speziellen Skala gleich häufig repräsentiert wären.

POSITIVE wie klein / gross; schmal / breit; leise / laut; wären demnach Komparative, die auf einem z w e i d i m e n s i o n a l e n System beruhen. Dadurch unterscheiden sie sich von den "echten" Komparativen wie kleiner / grösser; schmaler / breiter usw.

Die "echten" Komparative beruhen auf Eindimensionalität. Sie beziehen sich auf eine spezielle Skala und nur auf sie. Die Häufigkeiten werden ausser Acht gelassen. Demzufolge gibt es auf einer solchen Skala auch keinen NORMALWERT.

Die den Vergleich ermöglichende Setzung eines Skalenwertes gleich NULL erfolgt w i l l k ü r l i c h . Das erkennt man daran, dass jeder beliebige Skalenwert gleich NULL gesetzt werden kann. Beim NORMALWERT ist das nicht möglich. Er ist vorgegeben und bleibt, was er ist - sofern sich die ihn steuernden Umweltfaktoren nicht ändern.

So können wir den durch Bruno repräsentierten Wert für NULL setzen und sagen:

(3) Erna ist kleiner als Bruno

Umgekehrt können wir in derselben Weise mit dem von Erna repräsentierten Wert erfahren und sagen:

(4) Bruno ist grösser als Erna

Sagen wir aber:

(5) Bruno ist gross,

so beziehen wir uns auf das zweidimensionale System. Sagen wir schliesslich:

(6) Bruno ist grösser als normal,

so haben wir uns n i c h t auf das zweidimensionale, sondern abermals auf das eindimensionale bezogen, was soviel heisst, als dass normal hier w i l l k ü r l i c h gleich NULL gesetzt ist.

Man darf sich hier von diesem normal nicht täuschen lassen. Worauf es ankommt, ist allein der Umstand, dass in diesem Satz der "echte" Komparativ und nicht der Positiv gesetzt wird. Normal hat hier lediglich identifizierende Funktion: Er sagt uns, dass der hier willkürlich gleich NULL gesetzte Wert, im zweidimensionalen System ebenfalls gleich NULL ist.

Dass die Umschreibung

(7) Bruno ist grösser als normal

nicht gleich dem Befund

(8) Bruno ist gross

ist, sieht man auch an Folgendem:

Ist nach dem zweidimensionalen System irgendein Wert der Skala der NORMALWERT und hat er demgemäss die Eigenschaft der NULL, so kann er unter denselben Bedingungen nicht zugleich nicht der NORMALWERT sein.

Da er das nicht sein kann, kann er selbst auch nicht Gegenstand des Vergleiches sein, kann er also von sich aus nicht in einer der beiden Richtungen liegen und sind Kontrastivadjektive im Positiv auf ihn nicht anwendbar.

Weder ist der NORMALWERT klein, dann befände er sich in Minusrichtung von sich selbst, noch ist er gross, weil er sich dann in Plusrichtung von sich selbst befände.

Benutzen wir jedoch den "echten" Komparativ und sagen:

(9) Bruno ist grösser als normal,

so hindert uns nichts, den durch Bruno repräsentierten Wert dem anderen gegenüber als NULL zu setzen und zu sagen:

(10) Der Normalwert ist kleiner als der durch Bruno repräsentierte,

in welchem Falle demjenigen Wert, der nach dem zweidimensionalen System die Eigenschaft der NULL hat, diese selbe Eigenschaft abgesprochen wird. Das ist nur möglich, wenn das zweidimensionale System ausser Geltung gesetzt wird.

Die Aussergeltungsetzung wird signalisiert, und zwar durch den "echten" Komparativ.

Literatur

BARTSCH, R. und VENNEMANN, T. (1972):"Semantic Structures". Frankfurt/M.

BIERWISCH, M. (1967): "Semantic Universals of German Adjectivals". Foundations of Language 3, 1-36.

REITER, N. (1979): Komparative. Balkanologische Veröffentlichungen des Osteuropa-Instituts Berlin. I.

SAPIR, E. (1951): "Grading: A Study in Semantics", Selected Writings of Edward Sapir in Language, Culture and Personality. Hgg. von David G. Mandelbaum. Berkeley/Los Angeles, 122-249.

POUR UNE ETUDE CONTRASTIVE DES SYSTEMES PREPOSITIONNELS DANS LES LANGUES ROMANES

Eugeen Roegiest

1. Une analyse syntactico-sémantique de prépositions incolores donne lieu à un système dont les définitions sémantiques sont inévitablement fort abstraites. Si cette analyse a comme objet des langues diachroniquement très apparentées, comme l'espagnol et le portugais, les différences qu'on observe dans l'usage de ces langues, s'intègrent parfois difficilement dans la formulation de la signification générale (au sens jakobsonien) au niveau de la Langue. Le programme des recherches que nous proposons, - rendre compte de la diversité des systèmes prépositionnels développés à partir de la situation latine - implique nécessairement une démarche (synchronique) contrastive : sinon on risque d'aboutir dans une langue donnée à des définitions valables également pour d'autres langues.

Souvent en effet, les définitions formulées pour les mêmes prépositions dans des langues romanes différentes se couvrent, même si ces prépositions présentent de langue en langue des écarts d'emploi incontestables. C'est ainsi que la description de l'opposition sémantique qui caractérise le couple _a_/_en_ en espagnol et en portugais est identique dans les grammaires traditionnelles des deux langues, malgré quelques divergences évidentes dans l'usage. De même, les "schèmes représentatifs" établis par POTTIER (1962: 202sv.; 1963: 96sv.) négligent quelquefois ce qui particularise une langue par rapport à une autre. On trouve effectivement dans l'oeuvre de Pottier une série de définitions qui s'appliquent autant aux prépositions françaises qu'espagnoles, de sorte que les différences d'usage entre les deux langues ne se trouvent parfois guère explicitées au niveau formel de la Langue[1].

A nous d'examiner dans quelle mesure des descriptions de ce genre sont acceptables. Il est obvie que la dichotomie saussurienne Langue/Parole et la théorie guillaumienne de l'_in-posse_ jouent un rôle important dans cette problématique : les différences contextuelles entre les particules dans deux ou plusieurs langues reflètent-elles la structure du système sous-jacent ou sont-elles attribuables à l'action actualisatrice de la Parole ? Le caractère virtuel du système est unanimement accepté, mais on risque, en invoquant la Parole, comme dans les descriptions de Pottier, d'accorder trop de poids à l'actualisation des virtualités, de l'in-posse. Et même si on peut démontrer que les divergences de Discours, observées entre deux langues, correspondent à des Systèmes sous-jacents identiques, il reste qu'une description linguistique, compte tenu du principe hjelmslévien d'exhaustivité, doit pouvoir formaliser ces divergences, a fortiori si elles contribuent à la spécificité d'une langue donnée et qu'elles ne relèvent plus d'un comportement idiosynchrasique[2].

Nous rejoignons ainsi les auteurs qui considérent la bipartition saussurienne comme insuffisante. A COSERIU (1969: 11sv.) nous avons emprunté une subdivision tripartite où la notion de Norme s'intercale entre celles de Système et de Parole. Ce niveau intermédiaire est sensé rendre compte de ces faits de langue qui, sans affecter le Système, caractérisent toute une communauté linguistique. Tout en étant propres à une langue donnée (p. ex. variantes combinatoires, faits de fréquence), ces faits n'entrent pas dans les oppositions pertinentes. En outre, la Norme contribuera à écarter ce qui n'est pas pertinent pour la typologie des langues romanes.

2. Afin d'illustrer ces réflexions, nous avons choisi deux faits ibéro-romans examinés en vue de l'établissement des Systèmes prépositionnels de ces langues.

2.1. Prenons l'opposition a/en en espagnol et en portugais. Le parallélisme entre les emplois contextuels de ce couple dans les deux langues est frappant : les mêmes règles générales, les mêmes exceptions. La relation sémantique entre a et en est caractérisée dans les deux langues par l'opposition directif/non directif. Et dans un environnement statique, ce sont les mêmes classes de noms qui peuvent être précédés de a au lieu de en[3]. Ce parallélisme n'est cependant pas parfait. Si en espagnol les verbes du type habitar/residir s'accompagnent toujours de en (sauf dans les cas signalés ci-avant), en portugais a se substitue souvent à em avec les mêmes verbes, devant un inventaire limité de noms désignant des rues, places ou quartiers :

(1) Desejam morar à/na Rua dos Combatentes.

mais

(2) Moramos no número 2, Travessa de São Paulo.

De même, parmi les noms de lieu concrets "relatifs", le portugais présente quelques différences intéressantes: canto, centro, meio admettent a au lieu de em, contrairement aux équivalents espagnols (en el rincón, non al rincón), tout comme certains noms relatifs et inaliénables non spécifiquement locatifs tels que página et capítulo et quelques noms d'instrument susceptibles d'une interprétation spatiale:

(3) O cão dormia enroscado, a um canto, com o focinho entre as patas. (Andrade)

(4) A página vinte, lemos que...

(5) Depois, levantou-se, limpou as faces à manga da camisa.

Ces différences impliquent-elles une nouvelle opposition pertinente ? De la distribution du a statique, on déduit que la localisation exprimée par la préposition espagnole est toujours relative (/limite simple, coïncidence relative/)[4], contrairement au français et à l'italien (/limite simple, coïncidence absolue/) En portugais aussi, la commutation entre a et em n'est possible que dans ces conditions particulières, tandis qu'en français à se substitue à dans dans la plupart des contextes où la notion de limite double n'est pas pertinente. L'existence en français et en italien d'une autre relation pertinente entre à et dans est indiscutable.

Comment interpréter alors l'écart relativement réduit entre les a portugais et espagnol ? Il n'est pas probable qu'en portugais l'opposition a/em se caractérise par une notion de relativité différente. Cette hypothèse ne résiste pas aux faits. D'abord nous avons affaire à un nombre restreint de syntagmes prépositionnels, qui, étant fort usités, apparaissent comme plus ou moins conventionalisés[5]. Ensuite, l'argument d'une relativité de nature différente (non latérale, p.ex.) ne s'appliquerait qu'à quelques-unes des exceptions (centro, meio...): la structure sémique des autres syntagmes qui particularisent l'emploi du a statique portugais ne diffère pas fondamentalement de celle des noms espagnols. C'est ainsi que la présence de a devant certains compléments locatifs de viver, residir, etc. repose sur une interprétation spécifique de l'espace référé: dans l'énoncé (1), il y a coïncidence relative pour le portugais[6]; la préposition a n'est exclue que si cette interprétation n'est plus possible comme dans (2)[7]. Enfin certaines des tournures signalées semblent varier avec les générations: les types limpar ao lenço et à pàgina se retrouvent dans l'espagnol de l'époque classique.

Dans tous ces cas, il s'agit nettement de différences d'habitudes linguistiques qui n'ont rien de fonctionnel. La société linguistique portugaise, sur la base de modèles familiers et existants (ao lado de, ao pe de...) admet la réalisation de ao centro, etc., réalisation qui s'avère bloquée en espagnol. Ces faits ne sont pas attribuables non plus au niveau de la Parole, puisque celle-ci se définit comme un acte contingent concret et individuel. L'emploi de a derrière morar, habitar, devant centro, derrière limpar, relève d'une abstraction de l'acte de parole et appartient au côté social du langage. Ce sont des réalisations potentielles dans le Système et imposées comme n o r m a l e s par une société linguistique et culturelle donnée au sujet parlant qui est sensé les utiliser comme telles dans l'acte concret de la Parole. Les divergences mineures qui caractérisent l'usage du a statique exprimant une localisation relative en espagnol et en portugais relèvent donc toutes de la Norme des deux langues, Norme qui peut encore différer à l'intérieur de chacune d'elles, en fonction de l'extension géographique ou chronologique.

Du point de vue structural, le système relationnel sous-jacent aux emplois statiques n'apporte pas d'éléments nouveaux en portugais[8].

2.2. D'autres différences dans l'usage des prépositions s'expliquent par l'existence de structures morphémiques ou lexémiques différentes en dehors du champ prépositionnel. Nous justifions ainsi l'écart remarquable entre l'usage de l'objet direct prépositionnel en espagnol et dans d'autres langues romanes occidentales. La distribution du morphème de l'objet direct est beaucoup plus développée en espagnol que dans les autres langues et parlers (portugais, catalan, engadinois...) qui se servent de la même préposition. Dans ces dernières langues, les faits descriptifs, quoique non absolument identiques, se ressemblent fortement. Partout (a) les pronoms personnels toniques et le groupe um--outro s'accompagnent obligatoirement de la préposition a, (b) todos et ses équivalents dans les autres langues facultativement, (c) a apparaît fréquemment devant un nom propre animé, mais rarement - dans certaines langues même exceptionnellement - devant un nom commun qui se réfère à un objet animé humain, sauf dans les cas d'emphase, de risque d'ambiguïté ou de répétition du pronom objet par un nom, (d) jamais on ne trouvera a devant un nom peu déterminé (nom au pluriel, nom avec article indéfini, etc.). Tous ces emplois se laissent ramener à un dénominateur commun, qui est de nature syntaxique: partout a marque une cohésion moins forte avec le prédicat que celle qui caractérise habituellement l'objet direct. D'où corrélativement l'hésitation entre les pronoms datif et accusatif dans les seules situations syntaxiques, notamment dans les constructions où un infinitif avec objet direct dépend d'un verbe principal (ex. ptg.: mandei-o/-lhe escrever uma carta). Dans les mêmes constructions le locuteur portugais hésite entre présence et absence de la préposition devant l'objet nominal (sujet de l'infinitif) du verbe principal.

Les différences - d'importance secondaire - entre ces mêmes langues sont soit d'ordre fréquentiel (p.ex. devant un nom propre animé, a est quasi obligatoire en engadinois, mais tout à fait facultatif en portugais), soit idiomatiques (emploi grammaticalisé de a devant le pronom relatif quem en portugais; pronom relatif toujours non marqué en engadinois). Ces variations-ci sont à attribuer à la Norme.

Si dans toutes ces langues, la présence du morphème a répond, comme à son origine lors de la disparition du système casuel, à un besoin fonctionnel de marquer l'objet du verbe par rapport au sujet, il n'en est pas ainsi en espagnol moderne. Dans cette langue, l'objet direct prépositionnel dépasse de loin le rôle syntactico-stylistique de distanciation et de désambiguïsation qu'il assume dans les autres parlers romans. La fréquence plus élevée, mais surtout la distribution beaucoup plus complexe du morphème en témoignent. La présence de a est obligatoire non seulement devant un nom propre, mais devant tout nom animé déterminé. Et cette détermination couvre un champ très étendu; elle affecte même sous certaines conditions des référents inanimés. L'emploi de a ne devient facultatif que devant les animés non déterminés.

Il s'ensuit que les deux traits de l'animation et de la détermination ne sont pleinement pertinents qu'en espagnol. L'accusatif prépositionnel y est devenu un instrument sémantique dont le contenu est difficile à formuler dans un métalangage adéquat[10]. A part les conditions d'animation et de détermination d'autres arguments appuient d'ailleurs la thèse sémantique. Certains verbes opèrent une "personnification" de l'objet (querer, odiar,...) et se construisent presque exclusivement avec a, d'autres comme une "chosification" (p.ex. tous les verbes qui expriment une relation de possession) et rejettent a. La présence d'un infinitif - dont le sujet est objet du verbe principal - a, parce qu'il implique ainsi l' "activation" de l'objet, une forte influence sur l'emploi de a, que l'on observe même assez régulièrement devant des noms inanimés:

(6) A los chiquillos les gustaba ver pasar a los trenes.

Or, on constate que de toutes les langues concernées, l'espagnol est la seule à avoir rejeté, du moins partiellement, la structuration bicasuelle du système des pronoms personnels. Les pronoms lo/la et le ne couvrent plus nettement l'opposition syntaxique cohésion faible vs cohésion forte, à l'encontre des autres langues romanes où les hésitations sont plus rares. En réalité, l'espagnol a tenté, à partir de ce donné syntaxique, d'établir un nouveau binarisme. Ce phénomène a été bien analysé par S. FERNANDEZ (1950: 198sv.). Afin de distinguer neutre et masculin, comme dans d'autres pronoms (ese/esa/eso), l'espagnol recourt à la forme du datif, qui a, outre sa morphologie (forme en -e), l'avantage de se référer de préférence à un animé. La langue tend à créer ainsi un système unicasuel dont les trois unités le/la/lo présentent une opposition sémantique du type sexué vs non sexué. Ce nouveau trait pertinent, propre au système pronominal de l'espagnol, se retrouve dans le fonctionnement du morphème de l'objet direct: tout comme pour les pronoms, une catégorie sémantique conditionne l'apparition de l'accusatif prépositionnel.

3. A l'état actuel de nos recherches, nous n'avons nullement l'intention de fixer des règles permettant de distinguer entre faits de Langue et faits de Norme. Notre seul but est de fournir un instrument adéquat pour la description des Systèmes prépositionnels de langues génétiquement apparentées et pour la découverte de leurs oppositions distinctives. La méthode contrastive y joue un rôle très important. Nous avons essayé, au moyen de quelques problèmes représentatifs, d'illustrer comment on établit, en comparant les unités fonctionnelles (axe paradigmatique) et leur distribution (axe syntagmatique), un réseau d'oppositions distinctives, après suppression des différences relevant de la Norme, ensemble d'habitudes linguistiques, ou relevant d'autres propriétés structurales. Il en résulte en outre une contribution à la typologie des langues romanes.

Notes

1 Le traitement de en espagnol en comparaison avec dans français illustre l'embarras de POTTIER (1962: 339) à distinguer en Langue ce qui sépare ces deux prépositions.

2 Une construction comme integrado a - au lieu de integrado en - s'explique à travers le Système espagnol. Elle n'est cependant, dans l'usage littéraire contemporain, qu'une variante individuelle contingente qui, en tant que telle, n'entre pas dans la description linguistique de cette langue.

3 Rappelons-les à titre informatif: (a) les syntagmes nominaux quantitatifs indiquant la distance, (b) les noms de lieu concrets "relatifs" (lado, entrada,...) (c) des noms abstraits avec référence locative (sol, sombra, frio,...), (d) des noms qui désignent un objet inaliénable (mano, pie,...).

4 Pour une analyse détaillée de cette définition, nous renvoyons à ROEGIEST (sous presse).

5 La nature quasi idiomatique de ces syntagmes explique pourquoi une construction comme al fondo est possible en espagnol, alors que la même communauté linguistique rejette al centro pour des motifs lexicaux (M. SECO 1965: 25).

6 Tout comme dans les syntagmes statiques où a remplace en en espagnol, les grammaires portugaises observent intuitivement un "manque de précision" dans l'expression qui contient a (DUNN 1930: 572-573).

7 Dans l'usage des prépositions, les virtualités d'interprétation dans les limtes du même système sont fréquentes. Que l'on songe à la concurrence, différant selon les générations, entre dans et sur en français (GOUGENHEIM 1970: 44). Si les auteurs classiques écrivaient dans le trône, c'est une question d'interprétation de la réalité.

8 C'est dans les emplois directifs que le système portugais se développe différemment, notamment grâce à l'absence d'une unité (hacia) qui fait l'originalité de l'espagnol.

9 L'attitude du catalan littéraire contemporain est fort intéressant dans cette optique: son rejet presque absolu de l'accusatif prépositionnel (BADIA MARGARIT 1975: t.II,p. 57) s'explique comme une affirmation de son indépendance linguistique à l'égard du castillan.

10 A notre avis, l'animation et la détermination sont des aspects d'un trait plus fondamental: tous les objets directs prépositionnels se caractérisent par une activité potentielle. Le morphème a apparaît ainsi comme une marque de "désactivation", dès que le contexte virtualise dans l'objet un degré d'activité équivalant à celui du sujet.

Bibliographie

BADIA MARGARIT, Antonio (1975): Gramática catalana. Madrid: Gredos.

COSERIU, Eugenio (1969): Teoría del lenguaje y lingüística general. Madrid: Gredos.

DUNN, John (1930): Grammar of the portuguese language. London: Nutt.

FERNANDEZ, Salvador (1950): Gramática española. Madrid: Revista de Occidente.

GOUGENHEIM, Georges (1970): Etudes de grammaire et de vocabulaire français. Paris: Picard.

POTTIER, Bernard (1962): Systématique des éléments de relation. Paris: Klincksieck.

POTTIER, Bernard (1963): Introduction à l'étude de la morphosyntaxe espagnole. Paris: Ediciones hispanoamericanas.

ROEGIEST, Eugeen (sous presse): "Hacia una nueva definición semántica de las preposiciones espaciales 'a' y 'en' en español contemporáneo". Romanistisches Jahrbuch (1978).

SECO, Manuel (1965): Diccionario de dudas y dificultades de la lengua española. Madrid: Aguilar.

DUTCH NOG AND AL AS DEGREE PARTICLES

Jos Rombouts

1. Degree particles

The name Gradpartikeln[1] was coined by Hans Altmann to denote German particles such as nur, auch, sogar (only, also, even), their synonyms and negative counterparts. He regards these particles as operators in Seuren's sense of the word. Their main function consists in inducing certain scales and in allotting a place or value (Grad) on these scales to the constituents of the sentence that are in their scope. Altmann calls this the scalary interpretation. When a series is induced the elements of which are similar to the constituent within the scope of the particle, but no scale is suggested, Altmann speaks of a quantifying interpretation. (ALTMANN 1976: 1-2) KOENIG (1977b) was the first to apply the name Gradpartikeln to German noch and schon (still, already)[2]. In this paper I will discuss their Dutch counterparts, nog and al.

2. The quantifying interpretation

The durative use of nog and al seems to be the most frequent. It can be seen in sentences such as

(1a) Jan is al op kantoor. (Jan is already at the office.)
(1b) Jan is nog op kantoor. (Jan is still at the office.)

In both sentences the state-of-affairs (SoA) that is realized at the moment t_i is contrasted with comparable SoAs. (t_i is indicated by the tense operator or adverbs of time.) In this way a sequence of SoAs is created. Moreover, each SoA can be regarded as a sequence[3](of points in time at which the SoA is realized) in its own right. In this way we obtain for

(1a) Not at the office, t_{i-n} / at the office, t_i / at the office, a period immediately after t_i

(1b) At the office, a period immediately before t_i / at the office, t_i / not at the office, t_{i+n}

The claims concerning the three stages differ in pragmatic

status. The claims concerning t_i (i.e. the original sentences without nog or al) are assertions. The speaker assumes that the hearer knows or expects that the SoA is realized at some moment before t_i (when nog is used) or that it is realized some moment after t_i (when al is used). The new information the speaker gives is that the SoA is realized at t_i. There is some discussion about the status of the claims concerning the period before t_i. When al is used these claims are generalized conversational implicatures. They can be cancelled by adding expressions such as altijd or de hele tijd (always, the whole time), as in

(2) Hij heeft altijd al een zwakke gezondheid gehad. (His health has always been delicate.)

The claim that when nog is used, the SoA is also realized during a period immediately before t_i is regarded as a logical implication by ABRAHAM (1977) and KOENIG (1977a), but KOENIG (1978) calls it a conventional implicature[4]. The claims about the third stage are generalized conversational implicatures. The points in time at which the SoA is realized can be regarded as elements of a set V. Both nog and al conversationally implicate that t_i is a peripheral element of V, i.e. that the speaker expects the set V to end soon after or at t_i (nog) or begins shortly before or at t_i (al). In this way a succession of at least two sets is implicated: V is preceded or followed by a set of points in time at which the SoA is not realized. Although the speaker distinguishes between central and peripheral elements of V, no scale of values is induced. In the above examples nog and al are used in their quantifying interpretation.

The use of nog and al in the following sentences is similar to that in (1) and (2):

(3) Turnhout ligt nog in België, Tilburg ligt al in Nederland. (Turnhout is in Belgium, Tilburg is in the Netherlands.)

(4) Dat beeldje heeft mijn grootvader nog gemaakt; dit hier is al het werk van zijn broer. (That sculpture was made by my grandfather; this one is by his brother.)

(5) Peters vader leefde juist lang genoeg om nog mee te maken dat zijn zoon een drugverslaafde werd. (Peter's father lived just long enough to see his son become a

drug addict.)

The two contrasting sets are: for (3): the towns and villages in Belgium / those in the Netherlands; for (4): the sculptures made by my grandfather / those by his brother; for (5): the events during his life Peter's father knew about / those that took place after his death. The ordering of these sets with respect to each other is indicated by <u>nog</u> and <u>al</u>. The elements in connection with which <u>nog</u> is used belong to the first set. The ordering of the sets in (3) and (4) can be reversed by reversing the use of <u>nog</u> and <u>al</u>:

(6) Turnhout ligt al in België, Tilburg ligt nog in Nederland.

(7) Dat beeldje heeft mijn grootvader al gemaakt; dit hier is nog het werk van zijn broer.

The speaker assumes that the hearer knows a number of elements of the set V, but that the hearer does not know whether the element that is referred to (the reference point; henceforth: RP_i) is an element of V or not. This is why it is conversationally implicated that the RP_i is a peripheral element of V. A sentence like

(8) Brussel ligt nog in België. (Brussels is in Belgium.)

where Brussels is not a peripheral element is pragmatically deviant. The use of <u>nog</u> implicates that the speaker assumes that the hearer knows a number of Belgian places, whereas (8) contains new information only when the speaker assumes that the hearer does not know Brussels is in Belgium (so the hearer does not know Belgium at all). Again no scale of values is induced. The open formulas 'x is in Belgium', 'y is in the Netherlands', 'x was made by my grandfather'... are equally true for all elements of the respective V's.

A third type of quantifying <u>nog</u> can be seen in sentences such as

(9) Nog voor hij met het proletariaat in kontakt kwam, wist Marx welke rol dat proletariaat zou spelen. (Even before he came into contact with the proletariate, Marx knew what role would be played by that proletariate.)

(10) Nog voor ze geheel was uitgewerkt, kwam Chomsky's theorie ten grondslag te liggen aan een hernieuwde samenwerking met de psychologen. (Even before it had

been worked out in detail, Chomsky's theory became the foundation of a renewed co-operation with psychologists.)

The set V can be defined as follows: for

(9) V = {x | x (happened) before Marx came into contact with the proletariate}

(10) V = {x | x (happened) before Chomsky's theory had been worked out in detail}

V consists of all events that took place before a given point in time t_x. The speaker assumes that the hearer has some knowledge concerning the set V. The hearer does not know, however, that the RP_i (e.g. the fact that Marx knew what role would be played by the proletariate) is an element of V. The SoA denoted by the RP_i is realized earlier than the hearer is assumed to expect. The RP_i is a peripheral element because it refers to a SoA that was realized only shortly before t_x. The generalized conversational implicature that the RP_i is a peripheral element of V explains why (11) is pragmatically deviant:

(11) In 1830, nog voor de tweede Wereldoorlog, ontdekte men dat... (In 1830, even before the second World War, it was discovered that...)

This implicature can be cancelled, however:

(12) In 1830, nog lang voor de tweede Wereldoorlog, ontdekte men dat... (In 1830, long before the second World War, it was discovered that...)

For this type of quantifying <u>nog</u>, there is no corresponding use of <u>al</u>. It is possible to use <u>al</u> instead of <u>nog</u> in (9) and (10), but then its function is not similar to that in the previous types. If it had been, we would have expected t_x to refer to the beginning of the set V. The SoA that is denoted by the RP_i would have been realized after t_x; which is clearly not the case. When <u>al</u> is used in (9) and (10) it just evaluates as early a point in time at which a SoA is realized.

Thus far, I have discussed three types of quantifying <u>nog</u> and two types of quantifying <u>al</u>. They can be distinguished only on syntactic grounds. In sentences of the first type, the elements of V function as adverbial constituents. In sentences of the second type, the elements of V function as nominal constituents, i.e. they are NCs or embedded sentences with the same function. In sentences of the third type, they are sentences without an Adv C referring to a point in time or place (except when nog occurs in a constituent which is in apposition to such an Adv C, as in (11)). These syntactic differences do not result in semantic ones, however. In all examples the use of nog or al induces a series of similar elements. The speaker distinguishes between central and peripheral elements of this series, but this distinction does not influence the truth of the predication. The predication is either valid or invalid; no scale of validity is suggested.

3. The scalary interpretation

In this respect the quantifying interpretation differs from the scalary one. In this interpretation, the fact that the RP_i is a peripheral element of V does affect the validity of the predication. The scalary use of nog and al can be seen in

(13) Dit is nog een stoel, maar dat is al een fauteuil.
(This is still a chair, but that is already an easy-chair.)

The two contrasting sets in (13) are the set of chairs and that of easy-chairs. The ordering of these sets can be reversed by reversing the use of nog and al. The speaker who uses nog and al in sentences like (13) is faced with a problem. Language makes use of discrete labels such as e.g. the words chair and easy-chair, but very often no clear distinction can be made between the objects in the extra-linguistic reality those labels refer to. There are easy-chairlike chairs and chairlike easy-chairs. Thus before the speaker claims x is a chair or x is an easy-chair, he has to judge to what degree (Grad) these predications are valid. He does this by comparing the RP_i with similar objects. By using nog or al he indicates that the RP_i does not show all characteristics of e.g.a chair. The more marginal an element, the less likely it is that a given predicate will be applied to it. The speaker thus constructs scales. When a set V is compared with a preceding and a following set, we obtain the following result:

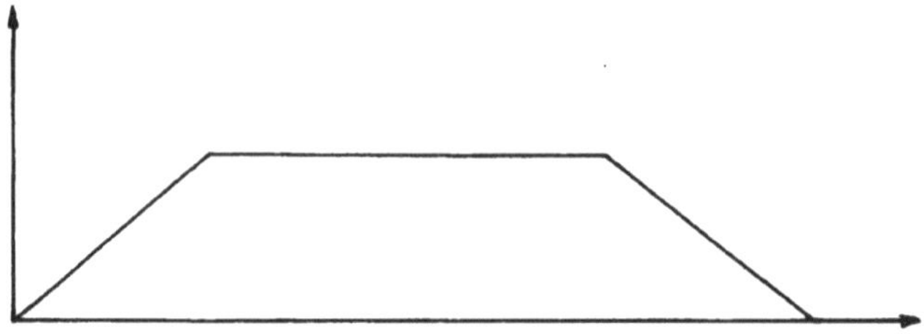

The horizontal axis represents chairlike objects, the vertical axis represents the degree in which these objects show the characteristics of chairs. The curve represents the set V. _Al_ is used only in connection with the rising part of the curve, i.e. that the RP_i is ranked higher on the scale than all preceding elements of V (if any). _Nog_ is used in connection with elements on the falling part of the curve. The RP_i is higher on the scale than all following elements of V (if any). The horizontal line represents the central, non-marginal elements. They are not ordered on a scale[5].

Another type of scalary _nog_ and _al_ is illustrated by sentences such as

(14) Er zitten al 53 reizigers in de bus.

(15) Er zijn nog 15 kippen in het hok.

Both sentences are ambiguous. This is caused by an ambiguity in the scope of the operators _nog_ and _al_, as is shown in the paraphrases:

(14a) Er zitten al reizigers in de bus. Ze zijn met 53. (There are already passengers in the bus. They are 53.)

(14b) Er zitten reizigers in de bus. Ze zijn al met 53. (There are passengers in the bus. They are already 53.)

(15a) Er zijn nog kippen in het hok. Ze zijn met 15. (There are still hens in the pen. They are 15.)

(15b) Er zijn kippen in het hok. Ze zijn nog met 15. (There are hens in the pen. There are 15 left.)

In the a-versions we have to do with the quantifying interpretation. The number (i.e. the predicate of the embedded sentence) is not in the scope of the operators _nog_ and _al_. In the b-versions _nog_ and _al_ are used in their scalary interpretation. In both versions of (14) it is implicated that at some point in time t_{i-n} there were no passengers and that at no point before t_i the number of passengers was greater than at t_i. But (14a) implicates only that there are at least 53 passengers during a period immediately after t_i, whereas (14b) expresses that the speaker evaluates the number of passengers as great. It is conversationally implicated that it will become even greater. The evaluating component of <u>al</u> in (14b) is made possible by the fact that in this version <u>al</u> is contrasted with <u>nog maar</u> (a combination of two degree particles):

(16) Er zijn nog maar 53 reizigers in de bus.
(There are only 53 passengers in the bus.)

However, (16) again is ambiguous. When <u>nog maar</u> is contrasted with <u>al</u>, it is used in connection with a number that the speaker expects to increase, but it can also be used in connection with a number that is expected to decrease, as in (17):

(17) Er zijn nog maar 15 kippen in het hok.
(There are only 15 hens left in the pen.)

(15a) and (15b) are similar in that in both interpretations a decrease of the number of hens, and thus an end of the SoA is expected. (15a) implicates that there were 15 hens during a period immediately before t_i, whereas (15b) implicates that at some point in time before t_i the number of hens was greater. To summarize these observations, let us represent the evolution of the number of passengers during a bus ride as follows:

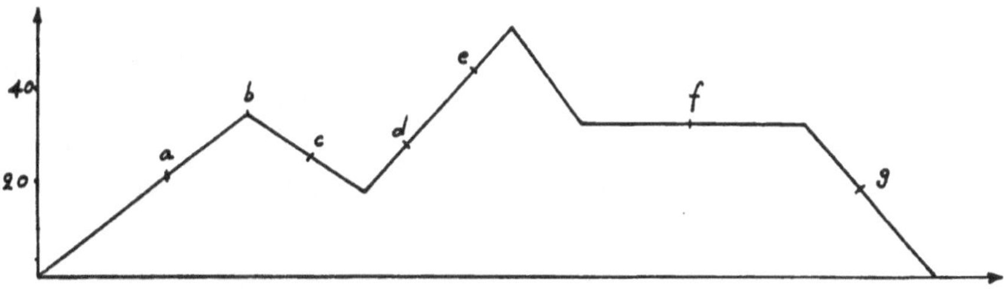

The horizontal axis represents points in time, the vertical axix represents numbers. V is represented by the curve. If degree particles are used in connection with elements a-g, then <u>al</u> can be used in connection with a, b and e. Though d is on a rising part of the curve, it cannot have <u>al</u>, since there are points in time before d at which the number of passengers is greater. This explains why nog is used with respect to d. Nog is also used in connection with f (quantifying) and c and g (scalary interpretation). The scalary use of nog and al in (14)-(17) is similar to that in (13), since al is used when the RP_i represents a stage in a rising gradation, whereas nog is used when the RP_i is a stage in a falling gradation. An important difference, however, is that in (13) only the peripheral elements are ranked on a scale, whereas in the scalary interpretation of (14)-(17) all elements are ordered on a scale. The use of nog and al in (14)-(17) does not implicate that the RP_i is a peripheral element. When the speaker thinks the RP_i is peripheral, he uses nog maar.

Sentences such as (14)-(17), in which the scale consists of quantities, are typical instances of a more general category. A few other examples:

(18) We zijn al in Parijs. (We are already in Paris.)

(19) Peeters is al bureauchef. (Peeters is already a head clerk.)

(18) and (19) are ambiguous. In the quantifying interpretation the set V consists of the points in time at which we are in Paris, respectively the points in time at which Peeters is a head clerk. It is implicated that only shortly before t_i, we were not so far as Paris and Peeters had a lower function than that of head clerk. This implicature occurs also in the scalary interpretation, but then it is conversationally implicated that at some later stage, we will be further than Paris and Peeters will have a higher function than that of head clerk. V is then for (18) the set of towns and villages a company passes on its way from Belgium to Spain, and for (19) V is the set of functions Peeters will be given during his career. The speaker evaluates the RP_i as fairly high on the scale. When he thinks the RP_i is rather low on the scale, he uses nog maar instead of al. When the speaker wants to implicate that Peeters has been degraded, he uses nog (in the scalary interpretation) or nog maar. Nog maar is generally preferred when the use of nog might result in an ambiguity.

4. Conclusions

The main purpose of this paper was to investigate whether the notions of quantifying and scalary interpretation, which were devised by ALTMANN (1976), can be used in a semantic description of nog and al. We have seen that the notion of scalary interpretation covers two semantically distinct ways of using these particles. We thus arrived at a threefold distinction. First, there are sentences in which nog and al are used only with respect to peripheral elements of the set V. No scale is suggested. Secondly, we have sentences in which nog and al are used with respect to peripheral elements which are ranked on a scale. The central elements are not. Finally, we have discussed examples in which the use of nog and al is not restricted to peripheral elements. All elements of V are ranked on a scale. Moreover, we have seen that there is an inclusive relation between the scalary and the quantifying interpretations: the quantifying interpretation is arrived at when a series of elements are all given the same rank on a scale (cf. the graphs). The uses of nog and al that are discussed in this paper do not constitute the full range of their uses as degree particles. For a more complete description of nog, al and a few related particles, I must refer to ROMBOUTS (1979) (forthcoming).

Notes

1 The term degree particles is a literal translation of the German term. Although the term particle in English is usually reserved for words such as for, up and with in to look for and to put up with, I prefer it to the more usual adverb in order to avoid confusion with the adverbs of degree.

2 For a more precise definition of the notion of sequence, see VANDEWEGHE (1979).

3 KOENIG (1978) distinguishes, however, between noch and schon as adverbs of time and noch and schon as degree particles. I reject this distinction, since there is a very close semantic similarity between both functions.

4 For the terms conventional and conversational implicature, see GRICE (1975).

5 Nog can also be used with respect to the non-peripheral elements on the horizontal parts of the curve. This is the so-called additive (quantifying) use. In this interpretation nog is translated as another. For more details, see ROMBOUTS (1979) (forthcoming).

References

ABRAHAM, Werner (1977): "Noch und schon als polare Satzfunktoren". SPRENGEL / BALD / VIETHEN (eds.) II: 3-19.

ALTMANN, Hans (1977): Die Gradpartikeln im Deutschen. Untersuchungen zu ihrer Syntax, Semantik und Pragmatik. Tübingen: Niemeyer.

GRICE, H. Paul (1975): "Logic and conversation". COLE, Peter / MORGAN, Jerry L. (eds.) (1975): Syntax and Semantics. III: Speech Acts. New York: Academic Press: 41-58.

KOENIG, Ekkehard (1977a): "Temporal and Non-temporal Uses of 'noch' and 'schon' in German". Linguistics and Philosophy I: 173-198.

--- (1977b): "Zur Syntax und Semantik von Gradpartikeln. Versuch einer allgemeinen Charakterisierung". SPRENGEL / BALD / VIETHEN (eds.) II: 63-69.

--- (1978): Direkte und indirekte Bewertung von Zeitintervallen durch Satzadverbien und Gradpartikeln im Deutschen und Englischen". WEYDT, H. (ed.) (1978): Partikeln des Deutschen. Berlin: De Gruyter.

ROMBOUTS, Jos (1979): Pragmatische aspecten van nog, al en ververwante graadpartikels. (working title) Wilrijk: Antwerp Papers in Linguistics (forthcoming).

SHETTER, William Z. (1966): "The meaning of german noch". Language 42: 42-66.

SPRENGEL, Konrad / BALD, Wolf-Dietrich / VIETHEN, Heinz W. (eds.) (1977): Akten des 11. Linguistischen Kolloquiums, Aachen 1976. II: Semantik und Pragmatik. Tübingen: Niemeyer.

VANDEWEGHE, Willy (1978): "Over de semantiek van het partikel al (= reeds)". De Nieuwe Taalgids 71: 193-205.

--- (1979): "Perspectivity Operators in Dutch: al, nog, nog.m". VANDEWEGHE, Willy / VAN DE VELDE, Marc (1979): Akten des 13. Linguistischen Kolloquiums, Gent 1978. II: Bedeutung, Sprechakte und Texte. Tübingen: Niemeyer.

PERSPECTIVITY OPERATORS IN DUTCH: AL,NOG,$\underline{NOG}_{n,m}$

Willy Vandeweghe

1. States-of-affairs (SoAs) and SoA-expressing predicates.

1.1. The "sequentialization" of SoAs.

In linguistics we traditionally speak of 'aspects', when a speaker uses grammatical means to situate a predication concerning an event or state of affairs temporally with respect to that event or state of affairs (henceforth: SoA) viewed as a whole. This implies that attention is paid to phases such as initial or terminal phase, to characteristics such as duration or iteration: "aspects are different ways of viewing the internal temporal constituency of a situation" (COMRIE 1976: 3), where by 'situation' is meant: " a state, an event, or a process" (COMRIE 1976: 13).

Language is also equipped with the means whereby it is possible to situate an SoA in a broader setting: it is possible to view an SoA occurring at a given moment, or obtaining with respect to a given individual, in contrast to comparable SoAs, obtaining at other moments, or with respect to other individuals. E.g., look at

(1) Hij is aangekomen. (He has arrived)

If we want to compare the SoA expressed in (1) and obtaining, say, at the moment t_i, with a preceding SoA, i.e., an SoA obtaining at t_{i-1}, we can have

(2) Hij is al aangekomen. (He has already arrived)

We have a SEQUENCE[1] of SoAs from 'not-arrived,t_{i-1}' to 'arrived, t_i'. Take as another example

(3) Hij is niet thuis. (He is not home)

There are two ways of "sequentializing" the SoA expressed in (3):

(4) a Hij is nog niet thuis. (He is not home yet)

b Hij is niet meer thuis. (He is no longer home)

The sequences are: for (4a): not home, t_i/home, t_{i+1}

(4b): home, t_{i-1}/not home, t_i

The same holds true for sentences in which a quantitative expression figures:

(5) Er zijn al 200 slachtoffers geëvakueerd. (200 victims have

already been evacuated)

The sequence: less (than 200) victims (have been evacuated), t_{i-1}/ 200 victims (have been evac.), t_i / more (than 200) victims (will - foreseeably - be evacuated), t_{i+1}[2]

In the preceding examples, an SoA obtaining at one point in time is compared with one obtaining at another temporal point. For the sake of completeness, note that it is equally possible to compare SoAs obtaining with respect to one entity - serving as Argument for the Predicate - with one obtaining with respect to another entity, as in the example

(6) a Eupen ligt nog in België; Monschau ligt al in Duitsland.
(Eupen is still in Belgium; Monschau is already in Germany)

b Eupen ligt nog in België;Monschau (ligt) niet meer (in B.).
(...; Monschau is no longer in Belgium)

As there are two predicates involved ('in Belgium', 'in Germany'), we have to deal with two (sub-)sequences:

(1°) in Belgium, Eupen / not in Belgium, Monschau

(2°) not in Germany, Eupen / in Germany, Monschau

We will come back to this special case later on[3].

1.2. Sorts of predicates.

Before engaging in a more thorough analysis of these matters, let us have a closer look at the relevant kinds of predicates which may appear in construction with the perspectivity particles al/nog/etc.:

(i) There are what we might call 'binary' predicates, i.e., predicates (mostly verbal) which are to be seen in contrast to their negative counterpart, equipped with the negative operator NEG, mostly appearing as 'not', and describing a complementary SoA: the SoAs are mutually exclusive. See for example the verbal predicates in (1) and (3).

(ii) We also have 'scalar' predicates, i.e. predicates appearing, e.g., as numerals, expressing a value on a scale, of which the most representative example - though not the only one - is the numerical scale, indicating quantitative values applying to entities. Each position on the scale is seen relative to every other position. See e.g. (5), but also

(7) a Hij is (al) ver gevorderd. (He has made a lot of progress - already)

b Hij is (nog maar) <u>een eindje</u> gevorderd.(...only a slight progress - as yet)

(8) a Het is al dinsdag. (It is Tuesday already)

b Het is nog maar dinsdag. (It is only Tuesday)

(iii) Then there are 'polar' predicates, that is, normally predicate pairs (mostly adjectives), related by the fact dat they express opposite polar values on a continuum established in terms of a dimension like SIZE, AGE, etc. Examples are <u>small</u>/<u>big</u>, <u>young</u>/ <u>old</u>. These polar predicates are very similar to scalar predicates in that they can be seen as defining values on a scale, but they are distinct from the scalar predicates in that they only allow us to express the <u>peripheral</u> values positively; the intermediate values can only be expressed negatively, e.g.: <u>not young</u>; <u>not old</u>.

We should note the special case of <u>weinig</u>/ <u>veel</u> (little, few/ much, many) and <u>zelden</u>/ <u>vaak</u> (seldom/ often) where we in fact have two antonyms, which however seem to behave like purely scalar predicates. We have to leave this topic undiscussed here, as we are not dealing with the semantics and syntax of the predicates (henceforth to be abbreviated as 'Pred(s)'), but rather with that of the perspectivity particles operating on them.

2. Perspectivity.

2.0. Thus far, we have only considered a more or less intuitive notion of 'Sequence'. In order to understand the nature of the Perspectivity operation, we have to proceed in a more analytic way.

2.1. With binary Preds.

Consider the sentence

(9) He is home.

There is first of all a deictic element in this sentence, namely the temporal point of reference (t_i, but rendered for the sake of generality by the notation 'RP_i'). This RP_i is not explicitly stated, but represented by the Tense operator PRESENT[4]. We can leave the entity denoted by the subject 'he' out of consideration, as it is not contrasted with comparable entities, when operators like 'already', 'not yet' etc. are introduced. So we can move to a second important part, the propositional part of the sentence, which we will render by the notation 'Pred' (= predicate). In (9), this is 'is home', which is predicated of 'he'. It takes on an assertory value (ASS)[5]: 'positive', rendered as 'ASS+'. ASS+ is not normally rendered in surface structure, unlike its explicitly expressed counterpart 'not', as in

(10) He is not home.

We can say that ASS+ is a member from a pair $\{$ASS+,ASS-$\}$, and we shall use 'ASS' as a neutral indication for one of these values. In 'ASS_{Pred}' the subscript 'Pred' indicates that ASS has the Pred from the sentence under consideration within its scope.

Now, we can equally well say that RP_i is a member from a set of RPs, which is a transitively, irreflexively and asymmetrically ordered set. If the set consists of temporal reference points, this order is normally fixed, as time moves only in one direction, but if they are Arguments for the Pred under consideration, as in (6), then the speaker may choose which of them he considers first. This in its turn has its influence on the Pred(s) chosen and the order in which they are applicable.

2.1.2. The operation to be carried out might be described as follows: with the chosen Pred as a standard, one considers in how far this Pred is applicable to the state of affairs with respect to a given RP. If the result is positive, the assertory value is ASS+, otherwise it is ASS-. But as we are comparing SoAs with respect to different RPs, these two values are placed in an ordered relationship with respect to each other, which results in what we will call a Sequence (Seq).

Seq $=_{def}$ an ordered, oriented set of values for ASS(Pred), with the ordering of the values corresponding to the order in which the RPs are mapped into it, and the orientation being dependent on what value is chosen by the speaker as the initial value.

As can be deduced from the definition of Seq, it consists of two differently specified zones - constituted by the mappings from the set of RPs for which the different values hold -, so it is evident that there must be a transition between them: we think of it as a transition point, namely p_x. This point is of vital importance for the operation under consideration, as the topological correspondent for RP_i on Seq, say p_i, will be alloted a relative position with respect to p_x. Now this assignment of a relative position for p_i is precisely what the Perspectivity operation consists of: if p_i is situated prior to p_x, we say that the perspective is PROSPECTIVE in nature; if it is located past p_x, we say that it is RETROSPECTIVE.

2.2. With scalar Preds.

The notion of 'Sequence', as aplied to binary Preds is, with certain adaptations, equally applicable to scalar Predicates. We should note first that it is not the ASS-value which is relevant here, as we are considering the use of scalar Preds where 'ASS+' is fixed[6]. As already stated, in the case of scalar Preds we are dealing with one of a set of values from a scale - mostly but not necessarily a quantitative scale - the ordering of which is again dependent on the ordering of the (set of) RPs. But the new thing here is that upon the asserted value - rendered by VAL_{scal_i} - is imposed an EVALUATION, that is, the speaker relates the VAL_{scal_i} to an evaluative scale which is a function of his own expectations with regard to what is 'much' or 'little' or neutral on the one hand and of the reference group of entities over one member of which the VAL_{scal_i} may quantify. E.g.

(11) Die boer heeft maar 3 koeien. (That peasant has only got three cows.)

By using the restriction operator maar (only), the Speaker implies that for cows, in his opinion, the number of 3 is little. (3 horses could, on the contrary, be 'much' in his opinion, and 3 cows could be a fair number for someone else.). The positive (non-restrictive) counterpart is normally not expressed, although in Dutch there is a lexeme by means of which it is possible to express positive appreciation, namely, 'wel':

(12) Die boer heeft (wel) drie koeien. (That peasant has got (as many as/no less than) 3 cows.)

'Maar' turns out to be the relative counterpart of absolute 'niet' we met in the case of the binary predicates. And, second analogy: the positive values, the absolute as well as the relative ones, may remain unexpressed. Now, if VAL_{scal_i} is assigned a relative position on the evaluative scale of the Speaker, then as a matter of course the comparable VAL_{scal}'s from the Sequence are assigned a position relative to this one, so that some place on the Sequence, the transition between $VAL_{scal}+$ and $VAL_{scal}-$ must be situated. Although this transition may be vague and consitute a zone, we shall for convenience also speak of a transition point p_x.

And again, as in the case of the binary Preds, the Seq may be oriented in two ways: positively or negatively. Here this means: from (relatively) 'less' to (relatively) 'more', or vice versa. In the following examples, the Speaker deals with a positively oriented Seq (13) and a negative one (14).

(13) a Die boer heeft nog maar 3 koeien. (That peasant only has 3 cows as yet)

b Die boer heeft al 3 koeien. (That peasant has got 3 cows already.)

(14) a Die boer heeft maar 3 koeien meer (nog maar 3 koeien).[7] (That peasant has only got 3 cows left.)

b Die boer heeft nog 3 koeien. (That peasant has got 3 cows left.)

2.3. With polar Preds.

The case of polar Preds is rather complicated, so we won't go into the details of it. The main thing to be said about the semantics of these Preds is that they may be viewed as located at the opposite ends of a lexically defined scale, which is defined by a common parameter such as AGE, SIZE etc. But as each of the Preds may be regarded as a binary Pred (e.g. old/not old) in its own right, the lexical relationship must be accounted for by a meaning postulate which

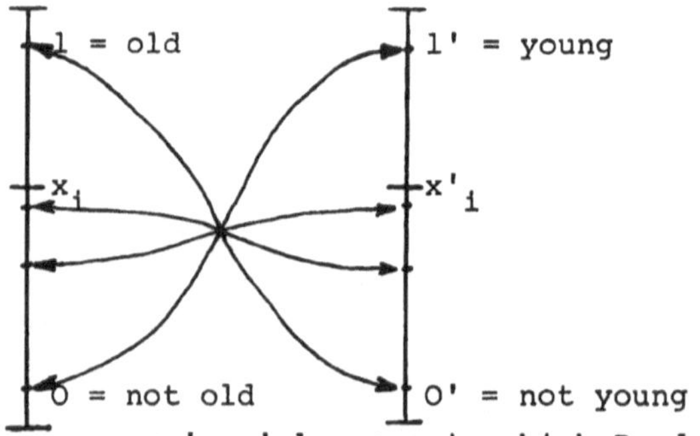

x_i'...' 1: zone in which Pred 'old' applies
x_i'...' : zone in which 'young' applies

- Table 1 -

says that the scales defined by the opposite polar predicates may be mapped onto each other, as is shown in Table 1.

It can be seen from this graphic representation that 'he is not young' may, for some instances, be mapped on 'he is old'. But it is far from us to say that these two expressions are equivalent, let alone synonymous.

As each of the polar Preds defines an own sequence, we may have oppositions like in (15) end (16):

(15) Hij is nog jong / Hij is niet jong meer. (He's still young/ He is no longer young)

(16) Het is nog niet laat/ Het is al laat. (It isn't late yet/ It's late already)

But as the scales may be mapped onto each other, and linked by a same standard, we may find nog/ al defining 'complementary' (polar) zones on a Sequence defined by this parameter:

(17) Hij is nog jong / Hij is al oud[8]. (He's still young / He's old already)

Now a problem - which is just mentioned in passing - is constituted by the fact that in some cases RESTR is possible, in others it is not:

(18) ?Hij is nog maar jong / Hij is al oud. (He's only young yet)

(19) *Het is nog maar vroeg/ Het is al laat.(It's only early yet)

But: nog maar klein (only small), nog maar weinig (only few). So, the polar Preds share all the properties of the binary ones,but in some cases, they share properties with the scalar Preds as well. Maybe it depends on how directly the idea of 'quantity' is present, but this matter must be left for further research. Henceforth, we will deal only with binary and scalar predicates.

2.4. Perspectivity and perspectivity particles.

2.4.1. We shall try to summarize in a diagram what has been said up to now. In the case of binary as well as scalar Sequences,we have

- the Sequence and its orientation: positive (- ⟶ +) or negative (+ ⟶ -); the order in which one 'goes through' the Seq is determined by the ordering of the set of RPs, and is represented in the diagram by the directionality from 'down' to 'up'.
- the transition (zone or) point p_x, being at the border of the 'positive' and the 'negative' zone.

Note, incidentally, that the (coming into)'mental' existence of a Sequence, and especially the relative location of p_x on it, is largely dependent on such pragmatic factors as presuppositions and speaker/hearer expectancy norms. (cf. VANDEWEGHE : 1978)

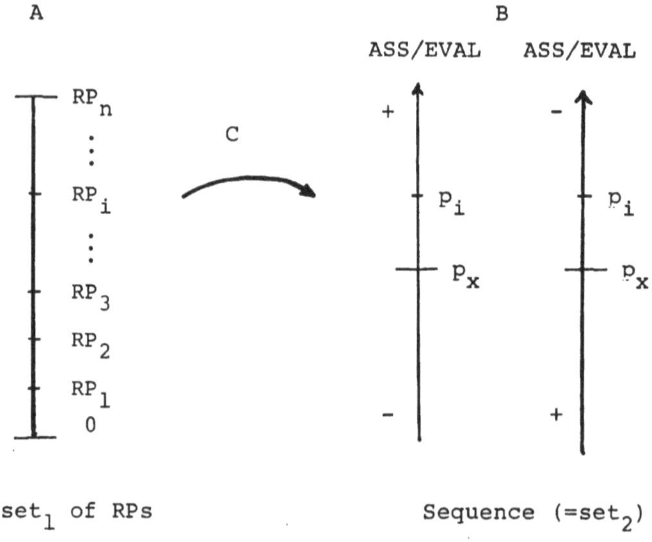

- Table 2 -

A is the set of RPs ($RP_1 < RP_2 < \ldots < RP_n$), all members of which may be connected with an (assertory or evaluative) value in Seq (B) through a functional relation C. The actual mapping from A into B in a given sentence is done for RP_i.

2.4.2. The role of the Perspectivity operator is now precisely to indicate that the SoA obtaining at RP_i is "put into perspective", which boils down to allotting p_i - the index of the value for RP_i - a <u>relative position</u> with respect to the transition point p_x.

The lexemes which are chosen in Dutch for the different situations are summarized in Table 3. Note that the subscript 'n','m' after <u>nog</u> or <u>meer</u> serves to differentiate these items from the 'positive polarity items' <u>al</u> and <u>nog</u> : 'n' stands for '<u>niet</u>' (not), 'm' for '<u>maar</u>'(only), i.e., the NEGATION- and the RESTRICTION-operator, with which they form a cluster.

Seq :	p_i prior to p_x		p_i past p_x	
	binary	scalar	binary	scalar
posit. orient.	NOG_n ($noch_n$, yet_n)	NOG_m(erst/ only)	AL/REEDS[9](schon/ already)	
negat. orient.	NOCH (noch/still)		$MEER_n$(n.mehr/ (not anymore/ no longer)	$MEER_m$/NOG_m (nur noch/ only (...left))

- Table 3 -

3. Al/nog: a suppletive set of perspectivity markers.

Leaving aside the case of $meer_{n,m}$ for the purpose of this paper, we are now able to bring out the systematic relationship between al and nog: they form a suppletive set of perspectivity markers.

3.1. Al appears to be used as a retrospective operator on the 'achievement'-line (= a positively oriented Seq): this is true for binary as well as for scalar Preds.

3.2. The lexeme nog appears in all prospective environments. It appears to be used whenever the zone in which p_i finds itself has a specification opposite to that of the Sequence (or, to be more precise, of the sequential orientation); this involves the foreseeability of an end-point to the SoA you find yourself in at RP_i. But note that p_x, the prospective end-point, need never be extensionalized in the later course of events! Only the possibility of there being an end-point at some point later on the Seq, is given[10].

The most striking exception to this seems to be the case of retrospective nog_m, combining with scalar Preds and RESTR: specification of zone and Seq seem to be the same! We may account for this, within the system outlined above, by assuming that for some speakers of Dutch, the relevant transition point for decreasing quantities is not situated between EVAL+ and EVAL-, but rather between the zone in which there is quantification, and the zone in

which there is not, i.e., the empty set: for those speakers, the relevant transition point is the zero-point.
The main virtue of the above-given account is again that it systematically relates the appearance of nog in binary sequences to that in scalar sequences. To give two contrasting examples:

(20) Hij is nog in het leger. (He is still in the army)

(21) Er zijn nog 30 mensen in de zaal. (There are 30 people left in de room)

As already suggested, it is fundamental to the understanding of (21) that the speaker situates his assertion against a background of DECREASING QUANTITY. In Dutch and German - unlike in English - this may in some cases be expressed by no other means than the adverbial nog resp. noch, although of course one can easily find means to express the same idea redundantly, as in

(22) Er zijn nog 30 mensen over. (German: übrig)

But the fact that nog is capable of expressing this all by itself, is significant. And as for the reading of (21) in which nog has the same meaning as in (20): the two readings each have to be assigned their own semantic representation, in which the sequence is defined differently:

(1°) 'normal', remainder-reading: the sequence is defined by the values + and - for EVAL (VAL_{scal}); the 'hyper'predicate which is in the scope of EVAL is: QUANTITY (or NUMBER).

(2°) continuative reading (as in 20): the sequence is defined by the value + and - for ASS ($Pred_{bin}$); in this case the $Pred_{bin}$ would be: 'there are 30 people in this room'.[11]

4. Concluding remarks.

We have given only a rough sketch of the semantics of the operators al, nog and $nog_{n,m}$ in order to bring out the main similarities between their use in different environments. Of course, we have covered only part of the range of use of these operators, especially of the operator nog: it has other uses which are not primarily perspectival like the 'additive' use in

(23) Geef me nog een kop koffie. (Give me another cup of coffee)

where German also uses noch. Similarities and differences between the different uses should form part of a large-scale investigation.[12]

The main purpose of this paper was an attempt to establish the notion of perspectivity, which should enable us to find our way in the sub-system of language into which these particles fit.

Notes

1 The notion is adapted from the one used in SHETTER (1966).

2 In this paper, I won't go into the pragmatic status of the different SoA-descriptions. One can find information about this in DOHERTY (1973), and especially in KÖNIG (1977).

3 For a discussion of German noch/schon (Eng. still/already) used in connection with PERIPHERAL values, see KÖNIG (1977 : 183 ff.).

4 For operators like the Tense operators and ASS, see: Seuren (1969).

5 ASS is only one - albeit the most important - of the performative operators. But I will only discuss cases with ASS, leaving QU(estion) and COMM(and) out of consideration.

6 A numerical Pred may of course function in a Sequence with a succession of ASS-values, but this only provides a sub-case of the binary Preds: see Section 3, esp. the discussion of example (21).

7 The distribution of the two clusters, maar...meer and nog maar used for decreasing quantity, seems, partly at least, to be dialectally determined: there is a central "wedge" from Brabant to Holland (nog maar) between the N-Eastern and S-Western areas of the Dutch-speaking area.

8 The fact that 'al jong'/'nog oud' (young already/still old) are very unlikely to appear is purely accidental, due to the semantics of these predicates and the 'model' in which they are interpreted, namely, the actual world we live in, where ageing is an irreversible process.(This need not be so in e.g. fairy tales).

9 Reeds in modern Dutch is a rather obsolete pendant of the much more frequently used al. Reeds is used mainly in formal speech.

10 KÖNIG (1977: 193) speaks about a "generalized conversational implicature" in Grice's sense of the word.

11 This is a slightly different interpretation than the one put forward in example (15a) of ROMBOUTS (1979)(in this volume).

12 Which I am undertaking at this moment, in preparation of my doctoral dissertation.

References

ABRAHAM, Werner (1977) : "Noch und schon als polare Satzfunktoren." SPRENGEL,K./BALD,W.-D./VIETHEN,H.W. (Hrsg.) : Semantik und Pragmatik. Akten des 11. Ling. Koll. Aachen 1976. Tübingen : Niemeyer : 3-19.

COMRIE, Bernard (1976) : Aspect. Cambridge: Cambr.Univ.Press.

DOHERTY, Monika (1973) : "'Noch' and 'schon' and their presuppositions." KIEFER,F./RUWET,N. (eds.) : Generative Grammar in Europe. Dordrecht : Reidel. 154-173.

KÖNIG, Ekkehard (1977) : "Temporal and non-temporal uses of 'noch' and 'schon' in German." Linguistics and Philosophy 1 : 173-198.

ROMBOUTS, Jos (1979) : "Dutch nog and al as degree particles." In this volume: VANDEWEGHE,Willy/VAN DE VELDE, Marc (Hrsg.) (1979) : Bedeutung, Sprechakte und Texte. Akten des 13.Ling. Koll. Gent 1978, Bd.2. Tübingen : Niemeyer.

SEUREN, Pieter A.M. (1969) : Operators and Nucleus. Cambridge : Cambr. Univ. Press.

SHETTER, William Z. (1966) : "The meaning of German noch." Language 42 : 42-66.

VANDEWEGHE, Willy (1978) : "Over de semantiek van het partikel al (= reeds)." De Nieuwe Taalgids 71 : 193-205.

DISKONTINUITÄT VERSUS DISAMBIGUIERUNG

Edda Weigand

1. Problemstellung

Disambiguierung wird vielfach definiert im Sinn von "Vereindeutigung für sich genommen mehrdeutiger sprachlicher Einheiten oder Relationen in einem weiteren Kontext" (WELTE 1974: 112 nach SPILLNER 1971: 247; ähnlich z.B. AGRICOLA 1970: 495f., LEECH 1970: 29 u. 269). Aufgrund der Vagheit des Begriffs "sprachliche Einheit" könnten somit auch Teile von Zeichen "disambiguiert" werden, in den Fällen, in denen für sich genommen mehrdeutige Ausdruckskomponenten eines diskontinuierlichen Zeichens durch andere Ausdruckskomponenten dieses Zeichens "vereindeutigt" werden. Danach wäre z.B. die Bildung des deutschen Passivs so zu beschreiben, daß das für sich genommen mehrdeutige werd- durch das Partizip Perfekt disambiguiert würde. Eine Definition von Disambiguierung in diesem Sinn läßt unberücksichtigt, daß sinnvollerweise nur solche sprachlichen Einheiten disambiguiert werden können, die als selbständige Zeichen für sich Bedeutung(en) tragen.

Damit aber ist das Problem nicht gelöst; denn es stellt sich sogleich die Frage nach der Abgrenzung eines Zeichens: Wann sind verschiedene Ausdruckseinheiten zu einem, als Ganzem eindeutigen, diskontinuierlichen Zeichen zusammenzunehmen, bzw. wann sind Ausdruckseinheiten für sich als mehrdeutige Zeichen anzusetzen, die durch weiteren Kontext disambiguiert werden? Diese Frage zielt auf ein Kriterium zur Unterscheidung von Diskontinuität und Disambiguierung. Während in den letzten Jahren zahlreiche Ambiguitätstests zur Unterscheidung von Ambiguität und Vagheit vorgeschlagen wurden (vgl. z.B. ZWICKY/SADOCK 1975), geht es hier also darum, einen Ambiguitätstest zu entwickeln, der es erlaubt, Fälle von Ambiguität/Disambiguierung und Diskontinuität zu differenzieren.

Im Rahmen einer größeren Arbeit zu den grammatischen Kategorien des Deutschen bin ich wiederholt auf diese Problemstellung eingegangen, ohne sie jedoch für sich geschlossen zu behandeln (vgl. WEIGAND 1978: 25ff. u. passim).

2. Kriterium zur Unterscheidung von Diskontinuität und Disambiguierung

Indem man klare Fälle von Diskontinuität und Disambiguierung einander gegenüberstellt, läßt sich ein Kriterium zu ihrer Unterscheidung gewinnen. Diskontinuität liegt vor, wenn zwei oder mehr in der Kette getrennte Elemente obligatorisch zum Ausdruck eines Zeichens gehören. Von Ambiguität eines Zeichens und Disambiguierung kann man dagegen nur sprechen, wenn die Elemente, die zur Eindeutigkeit beitragen, frei wählbar sind, also zufällige Kontexte bilden. In diesem Sinn stellen (1) und (2) klare Beispiele für Diskontinuität bzw. Disambiguierung dar:

(1) Sie haben ein Haus gebaut.
Perfekt Aktiv

(2) ein Satz der deutschen Sprache - ein Satz einer Sinfonie - ein Satz Briefmarken - ein Satz nach vorne

Mit diesem Kriterium der Klassifizierung des Kontexts als "frei wählbar" versus "obligatorisch" sind jedoch nicht alle Kontexte zu erfassen; vielmehr ist ein dritter, schwierigerer Kontexttyp anzusetzen, der die Auswahl unter verschiedenen Kombinationsmöglichkeiten erlaubt, die im Rahmen eines im Sprachsystem festgelegten Regelwerks bleiben. Diese "regelhaften" Kontexte sind somit weder frei wählbar, zufällig, noch obligatorisch in dem Sinn, daß das Vorkommen eines Elements obligatorisch mit einem bestimmten anderen gekoppelt sein muß. Da das Regelwerk jedoch die verschiedenen Kombinationsmöglichkeiten als Ganze obligatorisch vorschreibt und die Auswahl genau angebbaren Bedingungen unterliegt, sind regelhafte Kontexte in der Frage "Diskontinuität versus Disambiguierung" wie obligatorische zu behandeln. Ein Beispiel wird im nächsten Kapitel diskutiert.

3. Beispiele

In einer an Flexion und Kongruenz so reichen Sprache wie dem Deutschen ist die Frage "Diskontinuität oder Disambiguierung?" von grundlegender Bedeutung für die grammatische Beschreibung. An einigen Beispielen zur Kongruenz und zur Semantik grammatischer Kategorien soll gezeigt werden, wie dieses Beschreibungsproblem mit Hilfe des vorgeschlagenen Tests begründet im Sinn der einen oder anderen Darstellungsart entschieden werden kann.

3.1. Substantivflexion

Beschreibungen der deutschen Substantivflexion differieren u.a. in der Frage, ob die Substantivflexive für sich selbständige Zeichen (so z.B. RETTIG 1972, WEBER 1973) oder erst zusammen mit den kongruierenden Determinansflexiven diskontinuierliche Zeichen für den Kategorienverband Kasus/Numerus(/Genus) bilden (so z.B. WERNER 1975a: 49, AUGST 1975: 5ff.).

Die Frage der Trennbarkeit von Numerus und Kasus kann hier nicht erörtert werden (vgl. dazu z.B. RETTIG 1972: 16ff.). Einen Überblick über verschiedene Darstellungen des Systems der deutschen Substantivflexion geben RETTIG (1972: 41ff.) und BETTELHÄUSER (1976: 53ff.).

Betrachtet man die Substantivflexive als eigene Zeichen, so sind zahlreiche ambige Zeichen, darunter Zero-Elemente, anzusetzen, die großenteils durch die systematische Determinansflexion disambiguiert werden. Demgegenüber ergeben sich bei einer diskontinuierlichen Darstellung weit weniger ambige Zeichen, da auch bei homophonen Einzelelementen die Kombination aus phonologisch meist distinkten Komponenten vielfach eindeutig ist. So bildet z.B. in der NP <u>das Kind</u> das Zero-Element der Substantivflexion in Kombination mit dem Determinansflexiv ein eindeutiges diskontinuierliches Zeichen für den Kategorienverband Kasus/Numerus/Genus, während man bei einer nicht-diskontinuierlichen Darstellung von einem ambigen Nullzeichen sprechen müßte, das durch das Determinansflexiv disambiguiert wird.

Dieses Beschreibungsproblem läßt sich nur klären, indem man die Rolle des Determinans beim Bau von NPs im Deutschen klärt. Weder ist ein Determinans für ein appellativisches Nomen obligatorischer Kontext in dem Sinn, daß eine solche NP immer ein Determinans enthalten müßte, noch ist es frei wählbarer Kontext des Appellativums; sondern das Vorkommen eines Determinans ist obligatorisch an die Wahl eines bestimmten Nomens gebunden. Enthält die NP ein Nomen mit dem Merkmal [+zählbar], so ist es obligatorisch mit einem Determinans zu verbinden, allerdings mit der Besonderheit, daß der unbestimmte Artikel im Plural durch ein Zero-Element ausgedrückt wird. Enthält die NP ein Nomen mit dem Merkmal [-zählbar], so kann ein Determinans fehlen. Das heißt, die Regeln der deutschen Sprache für den Bau von NPs legen fest, wann eine NP ein Determinans enthalten muß bzw. wann es fehlen darf. Ein Determinans ist somit regelhafter Kontext des appellativischen Nomens, und für die Beschreibung der deutschen Substantivflexion ist folglich eine diskontinuierliche Darstellung zugrunde zu legen.

3.2. Verbflexion

Aufgrund der Kongruenz des finiten Verbums mit der Subjekts-NP stellt sich bei der Verbflexion im Deutschen ein ähnliches Beschreibungsproblem wie bei der Substantivflexion: Entweder man betrachtet - wie in der Regel üblich - die Verbflexive für sich als eigenständige Zeichen für den Kategorienverband Person/Numerus, die bei Ambiguität meist durch Pronomina bzw. Numerus(/Kasus/Genus)-Flexive der Subjekts-NP disambiguiert werden (vgl. z.B. AGRICOLA 1968: 35), oder man geht von diskontinuierlichen Zeichen aus, die als Ausdruckskomponenten die Verbflexive und Pronomina bzw. Numerusflexive der Subjekts-NP umfassen und durch die Kombination meist eindeutig sind (vgl. z.B. FOURQUET 1970: 21). So müßten z.B. in sie spielt nach der einen Darstellung zwei ambige Zeichen, das Pronomen sie und das Flexiv -t, angesetzt werden, die sich gegenseitig disambiguieren; nach der anderen Darstellung ergibt sich durch die Kombination ein eindeutiges diskontinuierliches Zeichen.

War bei der Beschreibung der Substantivflexion die Entscheidung "Diskontinuität versus Disambiguierung" dadurch erschwert, daß das Determinans regelhafter, nicht obligatorischer Kontext des appellativischen Nomens ist, so liegt hier unter Bezug auf das Testkriterium eindeutig Diskontinuität vor, da das finite Verbum im Deutschen obligatorisch mit einer Subjekts-NP zu verbinden ist, durch die die Kategorien Person und Numerus bestimmt werden. Allein bei Imperativsätzen kann das Verbum für sich bereits einen Satz bilden. Allerdings ergibt sich für eine diskontinuierliche Beschreibung der Verbflexion die Komplikation, daß die Ausdruckskomponenten des diskontinuierlichen Zeichens nur nach teilgleichen Kategorienverbänden zu klassifizieren sind So ist in unserem Beispiel sie spielt das Flexiv -t Ausdruckskomponente für den Kategorienverband Person/Numerus, während das

Pronomen sie zusätzlich die Kategorien Kasus und Genus enthält.

3.3. Kasus

Untersuchungen zur Semantik der Oberflächenkasus arbeiten vielfach mit der Annahme, daß Oberflächenkasus ambig sind. So hatte bereits die traditionelle Grammatik versucht, die verschiedenen Verwendungsweisen, also die Ambiguität der Oberflächenkasus aufzuzeigen und gegeneinander abzugrenzen. Nach FILLMOREs Kasustheorie werden Zuordnungsbeziehungen von Oberflächenkasus und Tiefenkasus angenommen derart, daß einem Oberflächenkasus mehrere Tiefenkasus entsprechen (ebenso z.B. HALLIDAY 1970: 148):

(3) Der Abiturient kämpft um einen Studienplatz.
Nominativ <=> Agentiv

(4) Der Gewinner bekommt einen VW.
Nominativ <=> Benefaktiv

(5) Der aufmerksame Leser merkt die Übertreibung.
Nominativ <=> Experiencer

<=> bezeichnet gegenseitige Zuordnung
Die Tiefenkasus Agentiv, Benefaktiv und Experiencer sind aus FILLMORE (1968 u. 1971) übernommen.

Mit dem Terminus Kasus beziehe ich mich, wie in der traditionellen Grammatik üblich, im Unterschied zu Fillmore, auf die Kasus der Oberfläche, also auf Klassifikationen der Kasus(/Numerus/Genus)-Flexive.

Auch ohne Annahme einheitlicher Kasusbedeutungen im Sinn von HJELMSLEV (1935) und JAKOBSON (1936) läßt sich u.a. mit dem Kriterium zur Unterscheidung von Diskontinuität und Disambiguierung begründen, daß man nicht sinnvoll von Ambiguität der Oberflächenkasus sprechen kann. Dazu ist es allerdings notwendig, zwei Probleme zu klären, die von Fillmores Kasustheorie nicht befriedigend gelöst wurden: die Frage der Definition von Tiefenkasus und die Frage nach der Funktion der Oberflächenkasus.

Einen wesentlichen Fortschritt in der Kasusdiskussion brachte die Erkenntnis Fillmores, daß die Zuordnung von Oberflächenkasus und Tiefenkasus nur über das jeweilige Verbum möglich ist. Nicht nur die Oberflächenkasus, sondern auch die ihnen zugrunde liegenden semantischen Relationen der Tiefenkasus seien durch das Verbum determiniert. Trotzdem kann Fillmore aufgrund seines lexikalistischen Standpunkts die Zuordnung einzelner Tiefenkasus zu einem bestimmten Verbum nicht explizieren, sondern muß zugeben, daß seine Entscheidungen intuitiv bleiben (vgl. FILLMORE 1971: 262). Wenn man aber davon ausgeht, daß die semantischen Relationen der Tiefenkasus im Verbum angelegt sind, so ist es nur naheliegend, auch den nächsten Schritt zu vollziehen und diese semantischen Relationen durch Dekomposition der Verben zu explizieren. Die einzelnen Tiefenkasus sind dann aufgrund ihrer Abhängigkeit von bestimmten atomaren Prädikaten zu definieren (zu einer Konzeption der Tiefenkasusstruktur, die von atomaren Prädikaten ausgeht, vgl. z.B. TEN CATE 1976).

Entscheidend für das Verständnis der Zuordnungsbeziehungen zwischen Oberflächenkasus und Tiefenkasus ist die zweite zu klärende Frage nach der Funktion der Oberflächenkasus. Wenn die Tiefenkasus als Teile der Verbsemantik aus den Oberflächenkasus allein nicht zu entnehmen sind, so kann ihnen - wie WERNER (1975b: 449f.) richtig erkannt hat - nur die Funktion von "Referenzindizes" zukommen, die die NPs der Oberfläche den im Verbum immanent enthaltenen und mit bestimmten Tiefenkasusrelationen verbundenen Argumenten zuordnen.

Vor dem Hintergrund einer so modifizierten Kasustheorie zeigt sich wiederum, daß die Frage der Ambiguität von der gewählten Sprachbeschreibung abhängig ist. Denn nun entsprechen einander nicht mehr Tiefenkasus und Oberflächenkasus, sondern den Tiefenkasus als Teilen der Verbsemantik sind komplexe Ausdrucksstrukturen zugeordnet, die diskontinuierlich Kasusflexive und das jeweilige Verbum umfassen. Bei gleichen Oberflächenkasus, aber jeweils verschiedenem Verbum, wie in den Beispielen (3)-(5), ist somit die Frage der Ambiguität überhaupt nicht zu stellen, sondern hier sind die Tiefenkasusrelationen zwischen den NPs und dem Verbum jeweils eindeutig den unterschiedlichen Ausdrucksstrukturen zu entnehmen.

Dagegen ist die Frage der Ambiguität der Oberflächenkasus zu diskutieren in den Fällen, in denen der gleiche Oberflächenkasus beim gleichen Verbum auf verschiedene Tiefenkasus zurückzuführen ist. Dazu ein aus FILLMORE (1968) bekannter Beispieltyp:

(6) Der Junge zerschlägt das Fenster mit einem Stein.

(7) Der Junge zerschlägt das Fenster.
Nominativ <=> Agentiv

(8) Der Stein zerschlägt das Fenster.
Nominativ <=> Instrumental

Nach neueren Vorschlägen zur semantischen Dekomposition von Handlungsverben und zur Ableitung der Tiefenkasus von atomaren Prädikaten lassen sich für Satz (7) und (8) Tiefenkasusstrukturen zugrunde legen, wie sie in Figur 1 und 2 dargestellt sind (vgl. DOWTY 1972, ROSS 1972, HUNDSNURSCHER 1975, TEN CATE 1976). Auf Einzelheiten der Darstellung braucht hier nicht näher eingegangen zu werden (einige Hinweise s.u.), da es in diesem Zusammenhang nur darauf ankommt, Tiefenkasus als Teile der Verbsemantik explizit zu machen.

(Fig. 1, zu Satz (7))

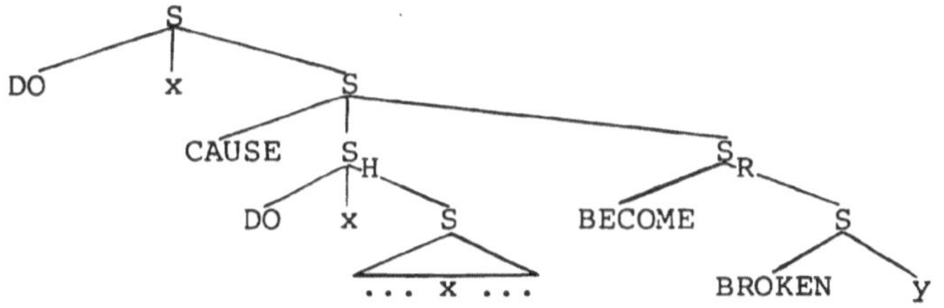

S_H: Handlungsbeschreibung, S_R: Resultatsbeschreibung

(Fig. 2, zu Satz (8))

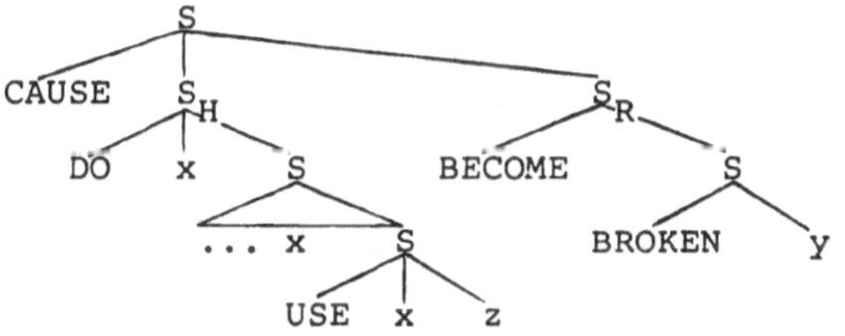

Die Darstellung in Figur 1 und 2 beruht im wesentlichen auf folgenden Grundannahmen: Agentive sind durch das begleitende Prädikat DO, Instrumentale durch USE definiert. Für Satz 7 (Fig. 1) wurde ein intentionaler Zusammenhang zwischen auslösender Handlung und Resultat angenommen, der durch das übergeordnete Prädikat DO, in das CAUSE eingebettet ist, repräsentiert werden kann (vgl. HUNDSNURSCHER 1975: 179). Der zugrunde liegende Agentiv in Figur 2 wird bei der Ableitung des Satzes (8) getilgt.

Auch in einem Phrasenstrukturmodell sind Tiefenkasus, die von atomaren Prädikaten abgeleitet werden, von diesen abhängig und können automatisch aus der Struktur abgelesen werden. Die Explizierung dieser Abhängigkeit in einer Dependenzdarstellung ist bei so definierten Tiefenkasus nur mehr eine Notationsfrage.

In beide Tiefenkasusstrukturen kann das Lexem zerschlag- eingesetzt werden, da dieses alternativ und fakultativ verschiedene Tiefenkasusrahmen zu sich nehmen kann. Die transformationellen Eigenschaften dieses Lexems legen die Zuordnung von Tiefenkasus und Oberflächenkasus in Abhängigkeit von dem jeweils gegebenen Tiefenkasusrahmen fest:

(Fig. 3)

<u>zerschlag-</u>

$$\left[\begin{array}{lll} \text{Ag} & <=> & \text{Nom} \\ \text{Instr} & <=> & \left\{\begin{array}{l} \text{Präp / [\_\_Ag+Instr+Objv]} \\ \text{Nom / [\_\_Instr+Objv]} \end{array}\right\} \\ \text{Objv} & <=> & \text{Akk} \end{array}\right]$$

Ag = Agentiv, Instr = Instrumental, Objv = Objektiv
Akk = Akkusativ, Nom = Nominativ, Präp = Präposition

Obgleich dem Nominativ hier beim gleichen Verbum einmal der Tiefenkasus Agentiv, einmal Instrumental zuzuordnen ist, kann diese Erscheinung nicht als Ambiguität des Nominativs beschrieben werden; denn Oberflächenkasus haben nicht für sich Bedeutung(en), sondern verweisen auf im Verbum festgelegte Bedeutungen. In einigen Fällen ergibt sich durch den Oberflächenkasus allein noch keine eindeutige Verweisung. Erst durch die Kombination mit einem semantischen Merkmal der NP kann teilweise Eindeutigkeit erzielt werden. Denn für das abstrakte Prädikat DO gilt, daß es Intentionalität impliziert (vgl. z.B. HUNDSNURSCHER 1975: 179, DOWTY 1972: 66, ROSS 1972: 105). Folglich kann nur eine NP mit dem Merkmal [+menschlich] Agentiv sein. In Satz (8) ist daher durch die Kombination des Oberflächenkasus Nominativ mit dem semantischen Merkmal [-belebt] eine eindeutige "Referenz" auf den Tiefenkasus Instrumental möglich. Im Unterschied zum Agentiv läßt sich jedoch der Instrumental nicht eindeutig einem semantischen Merkmal zuordnen. Satz (7) ist daher eigentlich ambig, wie folgender Kontext zeigt:

(9) Bei ihrem grausamen Spiel verwendeten sie den Zweijährigen als Ball. Der Junge flog durch das Zimmer, zerschlug das Fenster ...

Allein aufgrund einer "restriction on the world" liegt in Satz (7) ohne weiteren Kontext der Gedanke an ein menschliches Werkzeug fern (vgl. PLEINES 1975: 65, auch 1976: 88f.).

Auch durch das Kriterium zur Unterscheidung von Diskontinuität und Disambiguierung läßt sich begründen, daß Beispiele wie (7) und (8) nicht als Ambiguität, sondern als nicht eindeutige "Referenz" der Oberflächenkasus zu beschreiben sind. Wollte man von Ambiguität der Oberflächenkasus sprechen, so müßte die Disambiguierung durch frei wählbaren Kontext erfolgen. Das semantische Merkmal, das in Satz (8) eine eindeutige Zuordnung erlaubt bzw. in Satz (7) wahrscheinlich macht, ist jedoch Merkmal eines für die Kasuskategorie obligatorischen Kontexts.

3.4. Tempus

Charakteristisch für die semantische Analyse der Tempuskategorie ist die Abhängigkeit der Tempusbedeutung von verbinhärenten Merkmalen der Aktionsart (vgl. auch SALTVEIT 1970: 151). So kann z.B. das Präsens bei nicht eindeutig durativen Verben im allgemeinen als ambig, mit Gegenwarts- und Zukunftsbedeutung, beschrieben werden (vgl. (10a-c)); bei eindeutig durativen Verben hat es dagegen in der Regel nur Gegenwartsbedeutung (vgl. (11)):

(10a) Er holt seinen Bruder vom Bahnhof ab.

(10b) Er holt jetzt gerade seinen Bruder vom Bahnhof ab.

(10c) Er holt nachher seinen Bruder vom Bahnhof ab.

(11) Er schläft.

Nach WUNDERLICH (1970: 121) ist diese Erscheinung so zu erklären, daß die Tempuskategorie isoliert genommen ambig ist und erst durch inhärente Verbeigenschaften disambiguiert wird. Ähnlich bezieht z.B. auch SALTVEIT (1970: 144) Ambiguität offenbar auf die isolierte Kategorie, wenn er feststellt, daß sich die verschiedenen Bedeutungen einer "homonymen" Kategorie jeweils eindeutig bestimmten Verbklassen zuordnen ließen. Da jedoch das jeweilige Verblexem für die Tempuskategorie obligatorischen Kontext darstellt, ist es wenig sinnvoll, von Ambiguität der isolierten Tempuskategorie zu sprechen. Ambiguität einer Tempuskategorie ist erst dann zu erwägen, wenn sie beim gleichen Verbum mehrdeutig ist und durch frei wählbaren, zusätzlichen Kontext disambiguiert wird (wie in (10b) und (10c)).

4. Schlußbemerkung

Mit der Klassifikation des Kontexts als "obligatorisch/regelhaft" versus "frei wählbar" ist ein Kriterium gefunden, das eine begründete Entscheidung in der Frage "Diskontinuität versus Disambiguierung" erlaubt. Die Beispiele zur grammatischen Beschreibung des Deutschen haben ergeben, daß unter Bezug auf dieses Kriterium weit weniger Fälle von Ambiguität anzusetzen sind, als gemeinhin angenommen. Dieses Ergebnis trifft sich mit Untersuchungen zur Ambiguität im lexikalischen Bereich, nach denen viele Fälle, die bisher als ambig galten, als Fälle von Vagheit zu beschreiben sind (vgl. z.B. WEYDT 1975: 243). Indem auf diese Weise sowohl im grammatischen wie im lexikalischen Bereich das Vorkommen von Ambiguität eingeschränkt wird, nähert sich die Sprachbeschreibung nur der Alltagserfahrung, daß Ambiguität trotz vielfach homophoner Einzelelemente in der sich in Texten vollziehenden Sprachverwendung ein Ausnahmefall bleibt (vgl. auch AGRICOLA 1968: 13f., 1970: 493). Außerdem wird deutlich, daß Ambiguität in der natürlichen Sprache nichts absolut Vorgegebenes ist, sondern wesentlich von der gewählten Sprachbeschreibung abhängt (vgl. auch LYONS 1977: 401).

Literatur

AGRICOLA, Erhard (1968): Syntaktische Mehrdeutigkeit (Polysyntaktizität) bei der Analyse des Deutschen und des Englischen. Berlin: Akademie-Vlg.

--- (1970): "Relationen und Systeme der Disambiguierung". GRAUR, A. et al. (eds.) (1970): Actes du Xe congrès international des linguistes. Bucarest, 28 Août - 2 Septembre 1967. II. Bucarest: Éditions de l'Academie de la Rêpublique socialiste de Roumanie: 493-498.

AUGST, Gerhard (1975): Untersuchungen zum Morpheminventar der deutschen Gegenwartssprache. Tübingen: Narr.

BETTELHÄUSER, Hans-Jörg (1976): Studien zur Substantivflexion der deutschen Gegenwartssprache. Heidelberg: Winter.

DOWTY, David R. (1972): "On the syntax and semantics of the atomic predicate CAUSE". Papers from the 8th regional meeting of the Chicago Linguistic Society: 62-74.

FILLMORE, Charles J. (1968): "The case for case". BACH, Emmon / HARMS, Robert T. (eds.) (1968): Universals in linguistic theory. London etc.: Holt, Rinehart and Winston: 1-88. - Übers.: "Plädoyer für Kasus". ABRAHAM, Werner (ed.) (1971): Kasustheorie. Frankfurt/M.: Athenäum: 1-118.

--- (1971): "Some problems for case grammar". Working Papers in Linguistics. Ohio State University 10: 245-265.

FOURQUET, Jean (1970): Prolegomena zu einer deutschen Grammatik. Düsseldorf: Schwann.

HALLIDAY, M. A. K. (1970): "Language structure and language function". LYONS, John (ed.) (1970): New horizons in linguistics. Middlesex: Penguin Books: 140-165.

HJELMSLEV, Louis (1935): La catégorie des cas. Étude de grammaire générale. Aarhus: Universitets-forlaget.

HUNDSNURSCHER, Franz (1975): "Überlegungen zur Generativen Semantik". WERNER, Otmar / FRITZ, Gerd (eds.) (1975): Deutsch als Fremdsprache und neuere Linguistik. Referate eines Fortbildungskurses in Mannheim. München: Hueber: 166-183.

JAKOBSON, Roman (1936): "Beitrag zur allgemeinen Kasuslehre. Gesamtbedeutungen der russischen Kasus". Travaux du Cercle Linguistique de Prague 6: 240-288.

LEECH, Geoffrey N. (1970): Towards a semantic description of English. London: Longman.

LYONS, John (1977): Semantics. I/II. Cambridge etc.: Cambridge University Press.

PLEINES, Jochen (1975): "Kausale Relationen und Intentionalität". EHRICH, Veronika / FINKE, Peter (eds.) (1975): Beiträge zur Grammatik und Pragmatik. Kronberg/Ts.: Scriptor: 55-70.

--- (1976): Handlung - Kausalität - Intention. Probleme der Beschreibung semantischer Relationen. Tübingen: Narr.

RETTIG, Wolfgang (1972): Sprachsystem und Sprachnorm in der deutschen Substantivflexion. Tübingen: Narr.

ROSS, John R. (1972): "Act". DAVIDSON, Donald / HARMAN, Gilbert (eds.) (1972): Semantics of natural language. Dordrecht-Holland: Reidel: 70-126.

SALTVEIT, Laurits (1970): "Synonymik und Homonymie im deutschen Tempussystem". MOSER, Hugo (ed.) (1970): Studien zur Syntax des heutigen Deutsch. Paul Grebe zum 60. Geburtstag. Düsseldorf: Schwann: 137-153.

SPILLNER, Bernd (1971): "Polysemie und Disambiguierung". STECHOW, Arnim von (ed.) (1971): Beiträge zur generativen Grammatik. Referate des 5. Linguistischen Kolloquiums. Regensburg 1970. Braunschweig: Vieweg: 247-257.

TEN CATE, Abraham P. (1976): "Kausalstrukturen in der Kasusgrammatik". KERN, Rudolf (ed.) (1976): Löwen und Sprachtiger. Akten des 8. Linguistischen Kolloquiums. Löwen 1973. Löwen: Éditions Peeters: 191-212.

WEBER, Heinrich (1973): "Synpleremik II: Morphemik". ALTHAUS, Hans P. / HENNE, Helmut / WIEGAND, Herbert E. (eds.) (1973): Lexikon der Germanistischen Linguistik. Tübingen: Niemeyer: 163-175.

WEIGAND, Edda (1978): Die Zuordnung von Ausdruck und Inhalt bei den grammatischen Kategorien des Deutschen. Tübingen: Niemeyer.

WELTE, Werner (1974): moderne linguistik: terminologie / bibliographie. ein handbuch und nachschlagewerk auf der basis der generativ-transformationellen sprachtheorie. I/II. München: Hueber.

WERNER, Otmar (1975a): "Zum Genus im Deutschen". Deutsche Sprache 3: 35-58.

--- (1975b): "Zum Problem der Wortarten". ENGEL, Ulrich / GREBE, Paul (eds.) (1975): Sprachgebrauch und Sprachsystem. Festschrift für Hugo Moser zum 65. Geburtstag. II. Düsseldorf: Schwann: 432-471.

WEYDT, Harald (1975): "Mehrfachverständnis sprachlicher und nichtsprachlicher Zeichen. Systematische Analyse der Ambiguität". DRACHMAN, Gaberell (ed.) (1975): Akten der 1. Salzburger Frühlingstagung für Linguistik. Salzburg vom 24. bis 25. Mai 1974. Tübingen: Narr: 239-254.

WUNDERLICH, Dieter (1970): Tempus und Zeitreferenz im Deutschen. München: Hueber.

ZWICKY, Arnold / SADOCK, Jerrold M. (1975): "Ambiguity tests and how to fail them". KIMBALL, John P. (ed.) (1975): Syntax and semantics. IV. New York etc.: Academic Press: 1-36.

A PROPOS DE LA GRAVITATION EN LANGUE ET SON UTILISATION PRAGMATIQUE. LA MONTEE DE LA NEGATION ET AUTRES PHENOMENES EN FRANCAIS.

Dominique Willems

1. Nous tenterons dans la présente communication de formuler et de vérifier une hypothèse générale qui permettrait, d'une part, de rendre compte de certains faits difficiles à expliquer dans le cadre d'une théorie syntaxique ou sémantique standard, d'autre part de regrouperdes phénomènes épars, mais présentant néanmoins des caractéristiques communes, touchant la négation, l'interrogation, les temps, les constructions impersonnelles. L'hypothèse, si elle s'avère exacte, illustrerait par ailleurs le lien étroit entre syntaxe, sémantique et pragmatique.

L'hypothèse serait que, dans de nombreux cas, une distanciation syntaxique entraîne une atténuation et une ambiguïté sémantique. Le phénomène d'atténuation est ce que, à la suite de L. HORN (1975), nous appelons la loi de la gravitation en langue. Celle-ci, portant principalement sur deux types d'actes de parole, l'expression d'une <u>opinion</u> et celle d'une <u>volonté</u>, met en cause la relation locuteur-destinataire et est exploitée, au niveau pragmatique, dans la stratégie de la politesse. Nous voudrions illustrer ceci au moyen d'exemples touchant essentiellement la négation et l'interrogation.

1.1. Malgré le fait que la <u>montée de la négation</u> soulève de nombreux problèmes d'ordre syntaxique, lexical et sémantique, et a donné lieu à une bibliographie étendue, tant de la part des logiciens que des linguistes, surtout en ce qui concerne le statut transformationnel ou non de cette montée, il est certain que dans certains cas il y a montée de la négation de la subordonnée à la principale, sans raisons syntaxiques ou sémantiques apparentes, si ce n'est l'éloignement de la négation du terme qu'il affecte. Il s'agit d'un phénomène très général, qui dépasse largement le cadre des langues indo-européennes, l'ensemble des langues présentant ici une remarquable unité, même si les inventaires de verbes ne se superposent par toujours (HORN 1975: 285).

Cet éloignement s'opère, sans changement sémantique important, à l'intérieur de deux classes de verbes : les verbes d'opinion : croire, sembler, paraître, supposer (...) et les verbes de volonté : vouloir, désirer, souhaiter, falloir (...). Ainsi, les exemples (1) à (3), mettant en jeu les verbes croire et désirer, présentent à chaque fois deux variantes (a) et (b), la négation portant sur la subordonnée pouvant se trouver à la place normalement attendue ou "monter" jusqu'au verbe principal :

(1) (a) Je crois que Jean n' est p a s venu et je peux deviner pourquoi.
(b) Je n e crois p a s que Jean soit venu et je peux deviner pourquoi.

(2) (a) Je crois que Jean n' a p a s dormi de la nuit.
(b) Je n e crois p a s que Jean ait dormi de la nuit.

(3) (a) Je désire qu'elle n' ait p a s le moindre doute à ce sujet.
(b) Je n e désire p a s qu'elle ait le moindre doute à ce sujet.

Notons que la montée entraîne, pour les verbes d'opinion, l'emploi du subjonctif, signe explicite que le verbe de la subordonnée a été privé d'une information essentielle (DE POERCK 1966: 61). Ce point ne nous concernera toutefois pas dans la présente communication. Dans l'exemple (1), la montée est rendue évidente par l'adjonction de la proposition et je peux deviner pourquoi , qui renvoie sans ambiguïté au fait que Jean n'est pas venu. Dans l'exemple (2), l'agrammaticalité de l'expression positive *avoir dormi de la nuit rend l'interprétation par la montée nécessaire.

L'éloignement de la négation a pour conséquence une ambiguïté logique, mise en lumière et déplorée par bon nombre de logiciens, la négation pouvant s'interpréter comme portant sur le verbe principal, sans montée, ou sur le verbe de la subordonnée, dans le cas d'une montée.

L'éloignement entraîne d'autre part une atténuation, non seulement de l'effet négatif (POUTSMA, BOLINGER), mais aussi de l'acte de parole en général (l'opinion ou la volonté), ce qui explique sans doute qu'on préfère généralement la variante avec montée que sans. Ainsi, pour l'exemple (4),

(4) (a) Je veux que vous n e sortiez p a s
(b) Je n e veux p a s que vous sortiez.

P. Martinon nous dit : "A la phrase Je veux que vous sortiez correspond uniquement Je ne veux pas que vous sortiez. Cette tournure a en effet accaparé le sens de Je veux que vous ne sortiez pas, qui ne se dit pas." (MARTINON 1927: 536). Cette dernière phrase est en effet sentie comme "trop impérative" par les locuteurs interrogés sur ce point par B. DE CORNULIER (1973: 43). Notons d'ailleurs que cette différence d'acceptabilité disparaît lorsque la phrase sans montée est atténuée d'une autre façon, comme par l'emploi d'un conditionnel : Je voudrais que vous ne sortiez pas.

L'atténuation sémantique obtenue par l'éloignement de la négation peut présenter divers degrés. Dans l'exemple (5),

(5) (a) Il faut que ce mariage n e se fasse p a s.
(b) Il n e faut p a s que ce mariage se fasse.

la différence sémantique, toujours dans le sens d'une atténuation dans (b), est plus importante que dans l'exemple (4). Il est d'ailleurs remarquable que les deux variantes de l'exemple (5) de H. de Montherlant, cité par GAATONE (1971: 54), figurent à une page d'intervalle, la version (a) figurant en dernier lieu et reflétant la conviction accrue du héros, résultant d'une série de réflexions au sujet du mariage. Dans (6) (a) et (b), la différence sémantique est telle qu'on ne peut plus parler de montée (fait symbolisé par l'astérisque), le verbe savoir changeant de sens :

(6) (a) Je sais que Jean n' est p a s venu.
(b) *Je n e sais p a s que Jean soit (est) venu.
(c) Je n e sache p a s que Jean soit venu.
(d) Jean n' est p a s venu que je sache.

Dans (a), savoir indique une certitude, dans (b) un doute. Notons toutefois que le changement sémantique est également dans le sens d'un affaiblissement lexical du verbe. Dans les cas (c) et (d), (c) peut être décrit comme une montée par rapport à (d). Le sens du verbe savoir est alors proche de celui de croire. On note donc qu'à l'intérieur des champs des verbes d'opinion et de volonté, seuls les verbes relativement "faibles" admettent la montée sans changement sémantique important, pour les verbes "forts", l'éloignement de la négation s'accompagnant d'un affaiblissement sémantique plus important. Dans le champ sémantique de l'opinion, savoir exprime une certitude plus grande que croire. Lorsqu'il admet la montée, son sens s'affaiblit jusqu'à devenir un synonyme du verbe croire. Il en va de même dans le champ de la volonté, où falloir est plus "fort" que vouloir. Si le verbe que la négation peut rendre ambigu, possède sans elle un sens fort, il prend avec elle un sens faible ou neutre (cf. aussi CORNULIER 1973). Plus fort le sens du verbe, plus grande sera l'influence atténuante de la montée, jusqu'à devenir impossible : on ne peut en effet atténuer une conviction (savoir). Inversement, les verbes véhiculant un sens très faible (être possible, se pouvoir) ne semblent pas admettre la montée, la négation donnant à ces verbes un sens beaucoup plus fort. Seules les valeurs intermédiaires admettent donc la montée de la négation. HORN (1975) aboutit à une conclusion similaire pour l'anglais et établit une échelle allant du sens faible au sens fort en démontrant que seules les "midscale values" admettent, sans changement sémantique, la montée négative. A l'intérieur des notions d'opinion et de volonté, ce ne sont donc que les valeurs moyennes, les sens faibles de vouloir et de croire, qui admettent la montée.

Ceci nous amène à l'étude des contraintes sur la montée de la négation. En dehors des contraintes lexicales, que nous venons de discuter, nous remarquons que pour un même verbe, certains contextes admettent la montée sans changement sémantique important, d'autres non. Les contraintes exercées sont syntaxiques d'une part, présuppositionnelles de l'autre. Bien qu'à première vue fort éloignées des contraintes lexicales, nous croyons qu'en réalité elles les rejoignent. La comparaison des exemples (4) et (7) semble indiquer que, pour les verbes de volonté, l'infinitif se prête moins facilement que la complétive à la montée :

(7) (a) Je veux n e p a s sortir.

(b) Je n e veux p a s sortir.

Ceci semble également être le cas pour l'anglais (HORN 1975: 286). Bien qu'il s'agisse d'une contrainte peu claire et non générale (elle ne semble pas s'appliquer au verbe désirer par exemple), nous nous demandons si elle n'est pas liée au fait que la construction infinitive implique l'identité de sujet entre le verbe principal et l'infinitif, la construction complétive permettant deux sujets différents. Seule la construction complétive pouvant de ce fait exprimer un ordre, qui met nécessairement en relation deux personnes, on peut se demander si la contrainte syntaxique n'est pas en réalité d'ordre pragmatique, l'atténuation étant, dans le domaine de la volonté, surtout de mise dans l'expression d'un ordre.

L'étude de l'exemple (8) fait encore apparaître une autre difficulté :

(8) (a) Je voudrais n e p a s être Dieu

(b) Je n e voudrais p a s être Dieu.

(b) ne peut être décrit comme équivalent de (a), les présuppositions, entraînées par l'utilisation du conditionnel, étant distinctes : (a) présuppose que je suis Dieu, (b) que je ne le suis pas. Cet exemple nous oriente vers une contrainte plus générale, désignée par certains (POUTSMA, BOLINGER, LAKOFF) sous le noms de "uncertainty principle" (principe d'incertitude): pour qu'on puisse parler de montée, le fait exprimé dans la subordonnée doit être présenté comme incertain et ne peut être impliqué par le verbe principal. Ceci rejoint ce qui a été dit à propos du verbe savoir. En effet, la montée de la négation, entraînant une atténuation de l'acte de parole, donc une incertitude quant à la vérité du complément, sera exclue dans le cas de verbes où le complément est présupposé. L'équivalent de la certitude dans le champ de la volonté est l'obligation absolue (être nécessaire).

Il nous semble que l'analyse de ces contraintes nous oriente vers une interprétation pragmatique de la montée de la négation: la raison en serait essentiellement l'atténuation de l'acte de parole. Dans le contexte situationnel, deux actes se prêtent particulièrement à cette opération : l'affirmation d'une opinion et celle d'une volonté (ou plus précisément d'un ordre). Ce fait expliquerait d'une part l'impossibilité de montée pour les verbes forts : il y a en effet contradiction à vouloir atténuer une conviction ou un ordre absolu, d'autre part la même impossibilité de montée pour les verbes très faibles : l'atténuation est ici inutile.

1.2. Un même phénomène d'atténuation sémantique par distanciation, toujours dans le domaine de la négation, nous est offert sur le plan morphologique. Ainsi, dans les couples

être i m possible /v/ n e p a s être possible

d é courager /v/ n e p a s encourager

la négation intégrée est plus forte que la négation syntaxique, qui présente, par rapport à la première, un éloignement du terme à nier. Ici aussi, la distanciation entraîne une atténuation sémantique correspondante. Le phénomène semble également lié à un ensemble particulier d'adjectifs. L'étude des contraintes reste toutefois à faire.

2. Si ce phénomène d'atténuation par distanciation a déjà été mis en lumière pour la négation, nous pensons aussi pouvoir l'étendre à d'autres domaines. La <u>question rhétorique</u> présente sur ce point beaucoup d'affinités avec la négation. Sans entrer dans le détail des propriétés syntaxiques et sémantiques particulières qui permettent de caractériser la question rhétorique (notons seulement le caractère particulier du rapport question-réponse, la question positive entraînant une réponse négative, l'inverse se produisant pour la question négative, et le caractère souvent obligatoire de l'inversion, l'interrogation par intonation étant exclue), nous constatons avec A. BORILLO (1977) que la question rhétorique est dans de nombreux cas une négation atténuée. Ainsi, dans l'exemple (9),

(9) (a) Crois-tu que ce soit le moment ? [R. N o n]

(b) Tu crois que c'est le moment.

(c) Je crois que ce n' est p a s le moment.

(d) Ce n' est p a s le moment.

la question (a) n'est pas la forme interrogative de la phrase (b), mais bien celle de (c), contenant une négation. La phrase (c) n'est elle-même en réalité qu'une variante atténuée de (d). Dans le cas de la question rhétorique, la distanciation de la négation est maximale, celle-ci étant reléguée à une phrase réponse sous-entendue. L'atténuation consécutive est elle aussi maximale et l'ambiguïté joue à plein. Le lien avec la phrase négative dans la structure profonde de ces questions ressort clairement des exemples (10) et (11), où la question porte sur une locution verbale n'existant que sous sa forme négative :

(10) (a) Est-il besoin de le répéter ? [R. N o n]
(b) Il n' est p a s besoin de le répéter.
(c) *Il est besoin de le répéter.

(11) (a) A-t-on idée de faire tant de bruit ?
(b) On n' a p a s idée de faire tant de bruit.
(c) *On a idée de faire tant de bruit.

En réalité, dans la question rhétorique, plusieurs phénomènes de distanciation entrent en jeu : en dehors de la distanciation de la négation, il y a également une distanciation du locuteur, le j e étant remplacé par un t u (ex. 9), et l'affirmation étant remplacée par une question. L'introduction d'autre part de verbes modaux sans autre fonction sémantique que d'atténuer l'affirmation tout en rendant la phrase plus longue, plus complexe, contribue elle aussi à la distanciation. Notons d'ailleurs qu'une variante du type (9) (c) n'est pas possible dans l'exemple (12), preuve supplémentaire du caractère périphérique du verbe croire :

(12) (a) Crois-tu que cela m'amuse d'attendre ?
R. [Bien sûr que n o n]
(b) Cela n e m'amuse p a s d'attendre.
(c) *Je crois que cela n e m'amuse p a s d'attendre.

Dans le domaine de la question rhétorique, l'atténuation sémantique et l'ambiguïté jouent à plein et à plusieurs niveaux. A. BORILLO (1977) note l'importance de la modalité dans ce type de question : il est particulièrement fréquent dans le champ des prescriptions négatives (interdiction atténuée) et des jugements.

Nous retrouvons ici les deux domaines lexicaux de la montée de la négation.

Avant de quitter le domaine de la question, il faut noter les emplois très fréquents de la forme interrogative, là où la logique demandrait une forme affirmative ou impérative. En réalité, l'affirmation par rapport à l'impératif, la question par rapport à l'affirmation, constituent des éloignements successifs en ce qui concerne l'engagement du locuteur vis-à-vis du destinataire, entraînant des atténuations successives.

3. La distanciation par la personne, relevée pour la question rhétorique (ex. 9), se retrouve également dans l'emploi fréquent, pour certains verbes, d'une construction impersonnelle (éloignement de la personne) en relation d'équivalence avec une première personne. C'est là une autre forme de phénomène d'atténuation par distanciation :

(13) (a) J e p r é f è r e que tu ne viennes pas.
(b) I l e s t p r é f é r a b l e que tu ne viennes pas.

Il serait d'ailleurs intéressant de vérifier si les verbes souhaiter, préférer, en tant que radicaux des adjectifs préférable, souhaitable, figurant dans la construction impersonnelle, n'ont pas subi un affaiblissement sémantique.

4. La distanciation peut aussi être temporelle. Notons en français l'utilisation fréquente du temps futur là où un présent est logiquement attendu :

(14) (a) Je te f a i s remarquer que Jean est parti.
(b) Je te f e r a i remarquer que Jean est parti.
(15) (a) Ne r e v i e n s pas là-dessus.
(b) Tu ne r e v i e n d r a s pas là-dessus.

Le futur dans (15) prend même la valeur d'un impératif.

5. Tous les phénomènes discutés ici touchent deux domaines sémantiques, d'ailleurs apparentés : l'affirmation d'une opinion et celle d'une volonté ou d'un ordre. Les limites lexicales et les caractéristiques sémantiques nous orientent vers une

interprétation pragmatique unique. Ce qui est en cause, c'est la relation locuteur-destinataire. Par la distanciation, qu'elle soit spatiale, temporelle ou personnelle, on obtient à chaque fois une atténuation qui va de pair avec une ambiguïté croissante.

Sur le plan pragmatique, nous constatons l'exploitation de la loi de gravitation dans la stratégie de la politesse. Si la distanciation syntaxique entraîne une atténuation sémantique, celle-ci, en contexte, n'est que de surface. Dans la stratégie de la politesse, à travers l'atténuation, c'est le message non distancié qui est visé. Ceci est particulièrement clair dans le cas de la montée de la négation, où la forme atténuée a, dans certains cas, entièrement éliminé et remplacé la forme forte. Dans l'exemple (16), qui pourrait d'ailleurs être complété par d'autres tournures, l'éloignement progressif de l'aspect impératif va de pair avec un degré croissant de politesse. Celle-ci utilise donc des procédés syntaxiques précis de distanciation en vue d'une atténuation du message à donner.

(16) (a) Ne reviens pas là-dessus.
(b) Tu ne reviendras pas là-dessus.
(c) Il est inutile de revenir là-dessus.
(d) Il n'est pas utile de revenir là-dessus.
(e) Est-ce utile de revenir là-dessus ?
(f) Crois-tu qu'il soit utile de revenir là-dessus ?
(g) Ne crois-tu pas qu'il est inutile de revenir là-dessus ?
(h) Je veux que tu saches qu'il n'est pas utile de revenir là-dessus.
etc.

Références

BORILLO, Andrée (1977): "La question rhétorique". Communication faite au Colloque de Linguistique générale et de linguistique française. Paris 1977.

CORNULIER, Bernard de (1973) : "Sur une règle de déplacement de négation". Le Français Moderne 41: 43-57.

GAATONE, Daniel (1971) : Etude descriptive du système de la négation en français contemporain. Genève: Droz.

HORN, Laurence R. (1975): "Neg-Raising Predicates : toward an Explanation". Papers from the 11th regional meeting of the Chicago Linguistic Society: 279-295.

JESPERSEN, Otto (1908): "Negation in English and Other Languages". Selected Writings of Otto Jespersen. London 1917 : Allen and Unwin.

LAKOFF, George (1970): "Pronominalization, Negation, and the Analysis of Adverbs". JACOBS and ROSENBAUM (eds.) : Readings in Transformational Grammar. Waltham: Ginn and Company : 145-165.

MARTINON, Philippe (1927): Comment on parle en français. Paris: Larousse.

POERCK, Guy De (1966): "Syntaxe transformationnelle : le mode dans la complétive". Mélanges de grammaire française offerts à M. Maurice Grevisse. Gembloux: Duculot : 57-68.

POUTSMA, H. (1928) : A Grammar of Late Modern English. Groningen: Noordhoff.

2. PRAGMATIK

I N S I S T I E R E N

Ernst Apeltauer

INSISTIEREN ist ein komplexer Sprechhandlungstyp, der dazu dient, gegen den Widerstand eines anderen Agenten etwas durchzusetzen. [1] Diese Bestimmung gilt aber auch für andere komplexe Sprechhandlungsmuster wie z.B. SCHMEICHELN oder ARGUMENTIEREN. Im Unterschied zu diesen zielt INSISTIEREN aber auf schnelle Problemlösung, entweder mit dem anderen Agenten oder gegen den anderen Agenten. Man kann jemanden ANFLEHEN, nachdem er eine BITTE abgeschlagen hat. Man kann ihm aber auch DROHEN. Zwischen diesen beiden extremen Möglichkeiten gibt es eine Reihe von Alternativen, die sich durch Nachdruck voneinander unterscheiden. Im Folgenden sollen die für INSISTIEREN typischen Realisierungsmöglichkeiten am Beispiel eines Situationskontextes beschrieben werden. Dazu wird zunächst die Funktion des Handlungstyps herausgearbeitet 1). Im nächsten Schritt werden dann die für INSISTIEREN notwendigen Handlungsbedingungen - unter Einbeziehung der beiden vorausgehenden Sprechhandlungen - anhand eines einfachen Beispiels analysiert 2). Die Darstellung wird abgeschlossen durch eine systematische Beschreibung der Realisierungsmöglichkeiten von INSISTIEREN, sowohl auf der Handlungsebene 3) als auch auf der Äußerungsebene 4).

1) INSISTIEREN wird gewöhnlich mit folgenden Paraphrasen umschrieben: [2]

1. auf etwas beharren
2. auf etwas bestehen
3. bei etwas bleiben
4. etwas nachdrücklich fordern
5. etwas durchzusetzen versuchen

Hier zeigt sich bereits der Doppelcharakter von INSISTIEREN. Man kann aktiv INSISTIEREN, indem man z.B. 'etwas nachdrücklich fordert'. Man kann aber auch reaktiv 'auf etwas beharren' bzw. 'bei etwas bleiben'. Folglich sind zwei Formen von

INSISTIEREN zu unterscheiden:

INSISTIEREN$_1$, aktiv durch die Verwendung initialer Sprechhandlungstypen z.B. BITTEN oder AUFFORDERN

INSISTIEREN$_2$, reaktiv durch die Verwendung reaktiver Sprechhandlungstypen z.B. ABLEHNEN oder WEIGERN

Beide Formen können in Gesprächen wiederholt angewendet werden, wobei dieser Sprechhandlungstyp dann taktisch verwendet wird. [3] Ich möchte mich hier nur mit der aktiven Form von INSISTIEREN beschäftigen.

Worauf kann man INSISTIEREN? Hier bieten sich Substantive und Wendungen an, die mit 'insistieren' oder mit den in Wörterbüchern angeführten Paraphrasen von 'insistieren' verbunden werden können. INSISTIEREN kann man auf einem Standpunkt, einer Ansicht, einer Meinung, einem Vorsatz oder Entschluß, auf einer Frage oder der Ausführung einer Handlung, kurz: auf einem Anspruch oder Recht. Allerdings kann man das nicht, indem man eine Sprechhandlung oder gar eine Äußerungsform wiederholt. Auffallend ist vielmehr, daß zum INSISTIEREN sowohl Sprechhandlungstypen als auch Äußerungsformen verwendet werden, die sich durch das Maß an Nachdruck, das ein Sprecher mit ihnen zum Ausdruck bringen kann, unterscheiden. [4] Dazu ein Beispiel:

1 A: Werfen Sie die Zigarette weg!

2 B: Die Zigarette gehört zur Situation. Die Situation ist ungewöhnlich - für uns beide, Herr Professor.

3 A: Die Zigarette wegwerfen, sage ich!

4 B: Bedaure.

5 A: Sie sollten es wagen! Bursche! Sie werfen sie weg oder ich hemme Sie in Ihrer Laufbahn! [5]

Stellt man sich vor, daß A Satz 1 auch im 3. und 5. Zug verwendet, so denkt man entweder Intonationsvarianten mit, wodurch ebenfalls eine Nachdruckerhöhung erzielt werden kann, oder aber man tendiert dazu, A zu bemitleiden, weil er offensichtlich sein Ziel mit den falschen Mitteln verfolgt. Es wird zu zeigen sein, daß selbst da, wo ein Sprecher eine initiale Sprechhandlung scheinbar wiederholt, eine Variante dieses Sprechhandlungstyps verwendet wird, ein Untermuster, das Sequenzmerkmale aufweist und das darum als eine Möglichkeit zu INSISTIEREN aufzufassen ist.

Wann kann man INSISTIEREN? Formal betrachtet setzt INSISTIEREN immer zwei Züge mit Sprecherwechsel voraus, z.B. die AUFFORDERUNG von A und die WEIGERUNG von B. Allgemein kann man sagen, daß A einen initialen Sprechhandlungstyp verwendet, mit dem ein Aliokommitment eingeführt wird und daß B eine nicht-akkordierende Sprechhandlung gebraucht, mit der B A zu verstehen gibt, daß er das Aliokommitment nicht akzeptiert. [6] Abstrakt heißt das: A versucht Macht auszuüben, scheitert aber an B's Widerstand. Gibt A sein Interaktionsziel dennoch nicht auf und versucht in einem zweiten - vielleicht auch dritten oder n-ten - Anlauf doch noch an's Ziel zu gelangen, kann gesagt werden, A INSISTIERT. INSISTIEREN ist also ein sequenzabhängiger Sprechhandlungstyp, der von einem Sprecher vom 3. Zug an iterativ gebraucht werden kann.

2) Zu den Handlungsbedingungen für INSISTIEREN gehören folglich neben den für Sprechhandlungen typischen Kontext- und Situationsmerkmalen auch noch die beiden vorausgehenden Sprechhandlungen. [7] Sie sollen hier konstruktiv anhand eines einfachen Beispiels aufgebaut werden. Dazu möchte ich von folgenden Kontextbedingungen ausgehen:

- A und B sind Freunde
- die Beziehung ist symmetrisch

Als situative Bedingungen kommen hinzu:

- A hat einen Brief geschrieben, der noch heute zur Post muß
- A weiß, daß B heute noch bei der Post vorbeikommt
- A will, daß B H ('den Brief mit zur Post nehmen') tut

Das Interaktionsziel von A lautet: 'B dazu bringen H zu tun.' A wählt von den möglichen Handlungsalternativen BITTEN, das er durch folgende Äußerungsformen realisieren könnte:

(1) x-e doch für mich!

(2) Du x-t doch für mich?

(3) x-t du für mich?

(4) {Könntest / Möchtest} du für mich x-en?

(5) Würdest du für mich x-en?

(6) Wärst du {wohl / vielleicht} so {nett / freundlich} und würdest für mich x-en?

(7) Sei (doch) (bitte) so {nett / liebenswürdig} und x-e für mich!

(8) (Bitte) x-e (doch) (bitte) für mich, (bitte)!

(9) (Bitte) x-t du (bitte) für mich, (bitte)?

BITTEN kann unter diesen Bedingungen also mit Hilfe eines Aufforderungsatzes ((1) und (7)) oder eines Fragesatzes ((2) und (3)) realisiert werden. Zusätzlich stehen Formen mit Konjunktiv II + Modal- oder Hilfsverb zur Verfügung ((4) und (5)), sowie besondere Wendungen, die mit Frage- oder Aufforderungssätzen verbunden werden können ((6) und (7)). Schließlich kann der Handlungstyp durch Einfügen von 'bitte' an den durch Klammern markierten Stellen explizit gemacht werden ((8) und (9)).
Es soll davon ausgegangen werden, daß A zur Realisierung seiner BITTE (1) wählt. B realisiert daraufhin eine offene, nichtakkordierende Sprechhandlung, nämlich ABLEHNEN. [8] Als Äußerungsform wählt er dazu einen einfachen Aussagesatz:

(1o) Nein, ich werde nicht x-en.

Damit sind die Grundbedingungen festgelegt,und es stellt sich nun die Frage,wie (und womit) in einem solchen Falle INSISTIERT werden kann.

3) Die Annahme, daß A nun INSISTIEREN kann, indem er seine initiale Sprechhandlung einfach wiederholt, läßt sich darauf zurückführen, daß dieselbe Äußerungsform abermals verwendet werden kann. Eine solche Wiederholung setzt aber eine andere Betonung voraus. Sie ist neben dem Sequenzaspekt ein Indikator dafür, daß nicht dieselbe Sprechhandlung damit realisiert wird. [9]
Doch welche Möglichkeiten hat A im vorliegenden Fall, um seiner BITTE Nachdruck zu verleihen? Grundsätzlich stehen A zwei Möglichkeiten offen. Er kann sich für eine kooperative Form des INSISTIERENS entscheiden oder für eine kompetitive. Im ersten Fall wird er Handlungstypen wählen, durch die das ursprüngliche Aliokommitment abgeschwächt wird. Dazu gehören z.B. ERKLÄREN und APPELLIEREN. A könnte z.B. 'sagen, daß es nur eine

(kleine) Bitte ist'. Er könnte aber auch 'sagen, daß B kooperationsbereit sein soll'. Ein solcher APPELL an die Kooperationsbereitschaft hat zwar nicht notwendig eine Schwächung des Aliokommitments zur Folge, kann aber als positive intendiert gelten.[10] Wählt A kompetitive Formen, so muß er einen Sprechhandlungstyp wählen, durch den das ursprüngliche Aliokommitment verstärkt wird. Dazu gehören z.B. AUFFORDERN und DROHEN.

Neben dem strittigen Aliokommitment, das A, wie gezeigt wurde, erneut in schwächerer oder stärkerer Form einführen kann, spielen noch andere Handlungsbedingungen eine wesentliche Rolle. So wird durch die ERKLÄRUNG, daß es sich nur um eine (kleine) BITTE handle, das Gewicht der Handlung verringert und damit ein möglicher VORWURF vorbereitet. A kann das Handlungsgewicht aber auch erhöhen, indem er sagt, daß er die Handlung (sehr) hoch bewertet. Dadurch wird B gezwungen, seine Entscheidung zu überprüfen. Bleibt B bei seiner Entscheidung, wird er auf diese Weise genötigt, sich zu verteidigen. Natürlich kann A auch an B als Freund einen APPELL richten, indem er auf die für B günstigeren (oder die für A ungünstigeren) Handlungsbedingungen verweist, womit eine Handlungsmaxime folgender Art angesprochen wird:

> Wenn A und B befreundet sind und A den B bittet, für ihn zu x-en und nachweislich für A ungünstigere Handlungsbedingungen vorliegen, dann sollte B für A x-en.

Die ABLEHNUNG von B kann auch Gegenstand einer BEWERTUNG werden, dadurch daß A sagt, daß die ABLEHNUNG von B ein Akt der Unfreundlichkeit ist. Während 'sagen, daß die Handlungsbedingungen für B günstiger (bzw. für A ungünstiger) sind' B noch die Entscheidung überläßt, weshalb diese Form als positiv intendiert anzusehen ist, kann BEWERTEN in diesem Handlungskontext als kompetitiv klassifiziert werden. Schwieriger ist die Abgrenzung zwischen BITTE und AUFFORDERUNG, da für beide gilt: 'sagen, daß B x-en soll'. Für die Unterscheidung hier mag es jedoch genügen, AUFFORDERN von BITTEN durch den Zusatz 'nachdrücklicher sagen' zu unterscheiden. Für das DROH-GEBOT sind schließlich zwei Untermuster anzusetzen: 1. 'sagen, daß die

ABLEHNUNG Folgen haben werde' (Drohung mit unbestimmter Sanktion) und 2. 'sagen, daß A z-ten werde, wenn B nicht x-t' (Drohung mit bestimmter Sanktion).

4) Geht man davon aus, daß bereits auf der Ebene der Handlungen Alternativen existieren, die sich bezüglich des mit ihnen verbundenen Nachdrucks deutlich voneinander abgrenzen lassen, so erscheint es nur logisch, daß auf der Ebene der Äußerungsformen ähnliche Ausdrucksmöglichkeiten bestehen müssen. Darum werde ich nun die einzelnen Handlungsmuster und ihre Realisierungsmöglichkeiten, d.h. ihre Äußerungsformen, beschreiben.

Mit Hilfe welcher Äußerungsformen kann ERKLÄREN realisiert werden? Hierzu stehen eine Reihe von Aussage- und Aufforderungssätzen zur Verfügung.

(11) (Aber) das ist (doch) nur eine Bitte.

(12) (Aber) das ist (doch) wirklich nicht der Rede wert.

(13) (Aber) wenn ich (doch) ganz {lieb / herzlich} darum bitte!

(14) (Aber) das ist (doch) eine so kleine Bitte.

Durch ERKLÄREN wird also der Handlungstyp des 1. Zuges expliziert und das Gewicht der geforderten Handlung H als unerheblich klassifiziert. Mit Wendungen wie (12) wird nur auf den letzteren Aspekt abgehoben. Die Abtönungspartikel haben hier eindeutig verstärkende Funktion.[12] Zu den explizierenden Formen gehören auch die Äußerungsvarianten von (8), die an dieser Stelle verwendet werden können, weil sie im Kontrast zu (1) stehen. Die Fragesatzformen sind dagegen nicht mehr anwendbar. Durch ERKLÄREN kann aber auch die Bedeutung der Handlung H erhöht werden, wozu Konditional- und Aussagesätze verwendet werden können.

(15) Wenn du nicht x-t, komme ich in große Schwierigkeiten.

(16) (Aber) du würdest mir damit {einen (großen) Gefallen / aus einer Verlegenheit} {tun / helfen}.

Es wurde oben zwischen APPELL an die Kooperationsbereitschaft und Beziehungs-APPELL unterschieden. Für beide Formen werden Aufforderungssätze verwendet. Realisierungsmöglichkeiten der ersten Form thematisieren den Interaktionsmodus und gelten als abgeschwächte AUFFORDERUNGEN zur Kooperation.

(17) Sei doch lieb (bitte)!

(18) Sei doch nicht so unfreundlich!

(19) Stell dich doch nicht so (blöd) an!

Ohne die Abtönungspartikeln erhält man normale AUFFORDERUNGEN, weshalb sie in diesen Satzformen nicht fehlen dürfen. Äußerungsformen, die zur Realisierung des Beziehungs-APPELLs dienen können, wären folgende:

(2o) Das kannst du mir (doch) nicht antun!

(21) (Aber) das bist du mir (doch) schuldig!

(22) (Aber) das ist (doch) nicht dein Ernst!

(23) (Aber) das meinst du (doch) nicht ernst!

(24) (Aber) das ist (doch) wirklich $\left\{\begin{array}{l}\text{lächerlich}\\ \text{ärgerlich}\end{array}\right\}$, sich bei einer solchen Lapalie so anzustellen.

B wird mit diesen Formen in die Pflicht genommen, wobei (2o), (22) und (23) als schwächere Formen anzusehen sind, weil A mit ihnen B in seiner Entscheidungsfreiheit beläßt. Dagegen wird mit (21) B moralisch gefordert, mit (24) diese Forderung mit großem Nachdruck gestellt. Dennoch kann A mit keiner dieser Formen B AUFFORDERN zu x-en. Zu erwähnen sind noch die Formen, mit denen Handlungsbedingungen von A oder B angesprochen werden. Hierzu können sowohl Frage- als auch Aussagesätze verwendet werden.

(25) (Aber) du gehst doch sowieso bei der Post vorbei.

(26) (Aber) du kommst doch (heute) noch zur Post.

(27) Kommst du nicht gleich an der Post vorbei?

(28) Komme ich denn heute noch zur Post?

(29) Ich komme ($\left\{\begin{array}{l}\text{aber}\\ \text{doch}\end{array}\right\}$) (heute) nicht mehr zur Post.

(3o) (Aber) du weißt doch, daß ich heute nicht mehr zur Post komme.[13]

Der kompetitive Charakter von BEWERTEN wird an folgenden Äußerungsformen sichtbar.

(31) Ich finde das (wirklich) nicht gut von dir.

(32) Das ist ja (wirklich) das Letzte.

(33) Das ist ja (wirklich) {unglaublich / unerhört} .

(34) Das finde ich (aber) ganz schön beschissen.

Bis auf (31) haben diese Formen bereits den Charakter von festen Wendungen. In allen Formen kann durch die Partikeln der Nachdruck erhöht werden.

Mit AUFFORDERN kann sich A sowohl auf die Handlung H beziehen, als auch auf den Interaktionsmodus. Zur Realisierung können Aufforderungssätze verwendet werden. Aus (1) erhält man unter Weglassung der Abtönungspartikel (1'), aus (2) (2').

(1') x-e für mich!

(2') Du x-t für mich!

Fragesatzformen könnten nur dann verwendet werden, wenn der Eröffnungszug ebenfalls durch einen Fragesatz realisiert worden wäre. Aus den Formen (17), (18) und (19) erhält man unter Weglassung der Partikeln:

(17') Sei lieb!

(18') Sei nicht so unfreundlich!

(19') Stell dich nicht so an!

Für das Untermuster 'sagen, daß die Ablehnung Folgen haben wird' gibt es eine Reihe von Wendungen. Daneben besteht die Möglichkeit, den Aufforderungssätzen durch einen performativen Vorspann Nachdruck zu verleihen. Charakteristisch für beide Untermuster von DROHEN sind Bedingungssätze. [14]

(35) x-e (bloß)!

(36) Wehe, du x-t nicht!

(37) Los, x-e jetzt!

(38) Jetzt reiß dich aber zusammen, sonst vergess ich mich!

(39) Wenn du nicht (gleich) x-t, passiert was!

(4o) Ich warne dich: x-e (jetzt)!

(41) Ich rate dir: x-e (bloß)!

Für das Muster 'sagen, daß A z-ten wird, wenn B nicht x-t' sind folgende Formen wählbar:

(42) Entweder du x-t (jetzt) oder ich z-te.

(43) Wenn du nicht x-t, dann z-te ich.

(44) x-e (jetzt), sonst pfeife ich in Zukunft auf deine Freundschaft

(45) Wenn dir was an unserer Beziehung liegt, x-t du!

(46) Los, x-e (jetzt), sonst z-te ich.

Der angedrohte Beziehungsabbruch muß dabei nicht die größte Sanktionsmöglichkeit darstellen. So könnte es B z.B. mehr treffen, wenn A B's persönliche Geheimnisse ausplaudern würde. Mit den Äußerungsformen für DROHEN sind die sprachlichen Möglichkeiten der Nachdruckserhöhung im vorgegebenen Kontext erschöpft.

Zusammenfassend kann festgehalten werden, daß INSISTIEREN ein sequenzabhängiger Sprechhandlungstyp ist, zu dessen Realisierung eine Reihe von Handlungsalternativen existieren. Am Beispiel einer kleinen BITTE wurde exemplarisch vorgeführt, mit Hilfe welcher Sprechhandlungen A versuchen kann, sein Interaktionsziel ('B dazu bringen, daß er H tut') zu erreichen. Dabei kann nicht nur Nachdruckerhöhung sondern auch Nachdruckverminderung (vgl. z.B. ERKLÄREN) ein adäquates Mittel sein. Auf der Ebene der Äußerungsformen konnte gezeigt werden, daß die Abtönungspartikeln für manche Sprechhandlungen obligatorisch sind und folglich - unter bestimmten kontextuellen und situativen Voraussetzungen - als Indikatoren gewertet werden können. Weiter wurde festgestellt, daß die Äußerungsform des 1. Zuges die Wahlmöglichkeiten des 3. Zuges einschränkt. Konkret heißt das, daß in unserem Beispiel AUFFORDERN nicht mehr durch Fragesätze realisiert werden konnte.

Anmerkungen

1 Unter einem komplexen Sprechhandlungstyp wird (mit Ryle) ein Handlungstyp verstanden, dessen Beschreibung die Beschreibung anderer Handlungstypen voraussetzt. vgl. RYLE (1969: 258 ff.)

2 Vgl. dazu Duden 8

3 Zum Begriff der Gesprächstaktik vgl. APELTAUER (1978): 2.5. und 2.2.2.

4 Die Auffassung, dass die Wiederholung des Initial-Sprechaktes als die einfachste Form von INSISTIEREN angesehen werden kann, halte ich für falsch. vgl. dagegen HUNDSNURSCHER (1976: 261)

5 Der Gesprächsausschnitt ist eine etwas vereinfachte Form aus Heinrich Manns Prof. Unrat, TB-Ausgabe rororo 1970, S. 90

6 Zu den Begriffen Auto- und Aliokommitment vgl. WRIGHT (1968: 76 f.)

7 Die Unterscheidung in Kontext- und Situationsmerkmale übernehme ich von HALLIDAY (1973: 58 ff.)

8 Verdeckte nichtakkordierende Sprechhandlungen wären z.B. ABLENKEN oder SICH-NICHT-FESTLEGEN, welche ich beide zur Taktik AUSWEICHEN rechne.

9 Hundsnurscher schreibt: "dieselbe Äusserung, wiederholt, ist ein anderer Sprechakt, (...)" (S. 261) Eine Äusserung wird von einem Sprecher in einem Raum-Zeit-Punkt hervorgebracht. Folglich kann eine Äusserung nicht wiederholt werden. Nur eine Äusserungsform kann wiederholt gebraucht werden. Ebenso problematisch erscheint mir seine Gleichsetzung von 'Äusserung' und 'Sprechakt'.

10 Positiv intendiert darum, weil A mit einem APPELL B die Entscheidung nicht abnimmt.

11 Man nehme zur Verdeutlichung eine grosse BITTE: 'sagen, dass B A abschleppen kommen soll'. Darauf wird B nicht mit einfachem ABLEHNEN reagieren können. Er wird vielmehr eine Begründung nachliefern.

12 Durch die Klammern soll angedeutet werden, dass es sich hierbei um fakultative Elemente handelt.

13 Für 'Post' könnte man sich hier ebenfalls eine Variable denken, die durch entsprechende Bedingungsangaben zu belegen wäre.

14 Vgl. dazu APELTAUER (1977)

L i t e r a t u r

APELTAUER, E. (1977): "DROHEN": SPRENGER,K./BALD,W.D./VIETHEN, H.W. (Hg.): Semantik und Pragmatik, Akten des 11. Ling. Kolloquiums, Aachen (1976); Tübingen : Niemeyer. 187 - 198.

--- (1978): Elemente und Verlaufsformen von Streitgesprächen, Eine Analyse von Texten und Tonbandprotokollen unter Sprechhandlungstheoretischen Gesichtspunkten (Diss)

HALLIDAY, M.A.K. (1973): Explorations in the Functions of Language, London

HUNDSNURSCHER, F. (1976): "Insistieren." Wirkendes Wort, H. 4, 255 - 265

RYLE, G. (1969): Der Begriff des Geistes, Stuttgart

WRIGHT, G.H.v. (1968): An Essay in Deontic Logic and the General Theory of Action, Amsterdam

SPRACHGEBRAUCH UND REFLEXION
Kommunikationsversuche in einer anderen Welt

Bernd Ulrich Biere

Jemand kommt in eine andere Welt; sagen wir: der Südseehäuptling Tuivaii ins Land der Papalagi oder: der Ethnologe Evans-Pritchard zum Stamm der Azande oder: der Schriftsteller Hubert Fichte in die Welt der synkretistischen Religionen Südamerikas oder: Carlos Castaneda in die Welt des "mescalito", in die ihn der indianische Zauberer, den er Don Juan nennt, einführt oder: die Beatles kommen im gelben Unterseeboot nach Pepperland oder: das kleine, neugierige Mädchen Alice kommt ins Wunderland oder in eine Welt hinter den Spiegeln.
Manches wird ihnen dort vertraut sein, wird sie an "ihre" Welt erinnern. Sie werden manches zu verstehen glauben, was die Menschen in der fremden Welt tun, sie werden vielleicht sogar (bis zu einem gewissen Grad) an ihrem Leben teilnehmen können, werden vielleicht auch Vergleiche anstellen mit ihren eigenen Lebensformen usw. Aber vieles, was in der fremden Welt mit Selbstverständlichkeit getan wird, wird ihnen fremdartig und unverständlich erscheinen. Vielleicht wird der Wissenschaftler unter ihnen versuchen, einige der fremden Handlungsweisen zu beschreiben, aber das bereitet ihm einige Schwierigkeiten: was er beschreibt, sind nur die äusserlichen Bewegungen, die er sieht, er kennt nicht die Regel, nach der man in der fremden Welt handelt, die Regel, die unverständliches Handeln verständlich machen, ihm einen Sinn verleihen könnte.

Stellen wir uns einen Wissenschaftler aus einer "hochzivilisierten" westlichen Kultur vor. Die entscheidende Frage ist: W i e kommt er in diese andere Welt? Kommt er als Missionar, der sich im Besitz der Wahrheit wähnt und den sog. Fortschritt propagiert, als Eroberer, der sich einmal mehr die Überlegenheit seiner eigenen ("wissenschaftlichen") Lebensform unter Beweis stellen will?[1] Oder kommt er als Lernender, als jemand, der in Betracht zieht, seine eigenen Massstäbe (z.B. auch für das, was er Rationalität nennt) und Kriterien (z.B. für das, was er sinnvolles Verhalten nennt) relativieren zu müssen, als jemand, der verstehen will, kurz: als jemand, der eine fremde Lebensweise in dem Sinn ernsthaft studieren will, dass er seine eigene zu erweitern trachtet und "nicht einfach die fremde Lebensweise in die bereits bestehenden Grenzen seiner eigenen zu integrieren" versucht. (WINCH 1972/dt. 1975: 84)

Mit der ersten Haltung, der des Missionars oder Eroberers, will ich mich hier nicht weiter beschäftigen, mich interessiert die zweite Haltung, die des Lernenden und die Bedingungen, unter denen eine solche Haltung - Lernen - möglich ist.[2] Eine solche Haltung spielt jedoch nicht erst in der Konfrontation mit fremden Kulturen im Sinne des Ethnologen eine Rolle, sondern könnte sich als Voraussetzung sowohl des menschlichen Zusammenlebens (auch in einer scheinbar homogenen Gesellschaft) als auch für dessen Erforschung erweisen. Gleichwohl lässt sich das Problem, um das es mir hier geht, in einer in Richtung der Beispiele radikalisierten Form paradigmatisch aufzeigen und diskutieren. Ich habe dazu exemplarisch einen Textausschnitt aus Lewis Caroll, "Alice hinter den Spiegeln" gewählt.

In der fiktiven Welt "hinter den Spiegeln" lässt sich überdies eine zusätzliche Radikalisierung des Problems feststellen: es ist nämlich zunächst durchaus fraglich, was das denn für Wesen sind, denen Alice dort begegnet, ob oder in welchem Sinn es überhaupt Menschen sind. Damit werden wir sogleich vor die Frage gestellt, was denn "Menschen" sind und was nicht, anders ausgedrückt: welche Kriterien haben wir denn, um die u.U. folgenreiche Entscheidung zu treffen, ob einem Gegenstand das Prädikat "Mensch" zu- oder abzusprechen ist?

In der Begegnung mit Goggelmoggel[3] sieht sich Alice vor die - keineswegs triviale - Frage gestellt, ob etwas eine Krawatte oder ein Gürtel ist:

> "Was für einen schönen Gürtel Sie da anhaben!" bemerkte Alice plötzlich. (...) "Oder wenigstens" - verbesserte sie nach einiger Überlegung, "eine schöne Krawatte hätte ich vielleicht eher sagen sollen - nein Gürtel vielleicht - ach bitte, entschuldigen Sie vielmals!" schloss sie mit grosser Bestürzung, denn Goggelmoggel machte ein tiefbeleidigtes Gesicht,

> und sie wünschte langsam, sie hätte nicht gerade dieses Thema ausgesucht.
> "Wenn ich bloss wüsste",dachte sie im stillen, "wo bei ihm der Hals ist und wo der Bauch!" (CARROLL 1963/1975: 85 f)

Darauf erwidert Goggelmoggel nach einiger Zeit, "sichtlich erzürnt:

> "Es muss - überaus - schmerzlich - berühren, [..] wenn jemand eine Krawatte nicht von einem Gürtel unterscheiden kann." (CARROLL 1963/1975: 86)

Alice geht hier offensichtlich zunächst von der Annahme aus, dass sie diese Unterscheidung sehr wohl zu treffen imstande ist, bemerkt dann aber bald, dass ihre Kriterien für diese Unterscheidung nicht oder nur schwer anwendbar sind, d.h. es ist fraglich, ob diese oder jene Bedingungen, unter denen sie dieses oder jene Prädikat anwendet, hier tatsächlich vorliegen. Um so erstaunlicher ist es, mit welcher Sicherheit Goggelmoggel selbst eben diese Unterscheidung treffen kann. Kann er selbst trotz seiner Leibesform (Ei!) Hals und Bauch unterscheiden, hat er also die gleichen Kriterien wie Alice mit dem Unterschied, dass er sie auch in seinem Fall anwenden kann? Oder hat er andere Kriterien, eine Krawatte von einem Gürtel zu unterscheiden? Im Gegensatz zum Text, wo sich Goggelmoggel auf die Unterscheidungsmöglichkeit bzw. -notwendigkeit versteift, wäre noch eine andere Erklärung möglich, nämlich die, dass die betreffende Unterscheidung, die für Alice vielleicht wichtig ist, für Goggelmoggel einfach irrelevant sein könnte, so dass für ihn Krawatte und Gürtel einfach "ununterscheidbar" wären.

Alice erkennt ihr Problem: dass sie nämlich von Kriterien ausgeht, die hier offenbar nicht anwendbar sind oder dass s i e jedenfalls nicht in der Lage ist, diese hier "richtig" anzuwenden.

Was ich festhalten will, ist dies: Es scheint, dass wir in einer fremden Welt damit rechnen müssen, dass die Kriterien, aufgrund derer wir unsere Wahrnehmungen strukturieren und interpretieren, ihnen Sinn verleihen, möglicherweise nicht anwendbar sind, dass beispielweise geläufige, im allgemeinen unhinterfragte Unterscheidungen oder Einteilungen, Begriffe und Kategorien nicht ausreichen oder ganz versagen, wenn es darum geht, uns verstehend in dieser fremden Welt zurechtzufinden, d.h. diese Welt zunächst einmal so zu sehen zu versuchen, wie sie den Angehörigen der betreffenden Lebensform selbst als sinnhaft strukturierte Welt erscheint, in der Ereignissen und Handlungen vielleicht nur in einer anderen als der uns vertrauten Weise Sinn verliehen wird.[4]

Die seltsam anmutende Begegnung mit einem Ei, das mehr und mehr menschliche Züge annimmt, "Augen, Nase und Mund" bekommt, ja, einen Namen hat und schliesslich auch sprechen kann, mag auch für den Leser, der sich bis zum sechsten Kapitel schon an einige Überrachungen gewöhnt hat, immer noch einigermassen verwirrend sein. Nun ja, wenn jemand wie ein Ei aussieht, so dass man bei ihm nicht einmal Hals und Bauch unterscheiden kann, wie soll man mit dem überhaupt zurechtkommen, möchte man fragen, wie soll man mit dem gar reden können? Umso mehr überrascht es mich zunächst, dass dieser Typ nicht nur irgendeine Sprache spricht, sondern offensichtlich sogar "die gleiche" wie Alice, also Englisch bzw. im übersetzten Text Deutsch. Die Behauptung, Alice und Goggelmoggel sprächen die gleiche Sprache, erweist sich allerdings bei näherem Hinsehen als oberflächlich. Sie lässt sich - linguistisch gesehen - nur halten, wenn man von einem meiner Meinung nach reduzierten Sprachbegriff aus argumentiert. Denn die Regeln, nach denen Goggelmoggel hier (sprachlich) handelt, scheinen von den Alice vertrauten Handlungsmustern zumindest in bestimmten Bereichen doch sehr verschieden zu sein.

Sie versteht zwar, was Goggelmoggel s a g t , nach welchen Äusserungsmustern er handelt, aber die "Unterhaltung" mit ihm kommt ihr doch sehr eigenartig vor, ja: "Das Ganze kam ihr überhaupt nicht wie eine Unterhaltung vor." (CARROLL 1963/1975: 82) Wenn ich annehme, dass die Art, wie man eine Unterhaltung führt, beschreibbar ist als ein Teil der Regeln, die diese oder jene Sprache bzw. Lebensform ausmachen, kann offenbar nicht mehr davon die Rede sein, dass Alice und Goggelmoggel "die gleiche Sprache" sprechen.

Die Art des Handelns von Goggelmoggel ist für Alice zumindest bis zu einem gewissen Grad unverständlich. Aber sie konstatiert nicht einfach ihr Befremden, sondern versucht sich darüber klarzuwerden, w a r u m ihr das Ganze nicht wie eine Unterhaltung vorkommt, sie versucht ihren Eindruck ausdrücklich zu begründen, indem sie bzw. der Autor[5] äussert: "weil mit ihr ja eigentlich gar keiner sprach, die letzte Bemerkung, zum Beispiel, war offenbar an einen Baum gerichtet". (CARROLL 1963/1975: 82)

In welchem Sinn können wir Sätze dieser Art als Begründung ansehen? Was wird damit zu begründen versucht? Was tut Alice eigentlich, indem sie diesen Satz äussert? - Ich meine, sie begründet ihren Sprachgebrauch, indem sie Kriterien oder Bedingungen angibt, die für sie vorliegen müssen, damit sie etwas eine Unterhaltung nennt. Eine solche Bedingung gibt sie an, bzw. bezieht sich auf eine solche, indem sie behauptet, dass im vorliegenden Fall (ihrer Meinung nach) die Bedingung, dass man sich mit dem, was man sagt, an den Partner wendet, m i t i h m spricht, nicht erfüllt ist. Aber: eine Bedingung angeben ist eine Sache, und eine andere Sache ist, sich darüber zu verständigen, ob eine Bedingung in einem konkreten Fall tatsächlich vorliegt, "erfüllt" ist. (Dies ist eine genuin hermeneutische Frage oder, wenn man Kant bemühen will, eine Frage der Urteilskraft). Es geht hier, vergleichbar dem Problem der Krawatte-Gürtel-Unterscheidung genau darum, welche Kriterien wir dafür haben zu entscheiden, ob etwas eine Unterhaltung i s t oder nicht, bzw. darum, unter welchen Bedingungen wir etwas eine Unterhaltung n e n n e n . Die Kriterien, um die es hier geht, sind also letztlich Kriterien für einen bestimmten Sprachgebrauch und seinen Sprachgebrauch begründen heisst eben: Kriterien angeben, unter denen man diesen oder jenen sprachlichen Ausdruck, dieses oder jene Prädikat "normalerweise" verwendet.

Deshalb lässt sich eine Frage wie die, ob etwas eine Unterhaltung i s t , nicht so behandeln, als hätten wir es hier mit (Natur)gegenständen zu tun, über deren Existenz oder Nicht-Existenz wir ontologische Aussagen machen könnten, deren Wahrheit an einer "aussersprachlichen" Realität in einem empiristischen oder sensualistischen Sinn überprüfbar sei. Eine solche Frage lässt sich m.E. nur sprachanalytisch behandeln als Frage nach dem Gebrauch der betreffenden sprachlichen Ausdrücke oder Prädikate in unserer bzw. (vorsichtiger) in meiner Sprache.

Dies bedeutet wiederum, dass es hier gar nicht um eine Wahr-Falsch-Dichotomie geht, sondern um eine Explikation von Regeln, nach denen wir auf diese oder jene Weise sprachlich handeln. Einen Sprachgebrauch zu beschreiben, bedeutet zunächst nichts anders als zu behaupten: so handele ich eben und ich handele nach diesen oder jenen Regeln, die meinem Handeln zugrundeliegen. Ich fordere damit nicht unbedingt: So wie ich sollst du auch handeln! Folge den gleichen Regeln wie ich! Ich lege lediglich meine Kriterien offen, soweit mir das mit einiger reflexiver Anstrengung möglich ist, und fordere den Partner auf: jetzt expliziere du deine Kriterien! Dann werden wir weiter sehen.

An diesem Punkt wird deutlich, dass wir es hier in der Tat mit einem kommunikativen Problem zu tun haben, mit dem Problem nämlich, ob wir einen unterschiedlichen Sprachgebrauch der Partner akzeptieren und tolerieren können und wollen oder wie wir in der Lage sind, kommunikativ zu entscheiden, auf welchen Sprachgebrauch wir uns einigen wollen, welcher Sprachgebrauch uns zu welchem Zweck angemessen erscheint.[6]

Doch zurück in unser Land hinter den Spiegeln. Alice argumentiert ja viel vorsichtiger: "Das Ganze kam ihr überhaupt nicht wie eine Unterhaltung vor" (Hervorhebung von mir; B.U.B.), heisst es da. Und die Art, wie sie ihren eigenen Sprachgebrauch reflektiert, nämlich unter Bezug auf ein Gegenbeispiel, das Goggelmoggel ihr vor Augen führt, lässt sich m.E. zugleich als der Versuch verstehen, eine Hypothese zu machen über die Regeln, nach denen Goggelmoggel hier möglicherweise handelt. Vielleicht lässt sich sogar das anschliessende Aufsagen eines Gedichts nicht nur als Ausdruck der Hilflosigkeit deuten, sondern auch als ein Versuch, probeweise nach dieser Hypothese zu handeln, also sich etwa an einen Baum anstatt an den Partner zu wenden, denn dies könnte ja eine mögliche Regel sein, nach der Goggelmoggel kommuniziert. Anders ausgedrückt, scheint Alice im Unterschied zu manchem Wissenschaftler zumindest in Betracht zu ziehen, dass das, was für sie keine Unterhaltung ist, für den Partner doch eine sein könnte. Sprachanalytisch gewendet: dass Goggelmoggel einen anderen Sprachgebrauch haben könnte, der es ihm gestattet unter anderen Bedingungen als Alice etwas eine Unterhaltung zu nennen.

Betrachte ich den Verlauf der Unterhaltung oder Nicht-Unterhaltung zwischen Alice und Goggelmoggel weiter, so fällt mir auf dass ich, was der Text mir freilich nahelegte, bisher die Sache immer aus der Perspektive von Alice gesehen habe. Es ging um ihre Kriterien, um ihren Sprachgebrauch. Zwar habe ich eingeräumt, dass Alice durchaus bemerkt, dass Goggelmoggel einen anderen Sprachgebrauch haben könnte, nach anderen Regeln handeln könnte als sie; dabei habe ich jedoch immer unterstellt, dass sich Alice sozusagen normal verhält und Goggelmoggel demgegenüber irgendwie abweichend.

Wenn es zwischen den beiden zu dem oben beschriebenen Problem käme, kommunikativ entscheiden zu müssen, welcher Sprachgebrauch vorzuziehen, auf welche Regeln man sich zweckmässigerweise einigen sollte, so würde ich mich vermutlich für Alice starkmachen (vielleicht, weil ich Goggelmoggel für einen arroganten Typ halte? Ist er wirklich arrogant?).

Versuchen wir einmal, die Sache aus der Sicht Goggelmoggels zu sehen. Goggelmoggel ist in der Welt hinter den Spiegeln, in der Alice als Fremdling auftaucht, offensichtlich zu Hause, es ist seine Welt; eine Welt, in der man normalerweise gerade so zu handeln pflegt, wie er es tut. Nun kommt dieses kleine neugierige Mädchen daherspaziert und will mit ihm eine Unterhaltung anfangen, bzw. das, was sie dafür hält, was in ihrer Welt als Unterhaltung zählt. Dabei macht sie aus der Sicht Goggelmoggels jedoch offensichtlich einiges falsch. S i e ist diejenige, die sich abweichend verhält, die Fremde, die nicht dazu gehört, die sich bemühen muss, einen Zugang zu dieser Welt zu finden.

Ich meine, nicht zufällig ist Alice ein Kind. Als Kind ist sie ganz Lernende, und zwar in viel stärkerem Mass als ein Erwachsener, zumal, wenn dieser aus einer Kultur wie der unseren kommt, wo Wunder nicht vorkommen und auch nicht vorkommen dürfen. Wenn Alice sich in der Welt Goggelmoggels zurechtfinden und auch akzeptiert werden will, muss sie fast zwangsläufig die Haltung des Lernenden annehmen, sie muss wohl oder übel versuchen herauszubekommen, nach welchen Regeln man in dieser Welt handelt bzw. nach welchen Regeln sie handeln muss, wenn sie nicht ständig auffallen will.[7]

Das ist nicht leicht, aber hat Alice in Goggelmoggel nicht vielleicht einen guten "Führer" gefunden, der sie in diese Welt einführen kann?[8] Sicher: dieser Typ scheint uns zunächst eigenartig arrogant, aber das Argument lässt sich eben auch umkehren: Ist es in seinen Augen nicht vielleicht Alice, die sich zunächst ziemlich arrogant verhält? - Ich meine, dass Goggelmoggel anfangs recht energisch versucht, ihr die angemessene Rolle nahezulegen: die des lernenden Kindes oder - paradigmatisch gesehen - die eines Forschers, der sich als Lernender und nicht als Missionar und Eroberer begreifen lernen soll. Als Alice genau diese Rolle zu akzeptieren beginnt, wird Goggelmoggel schliesslich ganz verträglich: "Ununterscheidbarkeit! Das ist meine Meinung!"

> "Würden Sie bitte so gut sein und mir sagen", bat Alice, "was das heisst?"
> "So lässt sich schon eher mit dir reden," sagte Goggelmoggel mit sichtlicher Befriedigung. "Mit 'Ununterscheidbarkeit' meine ich, [..]" usw.
> (CARROLL 1963/1975: 88)

Ich muss die Analyse des Textbeispiels hier aus Platzgründen abbrechen. (Eine solche Analyse wird ohnehin immer bruchstückhaft bleiben müssen, immer nur unter bestimmten selektiven Aspekten durchgeführt werden können. Imgrunde scheint mir eine exemplarische Analyse letztlich die einzig sinnvolle, denn unsere Beschreibungen sollen ja nicht die Welt eins zu eins abbilden, sie sollen vielmehr (im Fall von Kommunikationsbeschreibungen) möglichen Kommunikationspartnern helfen, ihre kommunikativen Probleme besser zu verstehen und wenn möglich, eine i h n e n angemessen erscheinende "Lösung" zu finden. Wo es keine Probleme gibt, ist eine kommunikativ bzw. praktisch verstandene Beschreibung in gewissem Sinn überflüssig.)

Abschliessend will ich noch einmal zu unserer Ausgangsfrage zurückkehren und versuchen, das, was ich das Paradigmatische an unserm Textausschnitt genannt habe, klar herausstellen.

Der Versuch Alicens, sich in einer scheinbar verrückten Welt ("hinter den Spiegeln") zurechtzufinden, ihr Versuch, wenigstens einige der Regeln herauszubekommen, nach denen die Wesen in dieser fremden Welt kommunizieren und interagieren, scheint mir paradigmatisch im Hinblick auf bestimmte micht interessierende theoretisch-methodische Probleme der Beschreibung von Kommunikationen in den verschiedensten Bereichen. Er scheint mir paradigmatisch auch im Hinblick auf das Selbstverständnis des Wissenschaftlers, der damit beschäftigt ist, verstehende, exemplarische Analysevorschläge zu machen von Kommunikationen und Interaktionen in solchen sozialen Gruppen, deren Lebensformen von der je eigenen des Wissenschaftlers mehr oder weniger verschieden sind. Dabei muss ich gar nicht, wie unser Textbeispiel nahezulegen scheint, an völlig andersartige Kulturkreise denken, das Problem des verstehenden Zugangs zu den Sinnstrukturen uns nicht vertrauter sozialer Gruppen stellt sich in durchaus vergleichbarer Weise bei der Arbeit in der je eigenen Gesellschaft, die im Hinblick auf unterschiedliche Sprach- und Lebensformen eben nicht als homogen zu begreifen ist. Hier scheint mir immer noch die Argumentation Alfred SCHÜTZs in seiner "Einleitung in die verstehende Soziologie" relevant, dass wir zwar

> alle im schlichten Dahinleben unsere Handlungen als sinnvoll (erleben) und in natürlicher Weltanschauung davon "überzeugt" (sind), dass auch andere ihr Handeln als sinnvoll erleben, und zwar in genau der gleichen Weise sinnvoll, wie wir selbst ein solches Handeln erleben würden.
> [..]
> Wir sind weiters davon überzeugt, dass unsere Deutung des Sinns fremden Handelns im grossen ganzen zutrifft. Aber die unkritische Übernahme derartiger Vorstellungen in den Begriffsapparat einer Wissenschaft muss sich rächen [..]
> (SCHÜTZ 1932/1974: 16 f.)

Die Gefahr, eine solche Homogenitätsannahme dem Leser an einem fiktionalen Text anschaulich vor Augen zu führen und ihn dazu zu ermuntern, die Diskussion über die Frage eines angemessenen theoretisch und methodisch reflektierten Zugangs zur Beschreibung von Kommunikationen und Interaktionen nicht als abgeschlossen zu betrachten und dabei schliesslich auch auf die Rolle, oder wie ich es genannt habe: die Haltung des Wissenschaftlers gegenüber seinem sozialen Untersuchungs"gegenstand" kritisch zu reflektieren, war das Ziel dieses Beitrags.

Anmerkungen

1 Vgl. FEYERABEND (1977): "Kaufleute, Konquistadoren, Forscher, Missionare (und später, als alles schon ziemlich war, Wissenschaftler) - sie alle hielten es für selbstverständlich, dass sie die Wahrheit besassen, und sie hielten es für ihre Pflicht, diese Wahrheit 'mit allen Mitteln', wie wir an einer Bulle des

Papstes Alexander VI. lesen, zu verbreiten;" (25)

2 Es geht mir hier natürlich nicht um eine, sagen wir, lernpsychologische Begrifflichkeit, sondern um eine Art von Lernen, die Winch anspricht, wenn er schreibt: "Was wir aus der Erforschung anderer Kulturen lernen können, sind nicht nur andere Handlungsmöglichkeiten und andere Techniken. Wichtiger ist vielmehr, dass wir andere Möglichkeiten kennenlernen können, dem menschlichen Leben einen Sinn zu geben, andere Vorstellungen über die mögliche Bedeutung, welche die Ausführung bestimmter Tätigkeiten für Menschen annehmen kann, die den Sinn ihres Lebens als Ganzheit zu fassen versuchen." (WINCH 1972/dt. 1975: 93)

3 Ich berücksichtige nicht, dass es sich um einen aus dem Englischen übertragenen Text handelt und es ist mir gleichgültig, dass Goggelmoggel im Englischen Humpty-Dumpty heisst.

4 Vgl. das Zitat aus WINCH 1972 in Anm. 2.

5 Die Intervention des Autors muss ich hier ebenfalls unberücksichtigt lassen; ich interpretiere die folgende Textstelle so, als würde sich hier ein Kommunikationspartner tatsächlich Rechenschaft ablegen über seine jeweilige Interpretation sozialer und kommunikativer Phänomene.

6 Die "Lösung" letzterer Frage mündet de facto wohl meistens in den Versuch, den eigenen Sprachgebrauch (auch den von sog. Autoritäten, Institutionen usw.) als Norm zu setzen, für besser, richtiger, adäquater zu erklären und dies womöglich für alle nur denkbaren Zwecke. Das Problem der Normativität von Regelbeschreibungen ist also empirisch gewiss bedeutsam, auf der begrifflichen Ebene sollte man jedoch auf der Unterscheidung von Regel und Norm bestehen.

7 Die hier illustrierte methodische Problematik der teilnehmenden Beobachtung muss ich an dieser Stelle undiskutiert lassen. Möglichkeiten, das Konzept der teilnehmenden Beobachtung im Rahmen eines hermeneutisch orientierten Ansatzes zu spezifizieren habe ich an anderer Stelle exemplarisch aufgezeigt (siehe BIERE 1978: 12 - 60).

8 Vgl. zum Problem der "Einführung" in eine fremde Welt mithilfe eines "Führers" oder "Lehrers" CASTANEDA 1970. Die dort beschriebene Art des Lernens ist freilich nicht uneingeschränkt zu empfehlen. So stellt Castaneda nach dem Genuss von "Peyote" fest: "'Diese Art Lernen ist nichts für mich. Ich bin dafür nicht gemacht, Don Juan'" (CASTANEDA 1970/dt. 1973: 42).

Literatur

BIERE, Bernd Ulrich (1978): Kommunikation unter Kindern. Methodische Reflexion und exemplarische Beschreibung. Tübingen: Niemeyer.

CARROLL, Lewis (1963): Alice hinter den Spiegeln. Mit einundfünfzig Illustrationen von John Tenniel. Überse tzt von Christian Enzensberger. Frankfurt am Main: Insel-Taschenbuch, Insel 1975.

CASTANEDA, Carlos (1970): The teachings of Don Juan. Harmondsworth.-Übers: Die Lehren des Don Juan. Ein Yaqui-Weg des Wissens. Frankfurt am Main: Fischer, 1973.

FEYERABEND, Paul (1977): "Unterwegs zu einer dadaistischen Erkenntnistheorie". In: Unter dem Pflaster liegt der Strand, Bd. 4, hg. von H.P. DUERR, Berlin: Kramer: 9 - 88.

SCHÜTZ, Alfred (1932): Der sinnhafte Aufbau der sozialen Welt. Eine Einleitung in die verstehende Soziologie. Wien: Springer, Frankfurt am Main: Suhrkamp, 1974.

WINCH, Peter (1972): "Understanding a primitive society". In: WINCH, P.: Ethics and action, London, 8 - 49. - Übers: "Was heisst 'Eine primitive Gesellschaft verstehen'? In: WIGGERSHAUS, Rolf (ed.) (1975): Sprachanalyse und Soziologie. Die sozialwissenschaftliche Relevanz von Wittgensteins Spätphilosophie. Frankfurt am Main: 1975: 59 - 102.

RITEN DES ALLTAGS

Sequenzierungsbedingungen in präfigurierter Rede*

Florian Coulmas

1. Das Hauptmerkmal von Situationen alltäglicher Kommunikation ist ihre Komplexität. Der Grad der Komplexität bestimmt sich durch die Anzahl der möglichen Konfigurationen potentiell relevanter und als solche in die Definition von Kommunikationssituationen eingehender Faktoren. Für jeden Sprecher kommt es darauf an, ihre Komplexität zu ermessen, um seine eigenen Möglichkeiten, auf ihre Gestaltung bzw Fortführung Einfluß zu nehmen, beurteilen zu können. In der tatsächlichen Interaktion jede gegebene Situation für sich auf die sie konstituierenden Faktoren überprüfen zu müssen, wäre äußerst unökonomisch. Nun unterscheiden sich Kommunikationssituationen typischerweise hinsichtlich der Menge der dem Handelnden zur Verfügung stehenden Handlungsalternativen. Mit ihr variiert der Entscheidungsbereich des Handelnden. Vieles von dem, was täglich an sozialer Interaktion vollzogen wird, ist nicht einmalig, sondern wiederholt sich auf gleiche oder ähnliche Weise immer wieder, so daß der Entscheidungsbereich des Handelnden nicht jedesmal von neuem aufkonstruiert werden muß. Bestimmte Kommunikationssituationen existieren demgemäß als generische Typen. Als derartige generische Typen sind diese Situationen auf der Folie einer Kultur definiert, und abgesehen von naturgesetzlichen Beschränkungen richtet sich die Anzahl der jeweils möglichen Handlungsalternativen nach ihrer Maßgabe.

Wenn sich Handlungssituationen hinsichtlich der Menge von Handlungsalternativen der Partizipanten unterscheiden, so heißt das, daß der Grad der Fixiertheit ihrer Struktur variiert. Rekurrente Kommunikationssituationen neigen dazu, Muster zu ihrer Bewältigung auszubilden, durch deren Konventionalisierung der Entscheidungsbereich der Handelnden in den betreffenden Situationen u.U. drastisch eingeschränkt wird. Einschränkungen dieser Art rangieren auf einem Kontinuum zwischen S c h a b l o n e n für die äußere Absteckung von Art und ungefährer Folge von Handlungen resp. Teilhandlungen auf der einen Seite und a l t e r n a t i v e l o s e n H a n d -

l u n g s a b f o l g e n standardisierten Charakters auf der anderen . Hier sind sie unter dem Überbegriff "Riten des Alltags" zusammengefaßt. Riten, darüber herrscht weitgehend Einigkeit, sind Praktiken, die für eine Gesellschaft oder Kultur in besonderer Weise typisch sind. Sehr viel problematischer ist es, einheitliche Kriterien für die Identifizierung von Riten anzugeben, um sie von nicht ritualisierten Praktiken zu unterscheiden. Anthropologie und Soziologie suchten und fanden Riten vor allem in außeralltäglichen Bereichen im Zusammenhang mit individualgeschichtlich einmaligen Lebensab- resp. einschnitten (sog. Übergangsriten). Zudem wurde von Riten zumeist nur bezüglich präliterater Gesellschaften gesprochen. Hinter dem Vorurteil der Selbstverständlichkeit wurden die Idiosynkrasien des eigenen sozio-kulturellen Systems nur langsam und im Kontrast zu anderen Kulturen als solche sichtbar. Im wesentlichen hat die Soziologie seit DURKHEIM (1912) Riten im religiösen Leben exotischer resp. 'primitiver' Gesellschaften aufgespürt. Das alltägliche Leben in der industrialisierten Welt hat man erst in jüngerer Zeit begonnen, als einen vergleichbaren Untersuchungsgegenstand zu betrachten (cf. z.B. GOFFMAN 1972).

Verschiedene Merkmale springen als für rituelles Verhalten kennzeichnend besonders ins Auge: einmal initiiert, muß in einem Ritual eine Handlungsfolge auf ganz bestimmte Weise ohne Variation aublaufen. Die Ratio der Struktur einer solchen Handlungsfolge ist den Beteiligten dabei häufig nicht mehr völlig durchschaubar; ihre ursprüngliche Bedeutung ist m.a.W. verblaßt. Es ist deshalb nicht verwunderlich, daß Riten Bedeutung und Bedeutsamkeit im eigentlichen Sinn von denen, die sie praktizieren, oft abgesprochen wird, und daß sie als 'bloße Riten' oder 'Zwänge der Konvention' abgetan werden. Alltagsrituale sind, um es zusammenzufassen, Praktiken, in denen a) der Entscheidungsspielraum der Beteiligten reduziert ist und b) die Ratio der einzelnen Handlungszüge denen, die sie durchführen, nicht immer gänzlich transparent ist. Sie dienen der sozialen Integration. Der Übergang zwischen rituellem und nicht rituellem Verhalten ist ein kontinuierlicher, was nicht durch eine willkürliche strikte Trennlinie kaschiert werden sollte. Ein Ritual ist Ergebnis eines Prozesses: der Ritualisierung. Sie ist im wesentlichen als progressive Reduktion des Entscheidungsbereichs der

Partizipanten rekurrenter Kommunikationssituationen anzusehen.

2. In der Domäne sprachlichen Handelns manifestieren sich derartige Prozesse in der Routinisierung von Äußerungstypen, was wiederum zu einem direkten Niederschlag in der Sprache führen kann. Rekurrenz hat auch hier Konventionalisierung zur Folge: Sprachliche Einheiten, die sich als für die Erfüllung bestimmter kommunikativer Zwecke geeignet erwiesen haben, werden als Fertigteile in das Inventar einer Sprache aufgenommen und im Extremfall, bei Verlust jeglicher Produktivität, lexikalisiert. Die situationsadäquate Verwendung derartiger Ausdrücke, die als Versatzstücke bereitstehen, und zu denen es häufig keine Alternative gibt, stellt sich als p r ä f i g u r i e r t e R e d e dar. Insofern als präfigurierte Rede in starkem Maße an typifizierte Situationen gebunden ist und in ihren Vorkommensbedingungen den Erwartungen der Gruppenmitglieder entspricht, hat sie einen wohlbestimmten Platz im Normgefüge einer Gesellschaft. Ihre Hauptfunktion ist die Verringerung der Komplexität der Alltagskommunikation. Im wesentlichen wird das dadurch erreicht, daß präfigurierte Redemittel, d.h. Routineformeln im weitesten Sinne, relativ zu bestimmten Situationstypen hochmotiviert sind und daß die mit ihnen verbundenen Erwartungen und Wirkungen kollektiv definiert sind. Der Erfüllung dieser generellen Funktion sind verschiedene spezielle Funktionen subordiniert (cf. COULMAS 1978a).

Angesichts der starken situativen bzw. kontextuellen Konditioniertheit präfigurierter Rede stellt sich die Frage, ob derartige Sprachverwendung als Handlung im vollwertigen Sinne anzusehen ist (cf. REHBEIN 1977: 216). Sicher sind einige wesentliche Merkmale von Handlungen in präfigurierter Rede nicht gegeben, was sich vor allem in der starken Eingeschränktheit des Entscheidungsbereichs bemerkbar macht. Dies gilt besonders dort, wo präfigurierte Redemittel in Sequenzen vorkommen, so daß das Vorkommen einer Routineformel a die einzige notwendige und hinreichende Bedingung für die Verwendung einer anderen Routineformel b ist. Je enger die Verbindung zwischen Sequenzgliedern in präfigurierter Rede ist, desto größer ist die Berechtigung, sie als ritualisiert anzusehen. Unbeschadet ihres defizitären Handlungscharakters spielen derartige

"Riten des Alltags" als vorgeformte Ablaufmuster von Verhaltenseinheiten für die reibungslose Abwicklung der Alltagskommunikation eine überaus wichtige Rolle (cf. COULMAS 1978a,b).

3. Die Typen an die Verwendung präfigurierter Rede gebundener Alltagsrituale sind vielfältiger Natur und betreffen so unterschiedliche Verhaltensweisen wie Grüßen, Vorstellen, Danksagen, Glückwünschen, das Aufsagen von Kinderversen, Abzählreime (Eck, Speck, Dreck und Du bist weg!),Zeremonien im sakralen Bereich (Asche zu Asche, Staub zu Staub.) oder im profanen (Prosit Neujahr!), den Austausch verbaler Artigkeiten oder Unartigkeiten, Flüche, Drohungen, Komplimente sowie viele andere institutionsspezifische Sprechhandlungen. Es ist offensichtlich, daß sich Gesellschaften darin, in wie weit und auf welche Weise Interaktionen dieser Art ritualisiert sind, unterscheiden, und daß die Beschreibung solcher Muster somit eine selektive Beschreibung der Kultur der jeweiligen Sprechergemeinschaft ist. Typischerweise bestehen signifikante Unterschiede nicht nur zwischen den Strukturen einzelner ritualisierter Handlungsmuster verschiedener Gesellschaften, sondern auch hinsichtlich der generellen Einstellung gegenüber der Verwendung präfigurierter Rede und der Aktualisierung ritualisierter Handlungsmuster, die den Einzelnen in gegebenen Kontexten auf ganz spezifische Handlungszüge verpflichten (cf. GOFFMAN 1972, 1975: 63).

Sequenzierungsbedingungen spielen dabei eine doppelte Rolle bzw. werden auf zwei verschiedenen Ebenen relevant:

1. auf der Ebene der Organisation von Folge und Iteration abgeschlossener Alltagsrituale; und

2. auf der Ebene der internen Struktur von Alltagsritualen resp. präfigurierten Redeeinheiten.

4. Der erste Punkt betrifft die Fragen: (i) feststehende Interaktionsmuster welcher Art können aufeinander folgen bzw. können in welchen Kombinationen vorkommen? und (ii) unter welchen Bedingungen und in welchen Zyklen können welche Interaktionsmuster iteriert werden?

(i) Besonders ausgeprägt ist die Rigidität der Sequenzierungsreglementierung einzelner Muster in institutionellen Kommunikationsräumen: In kirchlichen und auch in standesamtlichen Trauungen ist der Platz jedes einzelnen Schrittes im Rahmen eines übergeordneten Ablaufschemas strikt festgelegt. Die generelle Bereitschaft zur Eheschließung kann nicht signalisiert werden, bevor die dazugehörigen peripheren Beschwörungen abgearbeitet sind. Mit den einzelnen Phasen, wie Erteilung von Lebensweisheiten, Ermahnungen und Gelöbnissen ist auch der Sprecherwechsel im Vorhinein eindeutig bestimmt. Längere Sequenzen, die in dieser Weise von Anfang bis Ende durchstrukturiert sind, finden sich nur selten. Vereidigungen, Ernennungen, Auszeichnungen, offizielle Sitzungen und andere Gelegenheiten mit Protokoll sind weitere Beispiele. In derartigen Interaktionszusammenhängen ist die Rederechtverteilung vielfach unabhängig von der normalen diesbezüglichen Funktionsweise der verwendeten sprachlichen Ausdrücke per Instanz eindeutig geregelt, was exemplarisch deutlich wird bei der uneigentlichen i.e. zitierten Verwendung von Ausdrücken z.B. beim Vorsprechen von Eidesformeln. Außerhalb institutioneller Kommunikationsräume kommt die vollständige Vorstrukturierung der Gesprächsorganisation nicht vor. Zu beobachten ist dort jedoch auch oft in einzelnen Bereichen eine geregelte Sequenzierung in variierender Rigidität. Typisch ist dabei, daß der höchste Grad diesbezüglicher Rigidität tendenziell in der Nähe von Interaktionsgrenzen auftritt. Relativ gut untersucht ist in dieser Hinsicht der terminale Bereich von Gesprächen von Sacks und Schegloff, die gezeigt haben, daß, bevor der Interaktionsabbruch vollzogen wird, in der Regel ein Angebot zum Abbruch gemacht und angenommen wird.

(ii) Interessant ist dabei, daß nicht nur zwischen Abbruchangebot und Abbruch beliebig viele Züge eingeschoben werden können, sondern daß auch der Abbruch selber rückgängig gemacht werden kann. Nach einem ausgetauschten Tschüß oder Wiedersehen kann der Faden sozusagen noch einmal aufgenommen werden, was häufig durch die Verwendung solcher Floskeln wie was ich noch sagen wollte... oder ach, da fällt mir noch ein... gekennzeichnet wird. Der Interaktionsabbruch kann also gewissermaßen wiederholt werden. Bei Interaktionseröffnungen ist dies nicht ganz trivialer Weise nicht der Fall. Die wiederholte Eröffnung einer Interaktion ist offenkundig redundant, was sich in den Verwendungsbedingungen von Grußformeln niederschlägt. Dies nicht nur in der Weise, daß Begrüßungsformeln nicht in unmittelbarer Folge oder in ein und demselben Gespräch wiederholt werden können. Offenbar haben Grußformeln einen zeitlichen Skopus ihrer Wirksamkeit. Wenn man sich begrüßt hat, was jedem anderen Interaktionszug vorausgehen muß, hält die Wirkung des Grußes normalerweise den ganzen Tag an. Wir ham' uns heute schon gesehen ist eine Floskel, die typischerweise verwendet wird, um einen wiederholten Gruß zu vermeiden und diese Vermeidung vor Dritten als nicht unhöflich zu kennzeichnen.

Ein Faktor, der in die Verwendungsbedingungen von Grußformeln und damit in die Bedingungen ihrer Iterierbarkeit eingeht, ist somit die Zeitspanne, die seit der letzten Begegnung der Kommunikanten vergangen ist. Daß gegenüber einer Person, wenn man ihr guten Tag gewünscht hat, das Soll für diesen Tag erfüllt ist, scheint in der Natur der Sache zu liegen. In anderen Gesellschaften haben Grüße jedoch einen anderen Effizienzskopus und folglich gelten für sie andere Iterierbarkeitsbedingungen. In südamerikanischen Gesellschaften erlischt die Obligation, bei der Interaktionseröffnung ein elaboriertes Grußmuster zu aktivieren, nicht damit, daß dies an dem betreffenden Tag schon einmal geschehen ist (CONDON/YOUSEF 1975). In unserem Kulturkreis erlischt eine derartige Obligation nicht nur: sie ist nicht einmal mehr als Option gegeben. Mehrfaches ausführliches Grüßen weicht von der Normalität des Verhaltens ab. Andererseits können etwa Lob, Dank, Entschuldigungen oder irgendwelche Beteuerungen durchaus mehrfach zum Ausdruck gebracht werden, was u. U. der Bekräftigung dient. Während also der Vollzug bestimmter Alltagsrituale wie Grüßen, Beglückwünschen, Danken, Entschuldigen etc. in die Interaktionsgeschichte der Beteiligten eingehende Fakten schafft, bringen diese es - anders als in institutionellen Kommunikationsräumen - nicht zwangsläufig mit sich, daß die sie konstituierenden Handlungen nicht wiederholt werden dürfen. Die diesbezüglichen Bedingungen sind in starkem Maße kulturspezifisch.

5. Der zweite oben genannte Punkt betrifft die interne Strukturierung von Alltagsritualen im Hinblick auf (i) Kombination und Folge routinisierter Äußerungstypen eines Sprechers; und (ii) durch

routinisierte Äußerungstypen bedingte Sprecherwechsel. Beide Aspekte hängen miteinander zusammen. Präfigurierte Redeeinheiten haben in ihrem Vorkommen oft gesprächsorganisierende Funktion, insofern als sie Bestandteile von Sequenzen sind, deren einzelne Teile Sprecherwechsel notwendig oder möglich machen. Durch die Verwendung vieler Formeln ist daher die Folge möglicher Äußerungen eines Sprechers notwendig begrenzt, da durch sie funktionale Leerstellen für Routinen des/eines Sprechers eröffnet werden. Dies ist so bei Formelpaaren (cf. SCHEGLOFF/SACKS 1973, "adjacency pairs"). Die beiden Elemente, aus denen Formelpaare bestehen, sind entweder symmetrisch oder komplementär aufeinander bezogen:

Formelpaar
- symmetrisch: Initialformel = Reaktivformel (Echo)
- komplementär: Initialformel ≠ Reaktivformel (Komplement)

In ersterem Fall ist die Reaktivformel eine genaue Kopie oder ein Echo der Initialformel, wie z.B. bei *hallo - hallo*; *bis dann - bis dann*; *Tag - Tag*; *Aufwiederhören - Aufwiederhören*;*Prost - Prost*; *Mahlzeit - Mahlzeit*; etc. Im zweiten Fall wird durch die Verwendung der Initialformel der Kontext für eine andere, von ihr verschiedene Formel geschaffen, so z.B. bei *Danke - Bitte*; *Machs gut - Machs besser*; *Herzlichen Glückwunsch - Vielen Dank*; *Darf ich vorstellen, N.N. - Freut mich sehr*; *Verzeihung - Macht nichts*; u.a. Dabei kann die Reaktivformel entweder polyfunktional sein und mit verschiedenen Initialformeln Formelpaare bilden, wie das etwa für die Formel *keine Ursache* gilt, die als komplementäre Reaktivformel von Dankes- und Entschuldigungsformeln vorkommen kann, oder sie ist an eine einzige Initialformel bzw. einen einzigen Initialformeltyp gebunden so daß deren Äußerung den einzigen Vorkommenskontext für sie darstellt. Ein Beispiel dafür ist die Reaktivformel *Gern geschehen*, die nur in Danksagungskontexten vorkommen kann. Allen angeführten Fällen gemeinsam ist, daß jeweils das erste Sequenzglied als Stichwort des zweiten Sprechers dient und somit einen Sprecherwechsel erzwingt oder zumindest ermöglicht. Daß die durch die Äußerung des ersten Sequenzgliedes eröffnete Leerstelle nicht ausgefüllt wird, kann als Ausnahme gelten, die nur selten kein Verstoß gegen die Norm ist.

Allerdings können Formeln, die Initialglieder von Formelpaaren sind, partiell auch so verwendet werden, daß das Rederecht nicht abgegeben wird bzw. daß die Äußerung der Reaktivformel vom ersten Sprecher inhibiert wird. Häufig zu beobachten ist dies bei der antizipatorischen oder uneigentlichen Verwendung von Entschuldigungsformeln, die gesprächsstrukturierend als Rahmenfloskeln für die Zurückweisung von Argumenten eingesetzt werden können: Entschuldigen Sie, aber.... Die sonst sehr feste Verbindung zwischen Gliedern von Formelpaaren ist ein Hinweis auf bestimmte allgemeine Kommunikationsnormen: Grüße und Wünsche der verschiedensten Art müssen erwiedert werden, Ausdrücke von Dank honoriert, Entschuldigungen akzeptiert resp. gewährt, Angebote gratifiziert werden etc., wenn sie in der gesellschaftlich geforderten Form vorgebracht werden. Diese ist wiederum Gegenstand kulturspezifischer Variation: Während in einer Sprache, z.B. dem Deutschen oder Englischen, präfigurierte Redeeinheiten vielfach als Bestandteile von Formelpaaren verwendet werden, stellen andere Sprachen, z.B. das Japanische, zahlreiche Formeln bereit, deren Verwendung nicht die Obligation einer Reaktion etabliert, die also eingliedrige Sequenzen konstituieren.

6. In alltäglichen Kommunikationsräumen scheinen in feste Muster gebundene präfigurierte Redeeinheiten in höchstens dreigliedrigen Sequenzen vorzukommen. Die Funktion einzelner Sequenzglieder kann jedoch komplex sein, wobei die Routinisierung der Kommunikationssituationen teilweise zu einem Höchstmaß an Ökonomie durch Verkürzung geführt hat: Mit der Reaktivformel Danke gleichfalls z.B. wird gleichzeitig ein Wunsch honoriert und erwiedert. In wie starkem Maße die Applikationsbedingungen solcher Formeln normativ bestimmt sind, wird hier sehr deutlich. Sie richten sich nicht nach Maßgabe der potentiellen Fungibilität. Danke gleichfalls kann nur als Reaktion auf ganz bestimmte Initialformeln folgen, obgleich sie sich auch in anderen Fällen als Mittel zum Ausdruck von Dank und Erwiederung anböte. Die Sequenz Guten Abend - Danke gleichfalls ist auf eine ganz bestimmte, diese Eigenschaft exemplifizierende Weise unakzeptabel. Die Ersetzung der Initialformel durch Schönen Abend gibt einen Hinweis auf den Grund. Schönen Abend kann als Grußformel verwendet werden, die gleichzeitig einen

Wunsch vermittelt. Diese Funktion hat die Formel Guten Abend, obgleich sie aus ihrer komponentiellen Bedeutung noch klar ableitbar wäre, durch Routinisierung vorloren. Sie ist zu einer reinen Grußformel geworden, deren wörtliche Bedeutung für ihre Funktion keine oder nur eine periphere Rolle spielt.

Hieraus läßt sich ein allgemeines Prinzip ableiten, das für die Sequenzierungsbedingungen vieler präfigurierter Redeeinheiten gilt: das Prinzip der Dominanz der Funktion über die wörtliche Bedeutung. Etwas präzisiert besagt es Folgendes: Gehört eine Initialformel zu einem Formelpaar, so steht der Angesprochene unter der Obligation, die entsprechende Reaktivformel zu verwenden. Gehört die Initialformel zu keinem Paar und bieten sich für die Reaktion verschiedene Möglichkeiten an, so darf mit der Reaktion nicht auf die wörtliche Bedeutung der Initialformel Bezug genommen werden.

Diese allgemeine, unmittelbar die Semantik präfigurierter Redeeinheiten betreffende Restriktion, die nicht für alle aber für viele von ihnen gilt, ist ein deutliches Indiz für den ritualisierten Charakter der Kommunikationszusammenhänge, in denen sie verwendet werden und für ihre Tendenz, zu idiomatischen Ausdrücken verfestigt zu werden.

Literaturauswahl

ADATO, Albert (1976): "Alltägliche Ereignisse - ungewöhnlich erfahren. Eine vergleichende Analyse von Erfahrungen des Abschiednehmens." WEINGARTEN, Elmar et al.(eds.)(1976): Ethnomethodologie. Beiträge zu einer Soziologie des Alltagshandelns. Frankfurt/M.: Suhrkamp: 179-202.

CONDON, John C. / Yousef, Fathi S. (1975): An Introduction to Intercultural Communication. New York: Holt.

COULMAS, Florian (1978a): "On the Sociolinguistic Relevance of Routineformulae." Erscheint in: Journal of Pragmatics 3/2.

--- (1978b) "Pragmatic Interferences in the Use of Routine Formulae. Their Role in Acculturation." Vortrag auf dem V. Kongreß der AILA 1978, Montréal. Erscheint demnächst.

DURKHEIM, Emile (1912): Les formes élémentaires de la vie religieuse. Paris: Bibliothèque de philosophie contemporaine.

GARFINKEL, Harold (1967): Studies in Ethnomethodology. Englewood Cliffs: Prentice-Hall.

GOFFMAN, Erving (1964): "The Neglected Situation." American Anthropologist vol. 66 No. 6 part 2: 133-136.

--- (1972): Interaction Ritual. Harmondsworth: Penguin Books.

--- (1974): Frame Analysis. Harmondsworth: Penguin Books.

REHBEIN, Jochen (1977): Komplexes Handeln. Stuttgart: Methler.

SCHEGLOFF, Emanuel A. / SACKS, Harvey (1973): "Opening up Closings." Semiotica 8: 289-327.

* Diese Arbeit entstand im Zusammenhang eines Forschungsprojekts der DFG (DFG-Az. Wu 86/4), deren Unterstützung hier dankbar vermerkt sei.

ÜBER DAS ABGIESSEN VON REIS

Ein Kommunikationsbedarfselement und Möglichkeiten seiner Deckung

Reinhard Fiehler

1. Aufgabenstellung

Jede Kooperation besitzt einen bestimmten Kommunikationsbedarf, und dieser Kommunikationsbedarf muß von den Beteiligten - durch Äußerungen - gedeckt werden, wenn die Kooperation gelingen soll. D.h. kooperative Praxis erfordert Kommunikation, sie muß kommunikativ organisiert werden.

Jede Kooperation läßt sich analytisch aufgliedern in vier Phasen: (1) die Konstitution des Teilnehmerkreises, (2) die Bildung einer identischen Zielvorstellung (die in der Kooperation durch die Beteiligten realisiert werden soll), (3) die Planung der Ausführung und (4) die Ausführung. Die Phasen wiederum können sich aus mehreren Sequenzen zusammensetzen. In diesen Sequenzen bzw. Phasen nun fallen einzelne Elemente des Kommunikationsbedarfs an und müssen befriedigt werden, damit die Kooperation weitergeführt werden kann.

Bei einigen Kooperationsformen gehört die Übermittlung der Zielvorstellung, die kooperativ realisiert werden soll, zu ihrem Kommunikationsbedarf. Ersichtlich ist dies ein Element des Kommunikationsbedarfs, das der Phase (2) Bildung der identischen Zielvorstellung zuzurechnen ist: Ein Individuum entwickelt eine Zielvorstellung, die es kooperativ realisieren will, und dieses Ziel muß es eben kommunikativ potentiellen Teilnehmern vermitteln, damit sie an seiner Realisierung mitwirken können.

Untersucht werden soll in diesem Beitrag, wie bei einer bestimmten faktischen Kooperation dieses Kommunikationsbedarfselement gedeckt wird und durch welche anderen Äußerungen es auch hätte gedeckt werden können. Dabei soll erklärt werden, wie es möglich ist, daß diese an der sprachlichen Oberfläche sehr heterogene Klasse von Äußerungen bei der betreffenden Kooperation das gleiche - eben die Übermittlung der Zielvorstellung - leistet, die Äußerungen also funktional äquivalent sind.

Sodann wird untersucht, welche sprachlichen Mittel zur Verfügung stehen müssen, damit diese Klasse funktional äquivalenter Äußerungen gebildet werden kann.

Aufgabenstellung dieses Beitrags ist es also, die Möglichkeiten und Bedingungen der kommunikativen Organisation kooperativer Praxis anhand eines bestimmten Kommunikationsbedarfselements zu untersuchen. Damit illustriert und konkretisiert diese Analyse an einem Beispiel die Kanngießer'sche Hypothese der Sprachfunktionalität[1], die besagt, daß als Folge der gesellschaftlichen Notwendigkeit, kooperative Prozesse zu organisieren, bestimmte Lagen des Kommunikationsbedarfs entstehen, die, damit sie durch Äußerungen befriedigt werden können, die Ausbildung bestimmter sprachlicher Mittel erforderlich machen.

2. Zur Erforderlichkeit der Zielübermittlung bei verschiedenen Kooperationsformen

Nicht bei allen Kooperationsformen ist die Zielübermittlung ein Element des Kommunikationsbedarfs. Einerseits ist sie bei Kooperationen nicht erforderlich, bei denen zwischen den Beteiligten keine symmetrischen sozialen Beziehungen bestehen: Aufgrund der Asymmetrie ist es hier möglich, daß ohne Zielübermittlung lediglich Teiltätigkeiten zugewiesen werden. Andererseits ist sie bei iterativen Kooperationen nicht erforderlich, weil hier als Folge der Wiederholung das Ziel allen Beteiligten aus vorhergehenden Durchführungen der entsprechenden Kooperation bekannt ist

Hingegen gehört die Zielübermittlung bei Kooperationen mit symmetrischen Beziehungen zwischen den Beteiligten in der Regel zum Kommunikationsbedarf. Sie hat u.a. die Funktion, einem potentiellen Teilnehmer die K e n n t n i s der (kooperativ zu realisierenden) Zielvorstellung zu vermitteln, damit er eine E n t s c h e i d u n g treffen kann, ob er diese Zielvorstellung übernimmt und sich an ihrer Realisierung beteiligen möchte. Eben diese Tatsache, daß bei symmetrischen Beziehungen jedem Beteiligten diese Entscheidung selbst vorbehalten ist, macht die Zielübermittlung zu einem Element des Kommunikationsbedarfs.

Betrachten wir nun eine Kooperation - unsere Beispielskooperation -, bei der die Zielübermittlung zum Kommunikationsbedarf gehört und dieses Kommunikationsbedarfselement auch - durch die

Äußerung (77) - befriedigt wird. Es handelt sich um eine Miniaturkooperation, die in eine größere Kooperation "Gemeinsam Kochen"[2] eingelagert ist: G will zusammen mit R den Reis abgießen.

(76) G laut zu R: Reini, kommst du mal?
Du mußt mir helfen.
R kommt in die Küche.
(77) G zu R: Ich will den Reis abgießen.
(78) R zu G: Soll ich?
(79) G zu R: Nein, nimm mal das Sieb.
R hält das Sieb.
G gießt ab.

Die Zielübermittlung ist insbesondere erforderlich, da für G und R keine gemeinsame Wahrnehmungssituation besteht, die eine kommunikationsfreie Alternative ermöglichen würde: R kann aufgrund der Wahrnehmungssituation beispielsweise nicht aus den (beobachteten) Tätigkeiten von G erschließen, daß sie zu einem bestimmten Zeitpunkt den Reis abgießen will.

3. Zur Explikation der Begriffe Zielvorstellung und Zielübermittlung

G bildet - in Abhängigkeit von der übergeordneten Kooperation "Gemeinsam Kochen" - eine Zielvorstellung aus, die sie kooperativ realisieren will, d.h. sie vollzieht eine Reihe von Bewußtseinstätigkeiten. Diese Zielvorstellung muß sie nun in eine Äußerung - also eine kommunikative Tätigkeit - umsetzen, oder anders gesagt: sie muß sie mittels einer Äußerung sprachlich abbilden. Diese sprachliche Abbildung versetzt R in die Lage, Bewußtseinstätigkeiten zu vollziehen, deren Resultat eine ähnliche Zielvorstellung bei ihm ist.

Was aber ist jetzt des genaueren eine Zielvorstellung und was heißt es, dieses Produkt von Bewußtseinstätigkeit mittels einer Äußerung sprachlich abzubilden?

Zielvorstellungen sind das Resultat einer Bewußtseinstätigkeit - der Antizipation - und somit mentale Phänomene.[3] Sie sind Antizipationen nicht-existenter (zukünftiger) Zustände. Aber nicht alle Zustandsantizipationen sind Ziele. Es handelt sich um Zustände, die durch Tätigkeiten realisiert werden sollen. Damit ist zweierlei klar: es handelt sich nicht um Zustände, die ohnehin (ohne daß jemand tätig wird) eintreten, und es sind gewollte

Zustände, d.h. den Zielen liegen bestimmte Bedürfnisse und Interessen zugrunde.

Zwischen dem gegenwärtigen Zustand, den ein Individuum mental repräsentiert, und dem zukünftigen Zustand, den es bewußtseinsmäßig antizipiert, besteht eine Differenz. Faßt man Zustände als Mengen von Zustandselementen auf, so ist diese Differenz beschreibbar als Unterschied dieser Mengen. Mit der Auffassung eines Zustandes als Menge von Zustandselementen ist zugleich eine Möglichkeit gegeben zu beschreiben, wie Antizipationen vorgenommen werden: nämlich durch die Schaffung einer neuen Kombination von Zustandselementen im Bewußtsein, oder anders gesagt: durch das Umbilden einer Menge von Zustandselementen, wobei einzelne Elemente eliminiert, durch andere ersetzt und neue Elemente hinzugefügt werden können. Wird beispielsweise ein Zustand bewußtseinsmäßig repräsentiert, in dem als ein Element ein Topf mit kochendem Reis vorkommt, so kann ein Ziel gebildet, d.h. ein anderer Zustand antizipiert werden, indem dieses Element 'ersetzt' wird durch ein Element "Topf leer und Reis in einem Sieb".

Zielvorstellungen können im Bewußtsein auftreten als Vorstellung des Zielzustandes oder als Vorstellung der zielrealisierenden Tätigkeiten. Diese Tatsache hat Konsequenzen für die sprachliche Abbildung von Zielen.

Ist eine Zielvorstellung gebildet, so muß sie, wenn sie einem anderen Individuum übermittelt werden soll, sprachlich durch eine Äußerung abgebildet werden. Die sprachliche Abbildung besteht in der Abbildung der Elemente des antizipierten Zustandes (wobei in der Regel nur die 'neuen' Elemente abgebildet werden müssen) oder in der Abbildung der zielrealisierenden Tätigkeit(en).

Nicht nur die Zielvorstellung muß relativ präzise sein, sondern auch ihre Abbildung. Stellt sich ein Hörer aufgrund der zielübermittelnden Äußerung ein anderes Ziel vor, so eröffnet dies nicht die Möglichkeit eines sinnvollen Zusammenwirkens.

Im Fall des Beispiels wird nicht der antizipierte Zustand abgebildet, sondern die zielrealisierende Tätigkeit. Nicht nur diese beiden Möglichkeiten, sondern auch die Tatsache, daß in beiden Fällen eine Klasse funktional äquivalenter Äußerungen diese Abbildung leistet, zeigt, daß ein Ziel die zielabbildende Äußerung nicht eindeutig festlegt.

4. Funktionale Äquivalente der zielübermittelnden Äußerung

Funktionale Äquivalente der Äußerung (77) Ich will den Reis abgießen. sind beispielsweise:[4]

Funktionale Äquivalente, die die zielrealisierende(n) Tätigkeit(en) abbilden

(1) Ich muß / möchte / beabsichtige, / habe vor, — den Reis / das Reiswasser — {abgießen. / ausschütten / abschütten} / abspülen / abschrecken

(2) (Reis) abgießen.
: :

(3) Es ist notwendig, / erforderlich — den Reis abzugießen.
: :

(4) Der Reis soll / muß / kann / müßte / sollte — abgegossen werden.
: :

(5) Wir wollen / müssen / können / Laß uns — den Reis abgießen.
: :

(6) (Los du Hund / Sei so nett) Hilf mir mal den Reis abgießen.
: :

(7) Ich kann den Reis nicht alleine abgießen.
: :

(8) Muß ich den Reis auch noch alleine abgießen?
: :

Funktionale Äquivalente, die den antizipierten Zustand abbilden

(9) Der Reis muß / kann / soll — ins Sieb. / auf den Tisch.

Funktionale Äquivalente, die weder Zielzustand noch die zielrealisierende Tätigkeit abbilden, aber die Möglichkeit eröffnen, diese zu erschließen

(1o) Der Reis ist fertig. / gar / weich / so weit

(11) (Der) Reis.

(12) [(a) Auf den Topf weisende Geste / (b) Den Topf abgießbereit vom Herd nehmen]

(13) (a) Halt mal das Sieb. [G nimmt den Reistopf]
[G nimmt das Sieb] Gieß mal den Reis ab.
(b) Willst du das Sieb halten oder soll ich es?

5. Zur Erklärung der funktionalen Äquivalenz dieser Äußerungen

Im folgenden soll erklärt werden, wie es möglich ist, daß (1) - (13) funktionale Äquivalente zu (77) sind, d.h. es soll geklärt werden, wie sie die Zielübermittlung gewährleisten. Für diese Erklärung müssen wir ansatzweise die Kette der Bewußtseinstätigkeiten rekonstruieren, die G vollzieht, bevor sie (76) äußert. Wie diese Kette auch immer faktisch genau ausgesehen haben mag, sie muß mindestens fünf Glieder/Zwischenresultate enthalten:

(A): Der Reis ist fertig. (Repräsentation des gegenwärtigen Zustandes (Ist-Zustand))

(B): Der Reis muß ins Sieb. (Antizipation des zukünftigen Zustandes (Soll-Zustand))

(C): Ich will den Reis abgießen. (Antizipation der zielrealisierenden Tätigkeit)

(D): Ich kann es nicht alleine tun. (Entschluß zur kooperativen Realisierung des Ziels)

(E): Jemand - R - muß mir helfen. (Planung des Teilnehmerkreises der Kooperation)

Resultat der Kette (A) - (E) ist die Äußerung (76), wobei (76) (D) und (E) im wesentlichen (für R erkennbar) enthält. (76) stellt die Sequenz "Kontaktieren der potentiellen Teilnehmer" der Phase (1) Konstitution des Teilnehmerkreises dar. Durch sein Kommen signalisiert R seine grundsätzliche Kooperationsbereitschaft, obwohl er das Ziel der Kooperation noch nicht kennt.

In dieser Situation könnte G alle fünf Zwischenresultate ihrer Bewußtseinstätigkeiten sprachlich abbilden, etwa mit folgender Äußerung: "Der Reis ist fertig. Er muß ins Sieb. Ich will ihn abgießen, kann es aber nicht alleine tun. Du mußt mir helfen." Dies ist sozusagen das 'größtmögliche' funktionale Äquivalent zu (77). Diese Äußerung würde bei R keinerlei Erschließungsleistungen erfordern. Jede weniger umfassende Wiedergabe dieser Kette erfordert, daß R Teile erschließen muß, und beruht damit auf einer bestimmten Hypothese von G über R's Fähigkeiten zur Erschließung und sein Wissen.

Faktisch bildet G aber nicht (A) - (E) ab.[5] Mit ihrer Äußerung (77) gibt sie lediglich (C) wieder: Sie übermittelt nur die zielrealisierende Tätigkeit. Sie setzt damit voraus, daß R (A) und (B) aus (C) erschließen kann. Geht man also davon aus, daß G an diesem Punkt der Kooperation nur die zielrealisierende Tätigkeit übermitteln will, so stellt man bei einer Durchsicht der funktionalen Äquivalente fest, daß unter dieser Voraussetzung nur (1) - (5) funktionale Äquivalente in einem engeren Sinne sind, weil nur bei ihnen die zielrealisierende Tätigkeit 'rein' abgebildet wird.

Auch die Fälle (6) - (8) bilden (C) ab, sie bilden aber zugleich mehr ab: (6) enthält zusätzlich - sprachlich explizit - (E), (7) und (8) enthalten (D). (8) und die Klammer in (6) zeigen, daß außer der zielrealisierenden Tätigkeit auch noch ganz andere Komponenten - hier: die soziale Beziehung - (sozusagen nebenbei) in einem funktionalen Äquivalent thematisiert werden können.

Statt (C) bzw. (C) + Komponenten der anderen Glieder abzubilden, könnte G auch (B) abbilden, also den antizipierten Zustand. Dies erfordert dann, daß R (A) und (C) erschließt. Insofern ist auch (9), das (B) abbildet, funktionales Äquivalent.

Ferner ist denkbar, daß G (A) abbildet, was erfordert, daß R (B) und (C) erschließt. Dieser Fall ergibt die funktionalen Äquivalente (1o) und (11), wobei (11) als elliptische Form von (1o) interpretiert werden kann, mit der R auf ein Element des Ist-Zustandes hingewiesen wird.

(12a) ist ein gestensprachliches - also nichtverbales - Äquivalent zu (11). (12b) ist demgegenüber expliziter: Es handelt sich um eine Tätigkeit, die - da R sie wahrnehmen kann - mit kommunikativer Funktion eingesetzt wird und R zur Erschließung von (C) veranlassen soll. (12b) springt - wie auch (13) - in die nächste Phase der Kooperation: die Planung der Ausführung. Dabei stellen (12b) und (13a) auf der einen Seite-hier wird R von der Entscheidung über die Tätigkeitsverteilung ausgeschlossen - und (13b) auf der anderen Seite verschiedene Formen der Planung der Tätigkeitsverteilung dar. In diesen Fällen (12b) - (13b) wird von der Kette (A) - (E) nichts abgebildet; sie ist zur Gänze von R zu erschließen.

Bei allen funktionalen Äquivalenten ist also (B) - der antizipierte Zustand - oder (C) - die zielrealisierende Tätigkeit - entweder sprachlich explizit abgebildet oder erschließbar. Damit ist in allen Fällen die Zielübermittlung - direkt oder indirekt - kommunikativ gewährleistet, die ihrerseits bei dieser Kooperationsform die Voraussetzung für die Weiterführung der Kooperation nach der Äußerung (76) ist.

Dieses höchst simple Beispiel zeigt hoffentlich in aller Deutlichkeit, wie komplex das Problem der Deckung eines Kommunikationsbedarfselements in der Kooperationspraxis gelöst werden kann und welche Vielfalt von Alternativen an diesem Punkt der Kooperation möglich ist. Es zeigt ferner, daß die bedarfsdeckende kommunikative Tätigkeit nur die Spitze eines 'Eisbergs' von Bewußtseinstätigkeiten ist, von Bewußtseinstätigkeiten, die auf der Grundlage von Annahmen über das Wissen der anderen Beteiligten vollzogen werden. Und es zeigt letztlich, wie automatisiert diese Bewußtseinstätigkeiten (in Form von Schlüssen, Entscheidungen etc.) sein müssen, da sie uns kaum mehr ins Bewußtsein kommen, wenn wir kooperieren.

6. Erforderliche sprachliche Mittel zur Bildung der funktionalen Äquivalente im engeren Sinne

Nach der Erklärung der funktionalen Äquivalente im weiteren Sinne soll nun auf die funktionalen Äquivalente im engeren Sinne eingegangen werden.

Faktisch wählt G zur Zielübermittlung die 'reine' Abbildung der zielrealisierenden Tätigkeit, sie gibt also (C) wieder. Äquivalent hierfür sind (1) - (5). Welche sprachlichen Mittel müssen nun zur Verfügung stehen, damit dies durch eine Äußerung geleistet werden kann?

Folgende vier Ausdrucksmöglichkeiten müssen zur Abbildung einer antizipierten zielrealisierenden Tätigkeit gegeben sein:

1. Möglichkeit zur Referenz auf eine Tätigkeit
2. Möglichkeit zur Referenz auf das Objekt der Tätigkeit
3. Möglichkeit zu indizieren, daß es sich um eine geplante, zukünftige Tätigkeit handelt
4. Möglichkeit zur Referenz auf den/die Tätigkeitsausführenden

Für unseren bestimmten Fall stehen zur Befriedigung dieser Erfordernisse bestimmte sprachliche Konventionen[6] zur Verfügung.

Z.B. kann zur Referenz auf die infragestehende Tätigkeit die Bedeutungskonvention "abgießen" benutzt werden. Es bestehen allerdings auch Bezeichnungsalternativen. Als Indikator für die Geplantheit bzw. Zukünftigkeit der Tätigkeit stehen das Tempussystem, Zeitadverbien oder auch - vermittelt über Schlußprozesse - (verschiedene) Modalverben bzw. eine Konstruktion wie "Es ist notwendig, daß ..." zur Verfügung.

Wie (2) - (4) zeigen, ist es nicht notwendig, alle Ausdruckserfordernisse sprachlich zu realisieren. Eine elliptische Minimalfassung ist (2), bei der nur die Tätigkeit (und das Tätigkeitsobjekt) abgebildet wird. Die fehlenden Teile können erschlossen werden.

Zur Bildung der verschiedenen funktionalen Äquivalente sind durchaus unterschiedliche Konventionen erforderlich. (1) - (5) stimmen aber darin überein, daß sie alle unter Verwendung z.B. der Benennungskonventionen "Reis" und "abgießen" für das Tätigkeitsobjekt und die Tätigkeit gebildet wurden.

Wie die Tatsache, daß (77) faktisch geäußert wird, zeigt, verfügt G über entsprechende sprachliche Mittel bzw. Bedeutungskonventionen, um das Kommunikationsbedarfselement "Zielübermittlung" durch eine Äußerung decken zu können. Das Problem der Deckung stellt sich also für sie dar als Auswahl geeigneter Konventionen aus dem Konventionensystem, über das sie verfügt. Sind die sprachlichen Mittel bzw. Konventionen nicht hinreichend, um den Bedarf decken zu können, so müßten entsprechende Mittel erst ausgebildet, eingeführt und konventionalisiert werden.[7]

Mit der Thematisierung der erforderlichen Konventionen ist der Durchgang durch das funktionalistische Schema

(FS) Kooperation ⟶ Kommunikationsbedarf ⟶ Äußerungen ⟷
Sprachliche Mittel/Konventionen

abgeschlossen. Für ein Kooperationssegment und das entsprechende Kommunikationsbedarfselement haben wir damit das funktionalistische Schema exemplifiziert und bis auf die Ebene konkreter Äußerungen und der ihnen zugrundeliegenden Konventionen verfolgt.

Wie die Analyse zeigt, sind die Beziehungen zwischen den einzelnen Gliedern des Schemas keineswegs determinstisch. Im Gegenteil, sie sind komplexer, als man es auf den ersten Blick vermuten möchte.

Anmerkungen

1 Cf. beispielsweise KANNGIESSER, Siegfried: "Spracherklärungen und Sprachbeschreibungen". WUNDERLICH, Dieter (ed.) (1976): Wissenschaftstheorie der Linguistik. Kronberg: Athenäum: 1o6-16o. Seziell: 131.

2 Cf. für die vollständige Wiedergabe der Kooperation "Gemeinsam Kochen" : BRÜNNER, Gisela / FIEHLER, Reinhard: "Gemeinsam Kochen. Rekonstruktion der kommunikativen und praktischen Tätigkeiten beim kooperativen Kochen einer Paella." CONTE, Maria-Elisabeth / GIACALONE RAMAT, Anna / RAMAT, Paolo (eds.) (1978): Sprache im Kontext. Akten des 12. Linguistischen Kolloquiums. Pavia 1977. Bd.2. Tübingen: Niemeyer: 151-153.

3 Als solche sind sie Gegenstand der Psychologie, die mir verzeihen möge, daß ich auf ihrem Gebiet dilettiere. Die folgenden Aussagen erheben nicht den Anspruch, den empirischen Prozeß der Zielbildung und -übermittlung zu beschreiben.

4 Die hier angegebenen funktionalen Äquivalente wurden intuitiv auf der Grundlage der eigenen Kooperaticnsfähigkeit und kommunikativen Kompetenz gewonnen.
Ich erhebe nicht den Anspruch, alle funktional äquivalenten Äußerungen, die in dieser Situation möglich sind (d.h. alle Äußerungen, die den Fortgang dieser Kooperation ermöglichen), erfaßt zu haben. Dies ist auch sicher nicht der Fall, da jederzeit Minimalvariationen möglich sind. Dennoch scheint es mir höchst wahrscheinlich, daß es sich um eine endliche Menge handelt.

5 Sie braucht dies - zumindest in Hinblick aud (D) und (E) - auch nicht zu tun, weil (D) und (E) - wie gesagt - in (76) "Du mußt mir helfen." schon ausgedrückt sind.

6 Ich verwende den Koventionsbegriff im Sinne von KUMMER, Werner (1975): Grundlagen der Texttheorie. Reinbek: Rowohlt.

7 In diesem Zusammenhang wäre das Problem der Sprachentwicklung bzw. Sprachveränderung zu thematisieren.

ICH WILL / KANN / MUSS / SOLL ... DEN REIS ABGIESSEN
Zur Funktion verschiedenen Modalverbgebrauchs

Angelika Redder

1. Fragestellung

Meine Überlegungen stehen in engem Zusammenhang mit den Untersuchungen von FIEHLER (1979). Das konkrete sprachliche Material, von dem wir ausgehen, ist identisch, nämlich die Teilkooperation "Abgiessen von Reis"; der transkribierte Text ist in Fiehlers Aufsatz abgedruckt. Untersuchungsgegenstand im engeren Sinn ist auch für mich die Äußerung (77), speziell der Gebrauch eines Modalverbs in dieser Äußerung:

(77) Ich will den Reis abgießen.

Fiehler arbeitet heraus, daß die Äußerung (77) der Übermittlung des Ziels der Kooperation dient; darüber hinaus diskutiert er eine Klasse von funktionalen Äquivalenten. An der sprachlichen Oberfläche dieser (funktionalen) Paraphrasen fällt auf, daß fast alle modale Ausdrücke enthalten. Ich will mich auf die Untersuchung der Modalverben beschränken und folgenden Fragen nachgehen:

- inwiefern tragen die verschiedenen Modalverben dazu bei, das Ziel der (Teil)kooperation zu übermitteln
- wie ist dieser Beitrag zur Kommunikation Funktion für die einzelnen Modalverben zu differenzieren und was bedeutet diese Differenzierung im Hinblick auf die Illokution.

Um diese Fragen behandeln zu können, will ich zunächst grob den Unterschied zwischen modaler und nicht-modaler Formulierung beschreiben und die Bedeutung der Modalverben 'möchten', 'können', 'müssen', 'dürfen', 'wollen', 'sollen' umreißen.

2. Modale vs. nicht-modale Formulierung

Vergleicht man die modale Formulierung in (77) (Ich will den Reis abgießen) mit der nicht-modalen

(77a) Ich gieße den Reis ab

so wird folgendes deutlich: mit der einfachen Verbalisierung rückt der Sprecher eine bestimmte Handlung (oder einen Handlungstyp) in den Aufmerksamkeitsbereich des Hörers, und zwar im Hinblick auf ihren Vollzug. Bezogen auf unser Beispiel heißt das, daß für den Hörer der Vollzug der Handlung Reis abgießen durch den Sprecher, der in diesem Fall Handlungsträger ist ('ich'), ins Blickfeld gerät. Demgegenüber interessiert bei der modalen Formulierung weniger der Handlungsvollzug, als die Vorstadien, in der sich der Handlungsträger im Bezug auf diese Handlungen befindet.

So verweist der Gebrauch von 'wollen' in (77) darauf, daß G ('ich') sich im Bezug auf die Handlung Reis abgießen in der Endphase der Planung befindet, nämlich der Absichtsbildung. D.h. daß der Vollzug der Handlung von nun an Ziel ist für G's weiteres Handeln bzw., im günstigsten Fall, daß G nun mit dem Vollzug ansetzt.

Die Weise, in der der Handlungsträger seine Aufmerksamkeit auf die Handlung richtet bzw. - von der Handlung her argumentiert - die Form, in der sie ihm mental präsent ist, wird mittels des Modalverbs bezeichnet und kommuniziert. Der Hörer erfährt dadurch, in welcher Form er die angesprochene Handlung repräsentieren soll.

3. Modalitäten und Modalverben

Zwischen beiden Kategorien besteht folgende Differenz: Modalverben sind sprachliche Ausdrücke. Insofern sind sie Mittel, bestimmte Modalitäten zu kommunizieren.

Diese Modalitäten sind demgegenüber mentale Phänomene, nämlich Präsentations- bzw. Repräsentationsformen von Handlungen oder von Sachverhalten, die Voraussetzung oder Ergebnis von Handlungen sind.

Eine solche Einschränkung des Inhalts dieser Modalitäten, der in der sogenannten Ergänzung des Modalverbs kommuniziert wird, entspricht meiner Annahme über die Verhältnisse, die vorliegen, wenn solche Modalitäten bewußt werden: ein Aktant richtet seine Aufmerksamkeit auf Handlungen oder deren Voraussetzungen/Ergebnisse im Bezug auf sich als einen Handelnden und positioniert sie innerhalb seines Handelns als Bedürfnis, Ziel/Absicht, Weg zu einem Ziel o.ä. Das Ergebnis dieser Positionierung besteht darin, daß diese Gegenstände seiner Aufmerksamkeit ihm nun in einer bestimmten Form präsent sind, daß er sie als Inhalte von Modalitäten begreift. Modalitäten sind demgemäß i.e.S. Repräsentationsformen von Handlungen oder Wirklichkeitskonstellationen, genau solche Formen, die deren handlungsmäßige Relevanz bzw. deren Position im gesamten Handeln des Handlungsträgers betreffen.

Dementsprechend gewinnen die Modalverben -als Kommunikationsmittel für solche Repräsentationsformen- für den Hörer die Funktion einer Vorschrift darüber, wie er den fokussierten Gegenstand in sein aktuelles Diskurswissen über den Handlungsträger einordnen soll. Damit erhält er wichtige Orientierungspunkte im gemeinsamen Interaktionsraum und einen Schlüssel für die Einschätzung der Handlungszusammenhänge, in denen der betreffende Handlungsträger steht.

Die Fähigkeit von Sprechern, die Perspektive ihrer Interaktanten oder auch von Aktanten einzunehmen, mit denen er nicht in einem direkten Handlungszusammenhang steht, drückt sich sprachlich dadurch aus, daß die finite Form des Modalverbs und der Subjektausdruck, mit dem der Handlungsträger bezeichnet wird, 2. oder 3. Personen markieren.

4. Modalitäten

Die Modalverben zeichnen nun folgende Modalitäten aus: das Mögen, das Wollen, das Sollen, das Können, das Müssen, Dürfen. Worin besteht die Spezifik der einzelnen Modalitäten?

Ist einem Aktanten z.B. ein Zustand in Form eines M ö g e n präsent, so bedeutet dies, daß dessen Realisierung ihm ein B e d ü r f n i s ist und daß er tendenziell eine Bedürfnisbefriedigung durch Handeln anvisiert.

Entwickelt sich diese Tendenz in der Weise, daß der Aktant Handlungswege aufsucht, deren Ziel bzw. Zweck in der Befriedigung dieses Bedürfnisses besteht, schätzt er also seinen Handlungsspielraum unter diesem Bedürfnis (als Ziel) stehend ein, so präsentieren sich ihm die infrage kommenden Handlungen (oder erreichbaren Zwischenziele) in der Form von K ö n n e n.

Werden solche Wege dadurch frei, daß eine Instanz, unter deren Kontrolle sich der Einschätzende befindet, sie ihm einräumt, so spiegelt sich das in der speziellen Form des D ü r f e n .

Ist sein Handlungsraum -wodurch auch immer- auf eine einzige Handlung beschränkt, so präsentiert sich ihm diese in der Form eines M ü s s e n .

Das zielgeleitete Einschätzen des Handlungsraums bildet für den Aktanten wiederum Motiv dafür, die ausgemachten Handlungsalternativen zu bewerten und sich zum Vollzug einer Handlung zu entschließen. Die Handlung, deren Vollzug er demgemäß beabsichtigt, die für sein gesamtes Handeln also die Position eines Z i e l s einnimmt, ist ihm dann in Form eines W o l l e n präsent.

Eine solche Absichtsbildung kann auch durch fremde Instanzen, besonders durch Interaktanten, initiiert werden, und zwar so, daß sie Ziele nahelegen oder -durch zielübermittelnde Handlungen- übertragen; solche Zielinhalte präsentieren sich dann für den Handlungsträger als ein S o l l e n .

5. Konkrete Analysen

Nun zurück zu (77) (Ich will den Reis abgießen).

Gemäß der Analyse von Fiehler entspricht hier der Modalverb-Gebrauch der Modalität: die Handlung Reis-abgießen ist Ziel der Aktivitäten von G; G steht zu ihr in der Relation der Absicht.

Bei der Interpretation ist zu beachten, daß 'Reis abgießen' einerseits eine Handlung ausdrückt, bestehend aus Topf-halten, Sieb-halten, Reis-abgießen und evtl. Abschrecken. Andererseits drückt es einen einfachen Teilakt im Rahmen der Handlung Reis-kochen aus. Für G erweist sich Abgießen gerade als eine Handlung, deren Teilakte sie nicht allein ausführen kann. Deshalb bildet sie die Absicht aus, den Vollzug kooperativ vorzunehmen und macht entsprechende (sprachliche) Handlungen. Innerhalb der Kooperation ist sie dann allerdings in der Lage, den einfachen Teilakt des Abgießens zu vollziehen und tut dies auch (79). Auf diesen Status des Handelns würde die eingangs benannte nicht-modale Formulierung abheben; mit (77a) (Ich gieße den Reis ab) würde die Arbeitsteilung vorgenommen; G übernähme klar den Vollzug des Teilakts Abgießen, während für R dann das Siebhalten bliebe. Wie an der Reaktion von R in (78) deutlich wird, referiert G (zumindest für sein Verständnis) in der modalen Formulierung auf Reis-abgießen als komplexe Handlung, die zu realisieren sie beabsichtigt ('wollen'). (77) ist an dieser sequentiellen Position Realisierungsform einer Begründung (die in (76) bereits als notwendig gekennzeichnet wird ('müssen')). Es handelt sich um den speziellen Typ einer Begründung, der dazu dient, dem Hörer/Interaktanten das Ziel bzw. den Zweck einer bereits begonnenen oder angekündigten Handlung klar zu machen.

Betrachten wir nun die funktionalen Äqivalente, die FIEHLER herausgearbeitet hat und erklären die Differenzierung auf der illokutiven Ebene, soweit sie durch den Gebrauch verschiedener Modalverben bzw. verschiedener Subjektausdrücke beeinflußt wird. Als erstes fällt dann Äquivalent (5 a) auf, in dem zwar das gleiche Modalverb, jedoch ein anderer Subjektausdruck verbalisiert wird:

(5 a) Wir wollen den Reis abgießen.

Worin besteht der Unterschied zu (77)? Durch das kollektive 'wir' wird R nicht erst mit in den geplanten Handlungsvollzug einbezogen, sondern bereits in die Absichtsbildung. Diese wird allerdings von G schon stellvertretend formuliert.

Gleiches gilt für das Äquivalent

(5 d) Laß uns den Reis abgießen,

das eine Variante darstellt, die an der sprachlichen Oberfläche dem Interaktanten etwas mehr Reaktionsspielraum einräumt. Beide Äquivalente kennzeichnen jedoch die momentane Situation als diejenige, in der eine Absichtsbildung stattfindet. Demgegenüber hat G bei der Äußerung von (77) berits eine Absicht ausgebildet und teilt sie R mit. Der Gebrauch des Verbs 'helfen' in (76) legt überdies nahe, daß G sich gemäß ihrer Absichtsbildung auch weiterhin als Hauptträger der angesprochenen Handlung versteht; es kann jedoch zwischen gleichberechtigten Partnern dennoch offen sein, wie groß die Anteile der Kooperanten am Vollzug sind.

Während (77) in erster Linie eine Begründung für die Kooperations-Bitte realisiert, würden die Äquivalente (5 a) und (5 d) eher einen Sprechakt vom Typ der Aufforderung, also einen initiierenden Typ, realisieren. (Das schließt nicht aus, daß auf einer dahinterliegenden Ebene der angesprochene R gleichzeitig ein Unverständnis über das Motiv der Bitte beseitigt und ihm daher die Aufforderung gleichzeitig eine Art von Begründung liefert. Kriterium für eine derartige Vordergrund-/Hintergrund-Klassifikation sollten Komplexität der sprecherseitigen Planung und hörerseitigen Verstehensprozesse sein.)

Versucht man die äquivalenten Formulierungen gemessen an der vorgegebenen Sequenz und der daran geknüpften Kommunikationssituation in näherliegende und fernerliegende Verbalisierungen zu unterteilen, so liegt die Wahl der Formulierungen (5 a) und (d) ferner; (5 a) ist geeigneter, dritten, nicht an der Kooperation beteiligten Aktanten das gemeinsame Ziel zu übermitteln, (5 d) liegt näher in einer Situation, in der G und R beide noch nicht in Aktionen verwickelt sind, die unmittelbare Voraussetzung für die fokussierte Handlung sind (G jedoch hat gerade den Reis abgedreht).

Die im Bezug auf das Modalverb schwächere Formulierung

(1 b) Ich möchte den Reis abgießen

ist ebenfalls eine fernliegendere Verbalisierungsform. Denn G würde gegenüber der progressiven Formulierung in (7 b) (du mußt mir helfen) ihre eigene Einschätzungskompetenz und damit auch die der Absichtsbildung zurücknehmen. D.h. (1 d) würde innerhalb des Diskurses eine Bitte um Rat realisieren, den gemeinsamen Handlungsvollzug also retardieren. Reis abgießen ist keine so schwierige Handlung, daß nach Ausbildung des Bedürfnisses erst noch eine Diskussion über das Können initiiert werden müßte; vielmehr kann das Mögen unproblematisch in ein Wollen umgesetzt werden. Insofern ist der Gebrauch von 'mögen' als Ergebnis verbaler Planung zu interpretieren, die auf eine sehr zurückhaltende, i.a. auf Höflichkeit bedachte Formulierung abzielt. Ein solcher Ton ist in unserem Fall der partnerschaftlichen sozialen Beziehung etwas gekünstelt, insofern fernerliegend. Wir haben bereits gesagt, daß (77) (Ich will den Reis abgießen) für die Bitte um Kooperation (76) eine Begründung bestimmten Typs realisiert. Nun ist auch das funktionale Äquivalent

(7) Ich kann den Reis nicht alleine abgießen

eine Begründung. Dadurch, daß hier eine Handlung in Form eines Nicht-Können fokussiert und kommuniziert wird, würde G unmittelbar den Grund für die Kooperation verbalisieren, den Schluß auf ihre Absicht dem R überlassen. In (77) verhält es sich demgegenüber umgekehrt, das Nicht-Können bleibt implizit; und das ist ökonomisch, da dies bereits durch (76) signalisiert wurde. D.h. mit den Verbalisierungen (77) bzw. (7) macht der Sprecher Gebrauch von verschiedenen Teilen des aktuellen Diskurs-/Handlungswissens von R. Das jeweils geäußerte Modalverb ist demgemäß nicht verschiedenen Modalitäten geschuldet, sondern der verbalen Planung, die an das Wissen des Hörers anknüpft. Mit den Äußerungen

(5 b) Wir müssen den Reis abgießen

(5 c) Wir können den Reis abgießen

würde G die Modalitätseinschätzung in einem größeren Rahmen vornehmen, und zwar würde dann Reis abgießen als Teilakt der Handlung Reis-kochen betrachtet, und Träger dieser übergeordneten Handlung sind -wegen gemeinsamen Kochens- in der Tat G und R ('wir'). Im Fall der Verbalisierung von 'müssen' ist stärker die Sequenzierung der Teilakte, d.h. das Schema vom zeitlichen Ablauf des Reis-kochens, maßgeblich dafür, wo der fokussierte Teilakt im Handeln der Kooperanten positioniert wird. Während beim Gebrauch von 'müssen' der momentane Zeitpunkt als dicht beim Maximum liegend eingeschätzt wird, fokussiert der Aktant mit 'können' einen frühest möglichen Zeitpunkt, an dem der Teilakt zu vollziehen ist. Dadurch gewinnt (5 c) eine positive Nuance, das Ziel, Reis gekocht zu haben, rückt näher; (5 b) dagegen hat einen eher mahnenden, erinnernden Charakter.

Da für die Realisierung der Handlung Reis-kochen (bei bestimmter Art ihres Vollzugs) das Abgießen einen sachlich notwendigen Teilakt bildet, spiegelt 'müssen' in (5 b) die sachlich vorliegende Modalität wider, während sowohl 'wollen' in (77) wie 'können' in (5 c) die subjektiven Bedingungen der Aktanten stärker hervorheben und sie als Entscheidungsinstanzen über den Vollzug ansprechen. Der Bezug der Modalität bzw. des Modalverbs in (5) auf den übergeordneten Zusammenhang bewirkt, daß diese

Äußerungen nicht in einem unmittelbaren Begründungszusammenhang mit der Kooperations-Aufforderung stehen: sie dienten vielmehr dazu, die Position des nun anstehenden Handelns im Gesamtablauf zu bestimmen, so daß sich Bedingungen für Hinweis und Aufforderung in ihnen überlagern zu einer Art Anweisung. Diese Überlagerung fehlt auf der verbalen Ebene, wenn kein Handlungsträger mehr benannt wird und lediglich auf die sachlich vorliegende Konstellation hingewiesen wird.

(4 b, c, d) Der Reis muß / kann / müßte abgegossen werden.

Die Kooperanten beim Reis-kochen wissen, daß mindestens einer von ihnen nun den Reis abgießen muß, wenn das Gesamtziel realisiert werden soll; in unserem Diskurs ist durch (76) bereits klar, daß G und R als Handlungsträger infrage kommen; allerdings bleibt im Fall von (4) noch offen, ob sich 'helfen', also das kooperative Handeln, auf Abgießen bezieht oder auf die letzten Teilakte vom gemeinsamen Kochen (G also z.B. sich weiter um die Erbsen und das Fleisch kümmert); (78) klärt dann aber die Art der Kooperation. 'müssen' und 'können' in (4) spiegeln die Modalität in der für (5) beschriebenen schwerpunktmäßig verschiedenen Weise wider; in der Konjunktivform 'müßte' überlagern sich beide Aspekte: einerseits drückt das Modalverb die sachlich vorliegende Modalität des Müssen aus, andererseits gesteht G durch den Konjunktivgebrauch dem R einen eigenen Entscheidungsspielraum zu, d.h. in dieser Form verzahnen sich objektive und subjektive Bedingungen für das fokussierte Handeln.

Das gilt auch für

(4 e) Der Reis sollte abgegossen werden.

Allerdings verschiebt sich das Gewicht etwas mehr in Richtung auf die subjektiven Bedingungen, ohne jedoch die externe, nämlich sachliche Dringlichkeit, auszublenden: 'sollen' leistet dies.

Das funktionale Äquivalent

(4 a) Der Reis soll abgegossen werden

ist im Rahmen der vorgegebenen Kommunikationssituation nur dann angebracht, wenn G etwa ein Kochbuch als entscheidende Instanz für die Zeitpunktbestimmung konsultiert. Im vorliegenden Fall wäre eine derartige Situation allerdings konstruiert.

Bleibt noch Äquivalent

(8) Muß ich den Reis auch noch allein abgießen?

Ein solcher Vorwurf mangelnder Kooperativität wäre, wie auch die Variante mit 'sollen', nur dann sequentiell akzeptabel, wenn er an (76) durch kontrastives 'oder' angeknüpft würde. 'müssen' entspricht dann der sachlichen Modalität, 'sollen' höbe ab auf die Absichten der Kooperanten.

6. Zusammenfassung

Wir sind ausgegangen von der authentischen Äußerung (77) (Ich will den Reis abgießen) innerhalb eines bestimmten Diskursabschnittes. Anschließend haben wir kommunikative Differenzierungen zwischen den von FIEHLER aufgeführten funktionalen Äquivalenten aufgesucht, soweit sie durch Modalverbgebrauch beeinflußt waren. Die Erklärung der Differenzen erfolgte im wesentlichen von zwei Perspektiven her:

- die Modalität derselben Handlung -in unserem Fall des Reisabgießens- stellt sich verschieden dar je nachdem, in welcher Weise auf übergeordnete oder untergeordnete Handlungszusammenhänge reflektiert wird.
- Modalverben sind nicht die 1:1-Übersetzungen von Modalitäten, sondern ihr Gebrauch hängt ab von der verbalen Planung des Sprechers; verschiedene Modalverben sprechen in dem Sinne verschiedenes Diskurswissen der Interaktanten an.

In den Bereich der verbalen Planung gehört auch die sprecherseitige Entscheidung darüber, ob stärker die sachlich vorliegende Modalität und damit die objektiven Bedingungen des Handelns hervorgehoben werden sollen oder aber den subjektiven Bedingungen der Interaktanten Rechnung getragen wird. Insofern waren die beiden Modalitäten Müssen und Wollen Formen, in denen die Handlung Reis-abgießen mental repräsentiert wurde. Da die sachliche Modalität ihrerseits ein zeitlich wachsendes Gewicht bekommt dadurch, daß der Vollzug der fokussierten Handlung innerhalb einer Zeitspanne mehr oder minder großer Ausdehnung ansteht, kann der Sprecher zu einem frühen Zeitpunkt darauf durch 'können' referieren, den Aktanten also noch einen Entscheidungsspielraum einräumen darüber, wann sie die Handlung ausführen. Daß sie die

Absicht ausbilden werden, den Reis abzugießen, ist bereits durch den größeren Handlungszusammenhang vorgezeichnet; sie wollen Reis-kochen, weil sie Paella (mit Reis) kochen wollen; zum Reis-kochen gehört auch der Teilakt Reis-abgießen, insofern haben sie diesen Handlungsvollzug in ihrem gesamten Handeln-Wollen mit eingeschlossen. Reis-abgießen als Inhalt des Wollen von G und auch R ist so nur zu verstehen auf dem Hintergrund des Handlungszusammenhangs, genauer: der über- und untergeordneten Struktur, die die einzelnen Handlungen zusammen bilden.

Literatur

FIEHLER, Reinhard (1979): "Über das Abgießen von Reis". VANDEWEGHE, Willy/VAN DE VELDE, Marc (eds.) (1979): Bedeutung, Sprechakte und Texte. Akten des 13. Linguistischen Kolloquiums, Gent 1978, Bd. 2. Tübingen: Niemeyer.

KONVERSATIONSPOSTULATE UND KOOPERATIVE TÄTIGKEIT

Gisela Brünner

In dieser Arbeit werden Überlegungen weiterentwickelt, die im Rahmen meiner Dissertation[1] entstanden sind. Dort wird versucht, den Zusammenhang zwischen der Form betrieblicher Kooperation, dem in solcher Kooperation auftretenden Kommunikationsbedarf und den in ihr vollzogenen kommunikativen Tätigkeiten zu beschreiben und zu erklären. Hier soll der Zusammenhang von Kooperation und Kommunikation speziell für bestimmte Interaktionsregeln dargestellt werden, nämlich die Griceschen Konversationspostulate.

In "Logic and Conversation"[2] versucht Grice ein Verfahren anzugeben, mit dem die Bedeutung von Äußerungen in Dialogen erschlossen bzw. rekonstruiert werden kann und die Anschließbarkeit von Äußerungen aneinander erklärbar ist. Als Basis eines solchen Verfahrens, nämlich der "conversational implicature", formuliert Grice ein "Kooperationsprinzip", das durch vier Konversationsmaximen (oder Konversationspostulate) derart spezifiziert wird, daß deren Befolgung im allgemeinen zu Ergebnissen führt, die mit dem Kooperationsprinzip übereinstimmen. Es handelt sich um die Maximen der Quantität, der Qualität, der Relation und der Art und Weise.

Grice weist seinen Postulaten einen doppelten Status zu: Als Bedingungen der Möglichkeit von Kommunikation überhaupt gelten sie *empirisch* universell, d.h. sie werden faktisch von fast allen Sprechern in fast allen Kommunikationen befolgt. Andererseits gelten sie *normativ* universell, d.h. die Befolgung der Postulate in jeder Kommunikation ist vernünftig ("reasonable") und sollte nicht aufgegeben werden.

Die Universalität der Postulate ist verschiedentlich bestritten worden[3], wenn auch mit sehr unterschiedlichen Argumenten. Auch ich halte den Universalitätsanspruch für falsch, glaube aber, daß die bisherigen Argumentationen zu kurz greifen.

Ich werde zunächst argumentieren, daß die Konversationspostulate nur für Kommunikation in bestimmten *Funktionen* gelten und insofern kontingent sind.

Grice selbst legt die Auffassung nahe, daß die Postulate nur für Kommunikation in bestimmten Funktionen Geltung haben, wenn er seinen Sprachbegriff erläutert:

> As one of my avowed aims is to see talking as a special case or variety of purposive, indeed rational, behaviour ... (GRICE 1968: 9)

Kommunikation hat aber - grob gesagt - nicht nur eine sachliche, sondern auch eine soziale Funktion. Legt man einmal (als Beispiel für viele ähnliche Klassifikationen) die von HARTUNG (1974: 3ooff.) unterschiedenen vier Grundfunktionen sprachlicher Kommunikation zugrunde, so läßt sich sagen, daß die Griceschen Postulate etwa nur für empraktisch eingebettete Kommunikation und für Kommunikation als Träger theoretisch-erkennender Tätigkeit gelten. Sie gelten nicht oder nur in einem höchst unbestimmten Sinne für Kommunikation in Kontaktfunktion und in der Funktion psychischer Entlastung.

Die Postulate erklären sich erst und haben ihre reale Basis und ihre Bedeutung dort, wo Kommunikation der Erreichung bestimmter sachlicher Ziele dient, insbesondere in K o o p e r a t i o n s p r o z e s s e n. Wo kooperativ auf die Realisierung konkreter, definierter Ziele hingearbeitet wird, entsteht ein Kommunikationsbedarf. Die Kommunikation, die diesen deckt, muß solcherart sein, daß sie den störungsfreien Ablauf und das Gelingen der Kooperation gewährleistet.

Besonders aber dort, wo in einer bestimmten F o r m kooperiert wird, nämlich unter kapitalistischen Prinzipien, werden an die (kooperationsbezogene) Kommunikation Anforderungen der Art gestellt, wie sie durch die Griceschen Postulate beschrieben werden. In der betrieblichen Kooperation bzw. Produktion, die den zentralen Fall solcher Kooperationsform darstellt und durch Bedingungen wie Verwertungsinteressen und Herrschaft gekennzeichnet ist, wird die Befolgung der Postulate wesentlich und entsprechend durch Sanktionen, formelle Reglungen etc. erzwungen.

Dazu, daß solche Kooperationsformen die Postulate erst erklären und den Stellenwert ihrer Geltung deutlich machen, einige genauere Bemerkungen.

Bezieht man das Gricesche K o o p e r a t i o n s p r i n z i p, das oberste Postulat, nicht nur auf die Erfordernisse des "talk-exchange", sondern auf die Kooperation selbst, der die Kommunikation dient, so wird sein Sinn deutlich: Für das Gelingen der (praktischen) K o o p e r a t i o n ist es notwendig, daß die Beteiligten dem Kooperationsprinzip Folge leisten, d.h. ihre kommunikativen Beiträge (aber nicht nur diese) in der Weise machen, wie es an der jeweiligen Stelle entsprechend dem Zweck (Ziel) der Kooperation erforderlich ist. Da die Beachtung dieses Prinzips in der betrieblichen Kommunikation und Kooperation von entscheidender Bedeutung ist, um den Kommunikationsbedarf zu decken, auf der anderen Seite aber die kommunikativen Notwendigkeiten wegen der Form der Zusammenarbeit nicht ohne weiteres überschaubar sind, sind die erforderlichen Beiträge teilweise formell vorgeschrieben; z.B. in Tätigkeitsbeschreibungen, die festlegen können, wann überhaupt ein Beitrag zu erfolgen hat, wie die Abfolge bestimmter Beiträge ist oder von welcher Art der Beitrag ist. Insofern kann gesagt werden, daß das Gricesche Kooperationsprinzip im Betrieb partiell kodifiziert ist und seine Nichtbefolgung mit härteren Sanktionen als in anderen Tätigkeitszusammenhängen verbunden ist. Die Absicherung seiner Befolgung ist ein Zeichen für seine besondere Bedeutung in Kooperationsprozessen und läßt die Annahme zu, daß das Prinzip auch faktisch in besonderem Maße eingehalten wird.

Ähnliches gilt für die Maximen der Q u a n t i t ä t , der R e l a t i o n und der A r t und W e i s e , die ja partielle Explikationen des Kooperationsprinzips darstellen.

Beiträge, die nicht hinreichend i n f o r m a t i v sind, gefährden das sachliche Gelingen der Kooperation. Der Kommunikationsbedarf wird nicht ausreichend gedeckt. Deshalb wird die Informativität von kommunikativen Tätigkeiten z.T. durch formelle Regelungen gesichert. Man denke z.B. an Formulare mit vorgegebenen Rubriken.

Überinformative Beiträge verletzen das Ökonomieprinzip verwertungsorientierter Kooperation. Sie gefährden das Verwertungsziel dadurch, daß der Nutzungsgrad der Arbeitskraft und der technischen Kommunikationsmedien sinkt. Darüberhinaus würde dem Bestreben der Betriebsleitung, den Informationsstand der Arbeiter möglichst gering zu halten, um sie leichter kontrollieren zu können, durch überinformative Beiträge entgegengehandelt.

Aus dem gleichen Grund, aus dem in Kooperationen informativ geredet werden muß, muß auch r e l e v a n t geredet werden. D.h. es muß nicht irgendwas gesagt werden, sondern das, was den speziellen Kommunikationsbedarf bzw. Teile davon befriedigt.

Das Postulat, k u r z (nicht weitschweifig) zu reden, ist teilweise für die Realisierung des Sachziels der Kooperation, teils für das Verwertungsziel von Wichtigkeit. Wenn z.B. Tätigkeiten, die in enger zeitlicher Abhängigkeit voneinander stehen, während ihrer Ausführung kommunikativ synchronisiert werden müssen, kann Weitschweifigkeit zu erheblichen Störungen führen. Sie verhindert ferner die optimale zeitliche Nutzung der Arbeitskraft und der technischen Kommunikationsmedien. In diesem Zusammenhang ist die Verwendung von Abkürzungen, die Terminologienormung etc. zu sehen.

Entsprechendes gilt für die Maximen, die die Vermeidung von A m b i g u i t ä t e n und U n k l a r h e i t der Sprechweise fordern. Ihre Einhaltung verringert die Wahrscheinlichkeit von Mißverständnissen bei der Kooperation und damit von Fehlern und Rückfragen. Auch hier haben Normierungen von Bedeutungskonventionen ihren Stellenwert. Sie sind Ausdruck der Tatsache, daß in Kooperationen, besonders in verwertungsorientierten, die Befolgung der Maximen besonders erforderlich ist.

Noch zentraler als die übrigen Maximen, die Grice nennt, ist die Maxime der Q u a l i t ä t in Kooperationen, und zwar in allen Kooperationen, nicht nur in solchen, die unter kapitalistischen Bedingungen stattfinden. Lügen kann hier haben und hat in der Regel auch gravierende Folgen. Damit überhaupt Zusammenarbeit möglich ist, muß man sich aufeinander verlassen können, gerade auch darauf, daß die Partner in dem, was sie sagen, zuverlässig und aufrichtig sind. Aufrichtigkeit ist schlechthin konstitutiv für jede Kooperation, ist unerläßliche Bedingung ihrer Möglichkeit. Deshalb wird unaufrichtiges Verhalten in der kooperationsbezogenen Kommunikation auch hart sanktioniert.

Weil kapitalistische Kooperation auf die Einhaltung der Postulate angewiesen ist, wird ihre Befolgung an bestimmten Punkten aus dem Entscheidungsspielraum des Einzelnen herausgenommen. Das geschieht z.T. dadurch, daß die durch die Postulate beschriebenen kommunikativen Anforderungen in formelle Regelungen gefaßt werden. Wo dies der Fall ist, ändert sich aber ihr Status und damit die mit ihnen verbundenen Implikationen.

Besonders deutlich sichtbar wird das im Fall bestimmter zentraler und sehr formaler Kommunikationsteile, nämlich der organisationsspezifischen (oder institutionellen) Sprechakte. Die Postulate verlieren ihren Charakter als vom einzelnen zu befolgende Normen. Für zwei Maximen soll dies im folgenden näher erläutert werden.

Bei organisationsspezifischen Sprechakten entfällt die Aufrichtigkeitsregel. D.h. der Vollzug solcher Sprechakte impliziert nicht notwendig, daß der Sprecher die befohlene Handlung wünscht, die vertragliche Verpflichtung einhalten will etc. Daraus ergibt sich, daß die Maxime der Qualität beim Vollzug solcher Sprechakte nicht einschlägig ist. Konversationelle Implikaturen, wie sie im Rekurs auf die Qualitätsmaxime für 'gewöhnliche' Aufforderungen möglich sind, sind für Befehle in Organisationen nicht möglich.

Es läßt sich zeigen, daß bei organisationsspezifischen Sprechakten die Searlesche Nichtoffensichlichkeitsregel entfällt. Das bedeutet, daß die Maxime der Quantität bei solchen Sprechakten außer Kraft ist. Die Äußerung z.B. eines Befehls in einer Situation, in der offensichtlich ist, daß der Angesprochene die befohlene Handlung ohnehin tun wollte, ist kein Verstoß gegen die Quantitätsmaxime und erlaubt keine konversationellen Implikaturen.

Ähnliche Überlegungen wie zu den Besonderheiten der Regeln organisationsspezifischer Sprechakte lassen sich auch für explizite performative Formeln anstellen, durch die solche Sprechakte ja weitaus häufiger als andere eingeleitet werden. Mit diesen Formeln sind nämlich Implikationen verbunden, die die Geltung der Maxime der Art und Weise betreffen.

In informalen Kontexten verletzt der Gebrauch expliziter performativer Formeln häufig die dritte Untermaxime ("Be brief (avoid unnecessary prolixity)"). Dies zeigt sich z.B. daran, daß Implikaturen möglich sind:

A und B seien Freunde, die zusammen wohnen und die Hausarbeit untereinander aufgeteilt haben. A sagt: "Wenn du diese Woche für mich abwäschst, mache ich für dich die Wäsche." B antwortet: "Hiermit verpflichte ich mich, für dich diese Woche die Wäsche zu waschen." B's Redeweise wirkt auf A sicherlich befremdlich. Gegenüber der normalerweise erwartbaren Antwort ("In Ordnung." o.ä.) verletzt sie in offensichtlicher Weise die Maxime "Be brief". A wird deshalb eine Implikatur vornehmen, die zum Ergebnis hat, daß B's Äußerung ein Scherz war oder bedeutet: "Mißtraue mir nicht immer."

Implikaturen dieser Art sind nicht möglich, wenn explizite performative Formeln in formalen organisationellen Kontexten gebraucht werden. Dies deutet darauf hin, daß die genannte Maxime hier nicht oder zumindest nicht in der gleichen Form gilt.

Wenn auf der einen Seite die v e r s c h ä r f t e G e l t u n g der Maximen in kapitalistischen Kooperationsprozessen betont wurde, und auf der anderen Seite gezeigt wurde, daß für organisationsspezifische Sprechakte die Maximen n i c h t e i n s c h l ä g i g sind, so scheint hier ein Widerspruch vorzuliegen. Es handelt sich aber lediglich um zwei Seiten derselben Medaille:

Betriebliche Kooperationen sind u.a. dadurch gekennzeichnet, daß zwischen den Mitgliedern Herrschaftsbeziehungen bestehen, auf der Seite der Arbeiter kein unmittelbares Interesse an der Zielrealisierung vorliegt und die Beziehungen der Mitglieder zueinander stark verrechtlicht sind. Unter solchen Bedingungen kann von der Seite der Betriebsleitung nicht ohne weiteres unterstellt werden, daß die Maximen aus eigenem Antrieb eingehalten werden, jedenfalls in der Strenge, wie es erforderlich ist. Ihre Einhaltung wird deshalb durch Sanktionen gesichert. Ein großer Teil der Sanktionen kann aufgrund der rechtlichen Einbindung der Kooperation nur durch Anrufung öffentlicher Gewalt erfolgen, z.B. durch die Einschaltung von Gerichten. Und hier muß die fragliche Situation nachträglich leicht beurteilbar sein.

Gerade die Sprechakte, die besonders folgenreich sind, werden nun als organisationsspezifische Sprechakte besonders ausgeformt

Was die Qualitätsmaxime betrifft, wird damit in der Sprechsituation die Möglichkeit, sie aufrichtig oder unaufrichtig zu äußern, nicht mehr zur Disposition des Sprechers gestellt. Sie werden vielmehr als "per definitionem" aufrichtig bzw. wahr gesetzt, soweit nur bestimmte äußere Bedingungen erfüllt sind. Auf diese Weise werden Zweifel des Angesprochenen an der Aufrichtigkeit des Sprechers und aus diesen resultierendes Fehlverhalten von vornherein ausgeschlossen, insofern alle Vermutungen des Hörers über die innerpsychischen Vorgänge im Sprecher überflüssig werden. Der Hörer kann sich auf den Sprecher verlassen, gerade weil der Sprecher beim Vollzug organisationsspezifischer Sprechakte keine Wahl hat zwischen Aufrichtigkeit und Unaufrichtigkeit. Das Als-wahr-Setzen formgerecht vollzogener organisationsspezifischer Sprechakte erlaubt es dem Angesprochenen, die Äußerung ohne weitere Bedeutungserschließung allein aufgrund formaler Kriterien für wahrhaftig oder aufrichtig zu halten und verpflichtet ihn gleichzeitig, entsprechend zu reagieren.

Gleichzeitig ermöglicht aber der Ersatz von Aufrichtigkeit durch formgerechten Vollzug, außerhalb der Sprechsituation, etwa in einer späteren rechtlichen Auseinandersetzung, die Bedeutung des Gesagten nachträglich eindeutiger zu klären.

Etwas anders liegen die Verhältnisse bezüglich der Maximen der Quantität und der Art und Weise beim Vollzug organisationsspezifischer Sprechakte und beim Gebrauch expliziter performativer Formeln. Betroffen sind in erster Linie die Submaximen, nicht überinformativ und kurz zu reden. Wenn, wie oben ausgeführt, ihre Beachtung für das Verwertungsziel erforderlich ist, sie aber andererseits im Falle organisationsspezifischer Sprechakte suspendiert sind, so ist dies ein Indiz für die Widersprüchlichkeit zwischen dem Bestreben, auch die kommunikative Tätigkeit zu ökonomisieren, und dem Versuch, gerade bei folgenreichen Sprechakten ein Sprachverhalten durchzusetzen, das auch abgelöst von der Sprechsituation eindeutig interpretierbar ist und über das in rechtlichen Auseinandersetzungen leicht geurteilt werden kann. Dieser Widerspruch zwischen Ökonomisierung und Kontrolle, die jener entgegenläuft, ist vom Bürokratieproblem her gut bekannt.

Um die letzten Überlegungen zusammenzufassen: Eben deshalb, weil die Maximen in kapitalistischer Kooperation in besonderem Maße eingehalten werden müssen, wird ihre Beachtung auf der einen Seite durch formelle Regelungen und Sanktionen erzwungen, auf der anderen Seite aus dem Entscheidungsspielraum der Individuen herausgenommen.

Wo aber gar nicht mehr die Rede davon sein kann, daß jemand die Wahl hat, den Maximen zu folgen oder nicht zu folgen, kann von einer Geltung der Maximen im Griceschen Sinne überhaupt nicht mehr gesprochen werden.

Anmerkungen

1 BRÜNNER, Gisela (1978): Kommunikation in betrieblichen Kooperationsprozessen. Theoretische Untersuchungen zur Form und Funktion kommunikativer Tätigkeit in der Produktion. (Masch. Diss. Osnabrück).

2 GRICE 1968: Teil II.

3 Cf. z.B. KANNGIESSER 1973, HINDELANG 1975, MAAS 1976 und OCHS KEENAN 1976.

Literatur

GRICE, H. Paul (1968): "Logic and Conversation". (Unveröffentl. Vorlesungsmanuskript).

HARTUNG, Wolfdietrich (ed.) (1974): Sprachliche Kommunikation und Gesellschaft. Berlin: Akademie.

HINDELANG, Götz (1975): "Äußerungskommentierende Gesprächsformeln. Offen gesagt, ein erster Schritt". EHRICH, Veronika FINKE, Peter (eds.): Beiträge zur Grammatik und Pragmatik. Kronberg: Scriptor: 253-263.

KANNGIESSER, Siegfried (1973): "Universelle und kontingente Bedingungen des sprachlichen Handelns". (Unveröffentl.).

MAAS, Utz (1976): Kann man Sprache lehren? Für einen anderen Sprachunterricht. Frankfurt a.M.: Syndikat.

OCHS KEENAN, Elinor (1976): "The Universality of Conversational Postulates". Language in Society 5: 67-8o.

WAS HAT SPRECHHANDLUNGSTHEORIE MIT LINGUISTIK ZU TUN ?

Götz Hindelang

1. Einleitung

In der Linguistik werden seit geraumer Zeit die Ergebnisse der Sprechhandlungstheorie rezipiert. Es ist den Linguisten jedoch bisher noch nicht gelungen, die bei Austin und Searle vorgezeichneten Bahnen einer sprachphilosophisch orientierten Betrachtungsweise zu verlassen und eine eigenständige linguistische Problemstellung bei der Beschreibung von Sprechhandlungen zu entwickeln.

Searle selbst betont in seinem Buch 'Speech Acts' den Unterschied zwischen seiner sprachphilosophischen Fragestellung und den in der Linguistik verfolgten Beschreibungszielen. Er charakterisiert sein Buch wie folgt:

> It is not an essay in linguistics. Linguistics attempts to describe the actual structures - phonological, syntactical, and semantic - of natural human languages. ... In that sense this essay is not in general about languages, French, English or Swahili, but is about language. (SEARLE 1969: 4)

Searle sieht also den wesentlichen Unterschied zwischen Linguistik und Sprachphilosophie darin, daß sich die Linguistik mit den konkreten Strukturen realer Einzelsprachen beschäftigt, während die Sprachphilosophie Sprache als solche, Sprache überhaupt, zu ihrem Gegenstand hat.

Es würde natürlich dem Selbstverständnis der Linguistik zuwiderlaufen, wollte man sie ausschließlich auf die Beschreibung einzelsprachlicher Strukturen festlegen. Gegen die Sprechhandlungstheorie als linguistische Disziplin kann man jedoch einwenden, daß sie bisher keinen nennenswerten Beitrag zur Beschreibung einzelsprachlicher Strukturen geleistet hat. Betrachtet man die große Zahl der Publikationen zur Sprechhandlungstheorie, die ja zu einem nicht geringen Teil auch von Linguisten stammen, so muß man feststellen, daß die einzelnen Sprechhandlungen hauptsächlich unter einer 'universalpragmatischen' Perspektive untersucht werden. Die systematische Analyse der speziellen sprachlichen Mittel, die in einer konkreten Einzelsprache zum Vollzug eines bestimmten Sprechhandlungstyps zur Verfügung stehen, und die grammatische Beschreibung dieser Strukturen wurde bisher weitgehend vernachlässigt. Die Antwort auf die Frage:"Was hat Sprechhandlungstheorie mit Linguistik zu tun?" muß also lauten:"Vorläufig leider noch zu wenig."

2. Drei Modelle für eine linguistische Sprechhandlungstheorie

Eine Sprechhandlungstheorie, die von einer linguistischen Fragestellung ausgeht, muß die Beziehung zwischen den einzelnen Sprechhandlungstypen und den in einer bestimmten Einzelsprache konventionellerweise zur Realisierung dieses Sprechhandlungstyps bereitstehenden Äußerungsformen aufzeigen können. Im folgenden möchte ich drei methodische Ansätze diskutieren, die man verfolgen kann, wenn man einzelsprachliche Strukturbeschreibung und Sprechhandlungsbeschreibung verbinden will. Die drei Ansätze sollen 'formorientiert', 'semantikorientiert' und 'illokutionsorientiert' genannt werden.

Das formorientierte Beschreibungsmodell setzt bei der Ausdrucksseite der Sprache an. Man geht aus von einem leicht abgrenzbaren grammatischen, morphologischen oder lexikalischen Phänomen F wie z. B. dem Konj. II, den Modalverben, einem bestimmten Tempusmorphem, einer Abtönungspartikel usw. Die Fragestellung des formorientierten Ansatzes lautet nun: "Welche Inhaltskategorien und welche kommunikativen Funktionen (=Sprechhandlungen H_1 ... H_n) kann man F zuordnen?"

Der semantikorientierte Ansatz geht aus von einer bestimmten Inhaltskategorie I und untersucht folgende Fragen: "Welche sprachlichen Mittel F_1 ... F_n können zum Ausdruck von I verwendet werden?" und "Welche kommunikativen Funktionen H_1 ... H_n können durch Äußerungen vollzogen werden, die der Inhaltskategorie I angehören?" Inhaltskategorien können am besten durch Ausdrücke wie (1) - (3) dargestellt werden.

(1) (sagen, was man tun wird)

(2) (sagen, was der Hörer tun wird)

(3) (sagen, was man will)

Innerhalb eines semantikorientierten Beschreibungsmodells würde man zu Aussagen der folgenden Art gelangen: Indem man sagt, was man tun wird, kann man ein Versprechen geben, eine Drohung aussprechen, eine Warnung äußern, einen Entschluß verbalisieren usw. Und: Die Inhaltskategorie (sagen, was man tun wird) kann man ausdrücken, indem man Äußerungsformen wie (4) oder (5) verwendet.

(4) Ich werde X-en.

(5) Morgen X-e ich.

Analog wird man für die Inhaltskategorie (2) festhalten, daß man, indem man sagt, was der Hörer tun wird, Bitten, Befehle, Ratschläge, Prophezeiungen usw. machen kann. Entsprechend müssen auch für (2) die Ausdrucksformen angegeben werden. Zu den Sprechhandlungstypen, die man vollziehen kann, indem man sagt, was man will, gehören Bitten, Befehle, Aufträge usw.[1]

Der illokutionsorientierte Ansatz geht von der Analyse einer bestimmten Sprechaktklasse wie z. B. den direktiven oder den kommissiven Sprechakten aus. In einem ersten Schritt werden die einzelnen Elemente dieser Klasse aufgrund ihrer sozialen und situativen Handlungsbedingungen abgegrenzt, oder, wie ich sagen möchte, ein kommunikatives Muster wie DIREKTIV oder KOMMISSIV wird in seine kommunikativen Untermuster aufgeteilt. Für die einzelnen kommunikativen Untermuster $H_1 \dots H_n$ einer Sprechaktklasse lautet die Fragestellung bei einem illokutionsorientierten Ansatz wie folgt: "Durch welche Inhaltskategorien $I_1 \dots I_n$ kann ein bestimmtes Untermuster H_i vollzogen werden?" und "Durch welche Ausdrucksmittel $F_1 \dots F_n$ können die Inhaltskategorien $I_1 \dots I_n$ bei H_i realisiert werden?"

Bevor ich das hier als illokutionsorientiert bezeichnete Vorgehen im dritten Abschnitt durch ein Beispiel verdeutlichen will, sollen die drei Beschreibungsmodelle zunächst nochmals in schematischer Form gegenübergestellt werden und die Vor- und Nachteile der drei Modelle gegeneinander abgewogen werden.

Formorientiert:

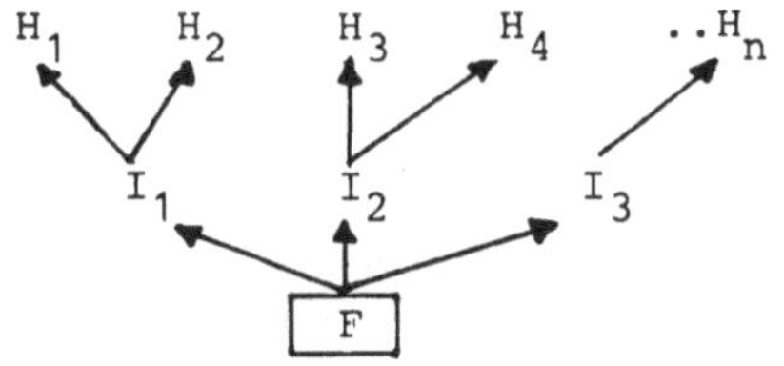

Semantikorientiert:

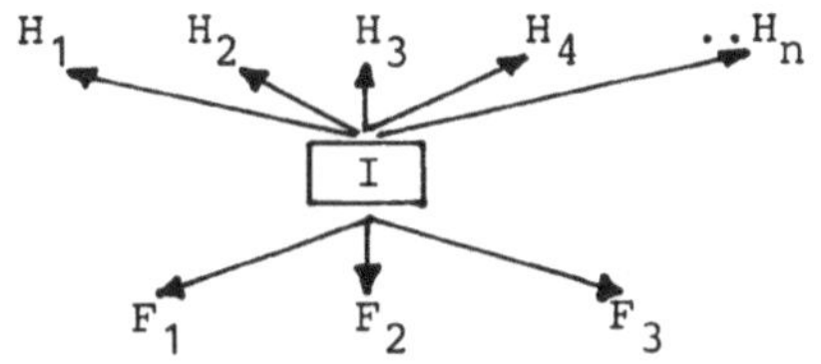

Illokutionsorientiert:

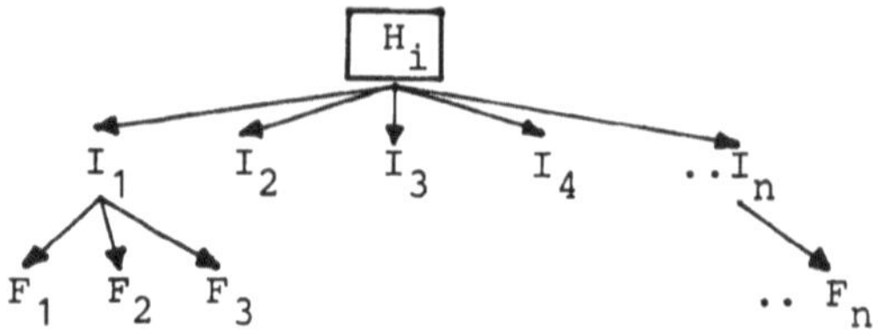

Der Vorteil eines formorientierten Vorgehens besteht darin, daß der Ausgangspunkt der Analyse unproblematisch zu bestimmen ist. Der Nachteil liegt jedoch, ähnlich wie bei einer semantikorientierten Analyse, darin, daß man es bei der Beschreibung der kommunikativen Funktionen der Form- bzw. Inhaltskategorien mit einer großen Zahl verschiedener kommunikativer Muster zu tun hat, die alle hinsichtlich ihrer relevanten Handlungsbedingungen charakterisiert werden müssen. Beim gegenwärtigen Stand der Forschung können semantik- und formorientierte Untersuchungen noch nicht auf explizite Beschreibungen der einzelnen Sprechaktklassen und ihrer Elemente zurückgreifen. Es empfiehlt sich also, zunächst bei der Analyse der einzelnen Sprechhandlungstypen anzusetzen und erst dann, wenn ein relativ gesicherter Überblick über das System der Sprechhandlungsmuster vorliegt, form- und semantikorientierte Beschreibungen zu machen. Bis dahin müssen bei solchen Versuchen die Beschreibungen der Sprechhandlungstypen zwangsläufig ad hoc erfolgen.

3. Skizze des illokutionsorientierten Beschreibungsansatzes

In HINDELANG (1978) habe ich die direktiven Sprechhandlungen untersucht. Es werden dort folgende Untermuster des AUFFORDERNS unterschieden: WEISUNG, AUFTRAG, BEFEHL, ANORDNUNG, FORDERUNG, GEBOT, ERPRESSUNG, KOMMANDIEREN, ANLEITUNG, RATSCHLAG, ANWEISUNG, VORSCHLAG, SYMMETRISCHE BITTE und ASYMMETRISCHE BITTE. Auf die Abgrenzung der Handlungsbedingungen dieser Aufforderungstypen soll hier nicht eingegangen werden. Es soll vielmehr versucht werden, für das Untermuster ANLEITUNG exemplarisch folgende Fragen zu beantworten:

- Durch welche Inhaltskategorien I_1 ... I_n können ANLEITUNGEN vollzogen werden?
- Durch welche Äußerungsformen F_1 ... F_n können die einzelnen Inhaltskategorien realisiert werden, und welche relevanten grammatischen Unterschiede bestehen zwischen diesen Äußerungsformen?

Zunächst ist es jedoch notwendig, eine kurze Beschreibung des Aufforderungstyps der ANLEITUNGEN zu geben. Als ANLEITUNGEN sollen solche Aufforderungen verstanden werden, bei denen Sp_1 seinem Hörer Sp_2 innerhalb einer Unterweisungsinteraktion eine bestimmte Handlungskompetenz beizubringen versucht und ihn deshalb zur Ausführung einzelner Teilhandlungen oder zu gewissen Ausführungsmodalitäten auffordert. Man kann sich hier Situationen in Fahrschulen, im Skikurs oder beim Sporttraining denken.[2]

Die Inhaltskategorien, die bei der Realisierung von ANLEITUNGEN eine Rolle spielen, sind durch die folgenden Ausdrücke (I 1) - (I 12) erfaßt.[3]

(I 1) (sagen, was positiv wäre)
(I 2) (fragen, ob Sp_2 X-en will)
(I 3) (fragen, ob Sp_2 nicht X-en will)
(I 4) (sagen, daß Sp_2 X-en wird)
(I 5) (fragen, ob Sp_2 X-en wird)
(I 6) (fragen, ob Sp_2 nicht X-en wird)
(I 7) (sagen, daß Sp_2 X-en muß)
(I 8) (fragen, ob Sp_2 nicht X-en muß)
(I 9) (fragen, warum Sp_2 nicht X-t)
(I 10) (sagen, daß Sp_2 X-en kann)
(I 11) (fragen, ob Sp_2 X-en kann)
(I 12) (fragen, ob Sp_2 nicht X-en kann)

Die durch Imperativsätze ausgedrückten ANLEITUNGEN stellen eine eigene semantische Gruppe dar; ebenso wären explizit performative Äußerungen einer eigenen semantischen Kategorie zuzurechnen. Es ist jedoch auffällig, daß ANLEITUNGEN kaum performativ formuliert werden können.

(6) ?Ich bitte Sie, den Talski zu belasten!
(7) *Ich fordere Sie auf, den Talski zu belasten!
(8) *Ich weise Sie an, den Talski zu belasten!

Die Beantwortung der zweiten Frage, d.h. die Spezifizierung der Äußerungsformen und ihre grammatische Charakterisierung kann hier auch wieder nur exemplarisch für (I 1) geleistet werden. Um alle in diesem Zusammenhang wichtigen Äußerungsformen unterscheiden zu können, ist es notwendig, drei Unterkategorien von (I 1) anzusetzen Sie lauten:

(I 1.1) (sagen, daß es positiv wäre, wenn Sp_2 X-en würde)

(I 1.2) (sagen, daß es positiv wäre, wenn ge-X-t würde)

(I 1.3) (sagen, daß es positiv wäre, wenn der Zustand p realisiert wäre)

Das folgende Schaubild (9) gibt eine Übersicht über die ANLEITUNGEN, die man machen kann, indem man sagt, was positiv wäre. Die entsprechenden Äußerungsformen selbst sind als (10a) - (12h) aufgeführt. In (9) sind die relevanten grammatischen und lexikalischen Unterschiede dargestellt, durch die sich diese Äußerungsformen unterscheiden.

(9)

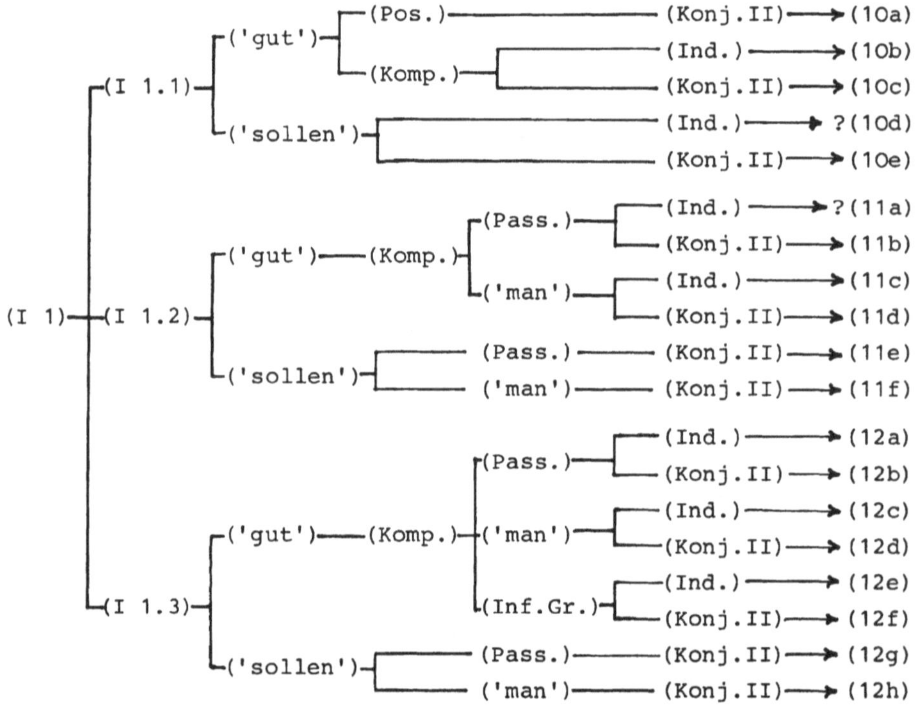

(10a) Es wäre gut, wenn Sie jetzt in den dritten Gang {schalten! / schalten würden!}

(10b) Es ist besser, wenn Sie jetzt in den dritten Gang {schalten! / schalten würden!}

(10c) Es wäre besser, wenn Sie jetzt in den dritten Gang {schalten! / schalten würden!}

(10d) ?Sie sollen in den dritten Gang schalten!

(10e) Sie sollten in den dritten Gang schalten!

(11a) ?Es ist besser, wenn jetzt in den dritten Gang geschaltet würde!

(11b) Es wäre besser, wenn jetzt in den dritten Gang geschaltet würde!

(11c) Es ist besser, wenn man jetzt in den dritten Gang schaltet!

(11d) Es wäre besser, wenn man jetzt in den dritten Gang schalten würde!

(11e) Jetzt sollte in den dritten Gang geschaltet werden!

(11f) Jetzt sollte man in den dritten Gang schalten!

(12a) Es ist besser, wenn hier im dritten Gang gefahren wird!

(12b) Es wäre besser, wenn hier im dritten Gang gefahren würde!

(12c) Es ist besser, wenn man hier im dritten Gang fährt!

(12d) Es wäre besser, wenn man hier im dritten Gang fahren würde!

(12e) Es ist besser, hier im dritten Gang zu fahren!

(12f) Es wäre besser, hier im dritten Gang zu fahren!

(12g) Hier sollte im dritten Gang gefahren werden!

(12h) Hier sollte man im dritten Gang fahren!

Das Schaubild (9) und die entsprechenden Äußerungsformen bedürfen noch der Erläuterung. Die Abkürzungen in (9) sind wie folgt aufzulösen: 'Pos.' = Positiv; 'Komp.' = Komparativ; 'Pass.' = Passiv; 'Inf.Gr.' = Infinitiv-Gruppe; 'Ind.' = Indikativ; 'Konj.' = Konjunktiv. 'Ind.' und 'Konj.' in (9) beziehen sich auf den Modus im übergeordneten Satz. Normalerweise sind Sätze wie (13), in denen der Modus von Haupt- und Nebensatz nicht übereinstimmt, ungrammatisch.

(13) *Es ist besser, wenn hier im dritten Gang gefahren würde!

In (10a) - (10c) ist der Modus des Nebensatzes jedoch nicht vom Modus des Hauptsatzes abhängig. So ist z.B. (10c') ohne weiteres akzeptabel:

(10c') Es ist besser, wenn Sie jetzt in den dritten Gang schalten würden!

Die Verwendung des 'würde'-Konjunktivs kann hier als eine Art Höflichkeitssignal gedeutet werden. (11a) ist m.E. nicht voll akzeptabel, jedoch auch nicht gänzlich abweichend wie (13). Die Äußerungsform wurde deshalb mit '?' markiert. In (10d) ist nicht die Äußerungsform als solche zweifelhaft, sondern ihre Zuordnung zum Untermuster ANLEITUNG. Äußerungen wie (10d) wird Sp_1 nur dann verwenden, wenn er seine Verärgerung darüber ausdrücken will, daß sein Schüler Sp_2 nicht von selbst, d.h. unaufgefordert ge-X-t hat, oder wenn Sp_2 eine zuvor ausgesprochene gleichlautende Aufforderung nicht befolgt hat. Diese Zusatzbedingungen sind in den Handlungsbedingungen des Aufforderungstyps ANLEITUNG jedoch nicht vorgesehen; solche Faktoren kommen erst in den Blick, wenn man das Sequenzmuster INSISTIEREN untersucht.

Ziel dieser exemplarischen Analyse eines Ausschnitts der Äußerungsformen der ANLEITUNGEN war es, zu zeigen, welche Form ein illokutionsorientierter Beschreibungsansatz haben könnte. Durch die Einbeziehung der Äußerungsformen der anderen Inhaltskategorien (I 2) - (I 12) entsteht ein Überblick über die sprachlichen Mittel, die im Deutschen zum Vollzug von ANLEITUNGEN gebräuchlich sind.

Eine systematische Ausweitung des hier skizzierten Ansatzes auf die Beschreibung aller Sprechakttypen ergäbe ein vollständiges Bild der sprachlichen Mittel des Deutschen und würde im Sinne DITTMANNS (1976) eine kommunikationsorientierte Grammatik darstellen. Eine so betriebene Sprechhandlungstheorie könnte sich als eine genuin linguistische Disziplin ausweisen.

Anmerkungen

1 Die Untersuchung des 'Aktes der Zeitreferenz', die DITTMANN (1976) vorgelegt hat, entspricht weitgehend dem hier als 'semantikorientiert' bezeichneten Ansatz.

2 Für eine genaue Beschreibung des Aufforderungstyps der ANLEITUNGEN siehe HINDELANG (1978: 378ff.)

3 In HINDELANG (1978) werden diese Inhaltskategorien als 'semantische Muster' bzw. als 'semantische Untermuster' bezeichnet.

Literatur

DITTMANN, Jürgen (1976) : Sprechhandlungstheorie und Tempusgrammatik. Futurformen und Zukunftsbezug in der gesprochenen deutschen Standardsprache. München: Hueber.

HINDELANG, Götz (1978) : Auffordern. Die Untertypen des Aufforderns und ihre sprachlichen Realisierungsformen. Göppingen: Kümmerle.

SEARLE, John R. (1969) : Speech acts. An essay in the philosophy of language. Cambridge: University Press.

PERMISSIVE

Ludger Hoffmann

Das Konzept der Erlaubnis hat bisher vor allem in den Systemen der deontischen Logik Beachtung gefunden. Eine Voraussetzung solcher Systeme ist eine Funktion, die einer Menge von Handlungen oder Handlungssätzen eine Bewertung durch die Operatoren des Gebots, Verbots oder der Erlaubnis zuordnet. Die deontische Logik unterscheidet drei Erlaubniskonzepte:

(1) Erlaubnis als Abwesenheit eines Verbots;

(2) Erlaubnis als Abwesenheit eines Gebots;

(3) Erlaubnis als Abwesenheit von Gebot und Verbot.

Damit ist bereits die Basis gegeben für den nächsten Schritt: die pragmatische Analyse von Interaktionsgeschichten, in denen Permissive vorkommen.

Durch die Handlung eines Individuums oder einer Institution wird eine Norm eingeführt, durch die ein Adressat oder eine Gruppe von Adressaten zu bestimmten Handlungen, Unterlassungen oder Arten von Handlungsausführungen verpflichtet werden. Die Geltung dieser Norm ist meist an einschränkende Bedingungen gebunden; für Normverletzungen werden Sanktionen in Aussicht gestellt. Normen sind also nur denkbar in einem sozialen System, das durch Machtbeziehungen gekennzeichnet ist. Die urtümlichste Form der Normsetzung ist die durch 'personale' Autorität, die aus der rein physischen oder intellektuellen Überlegenheit, psychischer Stärke, sachlicher Kompetenz etc. abgeleitet ist. Daneben findet sich in höher entwickelten sozialen Systemen stets auch die Herrschaftsform der 'positionalen' Autorität, einer an eine bestimmte soziale Position und spezielle Bereiche gebundenen Amtsbefugnis. Eine dritte Art von Autorität bildet das kodifizierte Normensystem, das neben Geboten und Verboten auch Erlaubnisse, gemeinhin als 'Rechte' bezeichnet, enthält. So bestimmt § 137 StPO im Absatz (1):

> "Der Beschuldigte kann sich in jeder Lage des Verfahrens des Beistands eines Verteidigers bedienen."

Dem Angeklagten ist es freigestellt, ob er von diesem Recht Ge-

brauch macht oder nicht; bei bestimmten Verfahrenstypen allerdings wird aus dem Recht eine Obligation.
Positionale Autoritäten haben eine Vermittlerposition, in der ihnen die Aufgabe zufällt festzustellen, ob die vom Normensystem her gegebenen Möglichkeiten im Einzelfall realisierbar sind, entsprechende Bedingungen vorliegen, ein Antrag formgerecht gestellt worden ist etc. In diesem Bereich ist dann eine fortlaufende Interaktion bis zu einem bestimmten Punkt hin vorgesehen, an dem eine Entscheidung, die in manchen Fällen durch andere positionale Autoritäten noch revidierbar ist, fällt. Eine zentrale Rolle kommt dabei jeweils den für die Interaktanten obligatorischen Begründungen zu; sie sollen nachweisen, daß der betreffende Spielzug innerhalb des jeweiligen Normensystems zulässig ist (formaler Aspekt) und daß er zur Entscheidung der strittigen Sache beitragen kann (substantieller Aspekt). Hiermit kommt bereits eine grundlegende Bedingung für die Vorgeschichte von Permissiven ins Spiel:

(4) Wer eine personale Autorität um eine Erlaubnis bittet oder bei einer positionalen Autorität einen entsprechenden Antrag stellt, muß den Status der normsetzenden Instanz und der normkontrollierenden Instanz sowie das Normensystem, zumindest aber die Bezugsnorm, präsupponieren.

Dies kann natürlich auch stillschweigend geschehen. Eine strategische Regel, der im allgemeinen gefolgt wird, ist:

(5) Die Bitte bzw. der Antrag sind zu begründen.

Die Geltung dieser Bedingungen ist besonders offensichtlich im Bereich personaler Autorität. Normaufhebungen werden hier argumentativ ausgehandelt; nur selten wird der Geber oder Nutznießer einer Norm seine Präferenzen aus freien Stücken ändern. Dies muß ihm dadurch schmackhaft gemacht werden, daß man seinen Status prinzipiell anerkennt, keine Zweifel am Weiterbestehen seiner Autorität aufkommen läßt und eine Begründung dafür liefert, daß die Norm zeitweilig oder - was seltener und schwieriger zu erreichen ist - generell aufgehoben werden soll.
Ich wende mich nun einigen typischen Formen des BITTENS um Normaufhebung im Bereich personaler Autorität zu.

Unter strategischem Aspekt ist daran zu erinnern, daß nach (4) der Übergang zu einem Autoritätsdiskurs nach Möglichkeit zu vermeiden ist. Zunächst muß unterschieden werden zwischen Versuchen, die eine generelle Normaufhebung anstreben, und solchen, die sich mit einer Normverschiebung zufrieden geben. Das Muster UM NORMVERSCHIEBUNG BITTEN kann z.B. realisiert werden durch:

(6) Ist es nicht besser, H morgen zu machen, weil nämlich p?

(7) Laß mich doch H machen, indem ich H'' mache (statt H'); das geht viel besser!

(8) Wenn Sie wirklich sicher sein wollen, daß H getan wird, beauftragen Sie doch Z damit. Z ist darin sehr geschickt.

Die Beispiele illustrieren, daß wir mit einer Verschiebung auf den Ebenen Zeit, Modalität der Handlungsausführung und Adressat rechnen können, wenn die Versuche erfolgreich sind.

Wer eine generelle Normaufhebung zu erreichen sucht, ohne in einen Autoritätsdiskurs eintreten zu wollen, wird in den meisten Fällen auf Unverträglichkeiten im betreffenden Normensystem hinweisen. Aus dem Bereich personaler Autorität kann als Beispiel angeführt werden:

(9) Niemand kann zwei Herren dienen!

Autorität und Verpflichtung werden akzeptiert; es wird aber die Möglichkeit eines Normenkonfliktes als Begründung angeführt.

Die normkontrollierende Instanz hat, wenn die konfligierenden Normen zu ihrem Regulationsfeld gehören, eine entsprechende Entscheidung zu fällen. Gehört eine der Normen zu einem anderen Kontrollbereich, kann der Normadressat bei Uneinsichtigkeit beider Seiten in eine Zwickmühle geraten, in der er nur noch zwischen zwei Sanktionen wählen kann.

Ein verwandter Argumentationstyp ist der Hinweis auf unerwünschte Handlungsfolgen, die sich einstellen, falls die Norm bestehen bleibt:

(1o) Wenn ich H mache, werde ich mir den Hals brechen.

Es geht also um logische Folgen der Normerfüllungshandlung, von denen vorausgesetzt wird, daß sie die normkontrollierende Instanz wirklich schädigen. Wir stoßen hier auf die Tatsache, daß Argumente, um zwingend sein zu können, auf gemeinsame Annahmen und Einschätzungen der Interaktanten gestützt sein müssen.

Ein anderer Typ, der ebenfalls meist als Mittel zur generellen Normaufhebung eingesetzt wird, ist der Hinweis auf fehlende Handlungskompetenz:

(11) Ich will H ja gern versuchen, aber ich schaffe das bestimmt nicht.

Wer sich dieses Musters bedient, muß gegebenenfalls nachweisen, daß ihm die entsprechende Fähigkeit tatsächlich abgeht; eine gängige Reaktion ist nämlich:

(12) Kannst du nicht oder w i l l s t du nicht?

Erst ein Versuch entscheidet, ob das Muster 'berechtigt' oder nur eine Ausrede war; es ist also strategisch weniger erfolgversprechend. Besonders im Bereich engerer Sozialbeziehungen - etwa in der Familie - finden sich Appelle auf der Beziehungsebene:

(13) Ich möchte doch so gerne etwas anderes machen!

(14) Sei so gut und laß mich H' machen!

Versuche dieser Art gehen davon aus, daß die Herstellung und Aufrechterhaltung emotional befriedigender Sozialbeziehungen einen hohen Rang auf der Präferenzskala des Partners einnehmen. Weitere konditionale Sprechakte werden realisiert durch Angebote und Drohungen:

(15) Wenn ich H nicht zu machen brauche, mache ich H'.

(16) Wenn du mich zu H zwingst, geht es dir schlecht!

Im ersten Fall ist Bedingung, daß die Kompensationshandlung auf der Präferenzskala des Partners positiv bewertet wird. Die Entscheidung hängt dann davon ab, ob sie positiver als H bewertet wird und interferierende Präferenzen (z.B. die Vermeidung von Autoritätsverlust) vorliegen. Präzedenzfälle sind unbeliebt.
Im Fall von (16) wird schon an der Autorität gerüttelt; wir müssen davon ausgehen, daß die Partner sich in verschiedenen Bereichen gegenseitig mit Sanktionen belegen können, so daß auch hier für die normkontrollierende Instanz ein Abwägen zwischen Präferenzen erforderlich ist.
Vielleicht sollte hier auch ein Spezialfall des Versuchs einer Normaufhebung erwähnt werden: die Bitte um Befreiung von Verpflichtungen, die durch sprachliche Handlungen übernommen worden sind, etwa durch ein Versprechen. Ist ein Versprechen akzeptiert worden - ohnehin wird es ja meist vom Partner initiiert -, be-

steht für den Geber die Verpflichtung zur Einlösung, während der Nutznießer die Handlung (nicht die Ausführungsmodalität) zu tolerieren hat. Die Verteilung der Präferenzen bei einem Versprechen macht deutlich, daß es nicht leicht ist, sich von einer solchen Verpflichtung entbinden zu lassen. Das zeigen auch die parallelen formalen Verfahren in Institutionen wie den Kirchen. Es werden regelmäßig komplexe Begründungsformen verlangt.
Ein weiterer Typ ist die Bitte um Handlungsfreiheit. Hier geht es nicht darum, die mit Sanktionen belegte Handlung ungestraft ausführen zu können, sondern sich beide Alternativen offenzuhalten:

(17) In diesem Fall möchte ich allein entscheiden, ob ich H mache oder nicht.

Es wird also versucht, eine Handlung vollständig aus dem Regulationsfeld des Normensystems herauszubekommen. Auch hier muß natürlich glaubhaft gemacht werden, daß Autorität und soziales System nicht gefährdet seien. Dabei wird oft von den Interessen und Handlungsmöglichkeiten des Normadressaten her argumentiert; es können sozial hoch geschätzte Werte wie Selbständigkeit, Verantwortungsbewußtsein, Entscheidungsfreudigkeit etc. ins Feld geführt werden. Entsprechend ist die Handlung des FREISTELLENS meist an eine Bewährungsbedingung geknüpft.
Wir kommen nun zu den Reaktionsformen auf die Versuche zur Normaufhebung oder Normverschiebung und verschaffen uns zunächst einen Überblick über die Sequenzen:

(18)

Muster des INSISTIERENS bekräftigen die Geltung der gesetzten Norm; manchmal wird der Druck noch verstärkt. Es sind ganz unterschiedliche Formulierungsweisen möglich:

(19) Ich bestehe darauf, daß du H machst. (performative Form)

(2o) Befehl ist Befehl! (Rekonstruktion der Ausgangshandlung)

(21) Darüber lasse ich mit mir nicht handeln! (Argumenta-

tionsvermeidung)

(22) Du machst H oder es setzt was! (Rekonstruktion und Drohung)

Die zentrale Handlungsbedingung für Permissive ist

(23) Nur die dazu legitimierte Instanz kann eine Norm aufheben, d.h. ein Permissiv erfolgreich äußern. Sie kann dies auch tun, ohne vorher darum gebeten worden zu sein.

Man kann davon ausgehen, daß ein Sprecher, der durch eine Bitte die Äußerung eines von der jeweiligen Verpflichtung befreienden Permissivs zu elizitieren sucht, die damit eröffnete Handlungsmöglichkeit auch wahrzunehmen intendiert. Ist dies nicht der Fall, so handelt es sich um eine Bitte um Freistellung oder die Einleitung eines Autoritätsdiskurses. Aus diesen Annahmen und Festlegungen ergeben sich folgende Handlungsbedingungen für Permissive:

(24) Die Bitte, eine bestimmte Handlung ausführen zu dürfen, verpflichtet bei gegebener Erlaubnis zu dieser Handlung.

(25) Die Bitte, einem Befehl nicht nachkommen zu müssen, verpflichtet bei Aufhebung des Befehls dazu, dies tatsächlich nicht zu tun.

(26) Wird auf eine Bitte hin eine alternative Handlungsmöglichkeit eröffnet, so besteht keine Verpflichtung zu einer der Alternativen.

(27) Eine nicht erbetene Erlaubnis verpflichtet nicht zu der nunmehr erlaubten Handlung.

Der Zweck einer Erlaubnishandlung wird also normalerweise durch das Ausnutzen der neuen Handlungsmöglichkeit erwiesen; wird eine solche Handlung unterlassen, kann eine Begründung verlangt oder ein Vorwurf erhoben werden:

(28) Wozu das ganze Theater, wenn du hinterher doch etwas anderes machst?

(29) Nun erlaubt man dir schon mal, H zu machen, und was machst du? Du sitzt hier herum!

Wie kommt es zu Obligationen dieser Art? Welche Funktion haben sie? Betrachten wir Verstöße gegen (24) und (25): in solchen Fällen wird der Schluß nahegelegt, daß der Partner ursprünglich eine ganz andere Intention verfolgt hat, z.B. PROVOZIEREN, AU-

TORITÄTSDISKURS EINLEITEN, EINEN PRÄZEDENZFALL SCHAFFEN. Und genau deshalb wird in der Folge eine Interaktion eingeleitet, die zu einer Handlungsbegründung oder Absichtserklärung führen soll. Es geht also um die Aufrechterhaltung der Autorität und Sanktionsgewalt. Das zeigt sich auch in den für Erlaubnishandlungen typischen Zusatzformeln:

(3o) Na schön, dann...

(31) Zwar habe ich starke Bedenken, aber (weil p) ...

(32) Ich bin nicht gerade begeistert, aber (weil p) ...

(33) Verdienst hast du es ja nicht, aber (weil p) ...

(34) Wenn du mir versprichst, daß p, bin ich bereit ...

(35) ... aber das muß eine Ausnahme bleiben.

(36) ... aber nur unter der Bedingung, daß p.

(37) ... aber das ist das allerletzte Mal!

Die Angst vor Autoritätsverlust wird hier ziemlich deutlich. (3o) und (37) lassen schon auf eine stark geschwächte Position schließen. (31) bis (33) stützen sich auf 'zwingende' äußere Gründe, während (34) bis (37) eine konditionale Erlaubnis markieren, bei der der für Permissive typische Prozeß des Aushandelns deutlich wird. Der Normadressat wird in jedem Fall in seine künftige Handlungsplanung Annahmen über die Glaubwürdigkeit, die Sanktionsbereitschaft und die Argumentationsstärke des Partners einbeziehen.

Die wesentlichen Realisierungstypen für ERLAUBEN und BEFEHL AUFHEBEN sind:

(38) Ich erlaube dir, H zu machen.
Ich bestehe nicht länger darauf, daß du H machst.

(39) Du darfst H machen.
Du brauchst H nicht zu machen.

(4o) Mach H!
Laß H bleiben!

(41) Von mir aus!

(42) Ja.

Wir finden also die performative Formel (38), die in normativen Kontexten wegen ihrer Deutlichkeit häufig ist, das Handlungsergebnis für den Adressaten angebende Modalverbkonstruktionen (39), Imperativsätze (4o), die sehr schön die Machtverhältnisse illu-

strieren, elliptische Konstruktionen (41), die auf eine der oben erwähnten Zusatzformeln reduziert sind, und Antworten auf als Entscheidungsfragen formulierte Bitten (42).
Bei Befehlen scheint der Normgeber hinsichtlich seiner Präferenzen etwas stärker situativ involviert zu sein, was dann in Sprecherhaltungen wie Unmut, Ärger, Frustration etc. zum Ausdruck kommt. Die Mehrzahl der Befehle wird zur kurzfristigen Lösung aktueller Handlungsprobleme verwendet, während die meisten Verbote zur länger andauernden Normeinführung dienen. In den überschaubareren Befehlssituationen ist eine Handlungsplanung vielfach leichter, so daß sich die Notwendigkeit seltener ergibt, einen Befehl aufgrund unzureichender Einschätzung der Lage wieder aufheben zu müssen. Während also Befehle meist vom Typ

(43) Mach H, und zwar sofort!

sind, lassen sich Verbote häufig als generalisierte Sätze wie

(44) Immer wenn du in einer Situation vom Typ C bist, unterlasse H!

darstellen.
Die Realisierungstypen für FREISTELLEN nehmen oft Bezug auf die damit eröffneten Handlungsmöglichkeiten:

(45) Hier habe ich einen interessanten Fall. Ich stelle Ihnen frei, ob Sie ihn bearbeiten wollen oder nicht.

(46) Mach, was du willst!

Auch hier gibt es verkürzte Formen, auf die wir nicht mehr eingehen wollen. Erfolgt die Handlung des FREISTELLENS als Reaktion auf eine Bitte um Aufhebung eines Befehls oder Verbots, gilt (26). Häufig finden sich Einschränkungen, die die Richtung der Entscheidung schon präformieren sollen:

(47) ... schließlich haben wir dich ja so erzogen, daß du weißt, was sich gehört.

Bei diesem Typ kann man von PSEUDO-FREISTELLEN sprechen. Er kommt oft in impliziter Form vor; dazu ein Beispiel aus einer in Arbeit befindlichen Untersuchung zur Sprache vor Gericht:

(48) (R= Richter; A= Angeklagter; S= Staatsanwalt)

R Ja, Herr H., Sie sind nich' verpflichtet, sich zu die-
R sem Vorwurf zu äußern. Woll'n Sie es dénnoch tun?
A [Nein, ich will mich nich' dazu äußern.
R [Sie woll'n sich
R [gar nich' dazu äußern? Überhaupt nich'? Nichts da-
A [Nein.
R [zu sagen? xxx Dann ham Sie heute natür-
A [(Kopfschütteln)
R lich 'n halben Tach umsonst gewartet, denn dann müssen
R wer die Verhandlung gleich wieder vertagen und Zéugen
R [laden. xxx
A [Warum?
S [Moment mal, ...

Gemäß § 243 StPO weist der Vorsitzende den Angeklagten auf sein Recht hin, zur Sache auszusagen oder sich nicht zu äußern. Bei der Vorbereitung des Prozesses war man offenbar davon ausgegangen, daß der Angeklagte aussagen würde; nicht nur, weil Verweigerungen selten sind, sondern auch, weil die zu verhandelnde Sache völlig klar schien. Nun also die Gefahr einer Vertagung, die zusätzlichen Zeit- und Arbeitsaufwand bedeutet. Nachdem der Richter dem Angeklagten demonstriert hat, daß er sich durchaus abweichend verhält, indem er gegen stillschweigende Regeln der Institution verstößt (und sich damit schadet), führt er ein erstes Argument an, das den Angeklagten zu einer Aussage ermuntern soll. Die Warum-Frage des Angeklagten ist ambig: sie könnte sich auf die Notwendigkeit einer Neuansetzung beziehen, sie könnte aber auch - nach meinem Eindruck als Beobachter ist dies die zutreffende Interpretation - die Berechtigung des gesamten Arguments zu hinterfragen suchen, etwa in der Form:

(49) Was kümmern mich die Probleme des Gerichts, wo ich doch nur mein mir zustehendes Recht wahrnehme?

Im Bereich des kodifizierten Normensystems finden sich Handlungen des FREISTELLENS primär da, wo Fehlschlüsse möglich erscheinen, man vergleiche dazu § 264 (2) StPO:

"Das Gericht ist an die Beurteilung der Tat, die dem Beschluß über die Eröffnung des Hauptverfahrens zugrunde liegt, nicht gebunden."

Die folgenden Beispiele zeigen, daß man sich von der Form nicht blenden lassen sollte:

(5o) Auf Ihren Antrag vom ... stellen wir Sie hiermit vom Wehrdienst frei.

(51) Frau N.N. erhält die Erlaubnis, nach Ablegung der Prüfung ein Kraftfahrzeug mit Antrieb durch Verbrennungsmaschine der Klasse drei zu führen.

Bei (5o) handelt es sich um eine Befehlsaufhebung; es wird nicht unterstellt, daß sich der Betroffene später doch noch für den Wehrdienst entscheidet. Im Falle von (51) ist es der Institution 'gleichgültig', ob Frau N.N. tatsächlich von der Aufhebung des zuvor bestehenden Verbots Gebrauch macht, es handelt sich also um eine Handlung des FREISTELLENS.

Abschließend ist anzumerken, daß zwischen Verpflichtungen verschiedener Art und Stärke zu differenzieren ist. Wenn etwa bei Befehlen hinter der Verpflichtung eine Sanktionsandrohung steht, so ist das schon etwas anderes, als wenn man zur Stützung seiner Behauptungen nicht bereit ist. Möglicherweise sind damit Extrempunkte auf einer Skala angedeutet, zwischen denen dann Fälle wie das NICHTEINLÖSEN eines Versprechens, das NICHTBEANTWORTEN von Fragen etc. anzusiedeln sind. Verpflichtungen, die aus der Verwendung von Permissiven resultieren, gehören - wie vermutlich die meisten Konversationsmaximen - eher an das untere Ende der Skala.

Literatur

HOFFMANN, Ludger (1978): Zur Sprache von Kindern im Vorschulalter, Köln und Wien: Böhlau

KATZ, Jerrold J. (1977): Propositional Structure and Illocutionary Force, Hassocks: Harvester Press

KUTSCHERA, Franz von (1973): Einführung in die Logik der Normen, Werte und Entscheidungen, München: Alber

LENK, Hans (ed.)(1974), Normenlogik, Pullach: UTB

LYONS, John (1977), Semantics, Vol.2, Cambridge: University Press

HANDLUNGSTHEORETISCHE ASPEKTE DES 'ERZÄHLENS' UND 'BERICHTENS'

Klaus-Peter Klein

1. 'Erzählen' und 'Berichten' als alltagssprachliche Kategorien

Eine Aufforderung wie:

(1) Erzähl' mal, wie es war!

ist geläufiger Bestandteil unserer alltagssprachlichen Kommunikation, ohne daß es für uns zum Problem werden könnte, ob der Aufforderung nun durch eine Handlung des ERZÄHLENs oder des BERICHTENs Folge geleistet werden muß. Ohne Schwierigkeit könnte (1) also auch substituiert werden durch:

(1') Berichte mal, wie es war!

Ganz undenkbar hingegen wäre es, wenn etwa der Verhandlungsleiter einer Delegation auf einer großen internationalen Konferenz eine bevorstehende Konsultation der heimischen Regierung mit dem Satz ankündigte:

(2) Ich werde meiner Regierung von dem gegenwärtigen Verhandlungsstand erzählen.

Vielmehr würde die Formulierung lauten:

(2') Ich werde meiner Regierung über den gegenwärtigen Verhandlungsstand berichten.

Noch evidenter wird die Differenz im Sprachgebrauch, wenn wir von den Textsorten ausgehen, die mit beiden Handlungsformen - zumindest begrifflich - jeweils in Verbindung gebracht werden:

(3) Der Major trug einen Bericht über die Manöverlage vor.

Mehr als Persiflage auf Militärisches würde die Formulierung verstanden werden:

(3') Der Major trug eine Erzählung über die Manöverlage vor.

Von der Vorstellung der fertigen Textprodukte, also der 'Erzählung' und des 'Berichts', war das Verständnis von ERZÄHLEN und BERICHTEN über Jahrzehnte hinweg fast ausschließlich bestimmt. Dabei spielten die Kategorien der Wahrhaftigkeit und Überparteilichkeit die entscheidende Rolle bei der Zuordnung eines Textes zu einer der beiden Textsorten. Damit thematisiert ist die Frage nach dem Verhältnis einerseits von Text und Wirklichkeit und andererseits von Text und Textproduzent.

Um beide Relationen kristallisiert sich die Diskussion in der erzähltheoretischen Literatur im Zusammenhang mit einer Definition von 'Erzählung' und 'Bericht' auch heute noch. Jenseits der Frage, inwieweit Sprache Wirklichkeit überhaupt abzubilden vermag, wird etwa von KANZOG (1976) der Wahrheitsbegriff zum Tertium comparationis der Unterscheidung zwischen ERZÄHLEN und BERICHTEN gemacht. Während dem ERZÄHLEN ein universeller Wahrheitsbegriff zugrunde liege, der die Darstellung fiktiver Ereignisse und Vorgänge ermöglicht, gelte für das BERICHTEN der Wahrheitsbegriff der zweiwertigen Logik, derzufolge jeder Aussage das eindeutige Merkmal 'wahr' oder 'falsch' (bzw. 'wirklich' oder 'unwirklich') zugeordnet wird: "Um über einen Vorgang b e r i c h t e n zu können, muß ihm das Merkmal 'w a h r' inhärent sein" (KANZOG 1976: 12, Hervorhebungen ebda). WERLICH (1975) stellt den fiktionalen Texten sog. 'faktische Texte' gegenüber, ohne dabei allerdings die begrifflichen Kategorien des ERZÄHLENS und BERICHTENS zu bemühen.

Die traditionell vor allem mit dem BERICHTEN verbundene Frage nach der 'Objektivität' von sprachlichen Wirklichkeitsdarstellungen hängt sowohl mit dem Wirklichkeitsbezug und Wahrhaftigkeitsanspruch eines Textes zusammen als auch mit dem Kriterium der 'Distanz' zwischen Text und Textproduzent bzw. mit dem Grad der subjektiven Perspektivierung bei der sprachlichen Abbildung von Wirklichkeit. Unter handlungstheoretischen Gesichtspunkten freilich ist diese Kategorie weitgehend obsolet, da jede sprachliche Äußerung eine geistige Tätigkeit voraussetzt, die auf der Grundlage der jeweils individuellen Intentionalität einer Persönlichkeitsstruktur und der diese mit beeinflussenden situativen Rahmenbedingungen vorgenommen wird. Überparteilichkeit im Sinne des völligen Fehlens subjektiver Elemente kann von daher durch den Berichtenden zwar angestrebt sein, ist aber letztlich nicht völlig realisierbar. Auf diesen Umstand weisen etwa auch LUDWIG/WOLFF (1978) hin. Die sich daran anschließende Frage, ob und inwieweit dem ERZÄHLEN und BERICHTEN identische Handlungsmuster zugeordnet werden können und müssen, soll im folgenden erörtert werden.

2. Die Textsortenspezifikation

Die textsortenspezifische Differenzierung zwischen 'Erzählung' und 'Bericht' ist nicht unerheblich beeinflußt und determiniert durch die spracherzieherische Übung, von beiden Textsorten jeweils entsprechende Aufsatztypen abzuleiten. Diese - alle Richtungen innerhalb der Sprachdidaktik ziemlich einheitlich erfassende - Tradition fixiert beide Textsorten allerdings auch weitgehend auf den schriftlichen Sprachgebrauch und weist ihnen eindeutige, teilweise kontrastierende stilistische Merkmale zu; so ist der Bericht demnach gekennzeichnet durch das Fehlen bestimmter grammatischer Mittel wie etwa der direkten Rede, des Präsens und des Konjunktivs.

Der allgemeine Sprachgebrauch ist von diesem spracherzieherischen Impetus deutlich geprägt. Kennzeichnend ist darüber hinaus, daß beide Textsorten ganz bestimmten Lebensbereichen zugeordnet werden: Der Bericht ist in Institutionen bzw. in institutionalisierten Kommunikationsverhältnissen angesiedelt, die Erzählung ist eine Kategorie des eher privaten Umgangs sowie des Literarischen. Dabei muß man nicht, wie LUDWIG/WOLFF (1978) es tun, die literarische Erzählung auf die Wiedergabe von Fiktivem begrenzen; auch reale Erlebnisse eines Autors oder anderer Personen können zum Gegenstand literarischer Bemühungen werden.

Die textsortenspezifische Differenzierung manifestiert sich zudem in den Handlungsverben, die beiden Substantiven im Alltagssprachgebrauch zugeordnet werden:

(4) einen Bericht (ab)geben
(5) Bericht über etwas erstatten
(6) eine Erzählung vortragen

Die Verben sind nicht beliebig austauschbar; jene in (4) und (5) verwendeten sind nicht nach (6) übertragbar, wohl aber das in (6) enthaltene nach (4) und (5). Die dem 'Bericht' zugeordneten Verbalausdrücke unterstreichen, daß die ihn erzeugenden Handlungen Bestandteil institutioneller oder quasi-institutioneller Abläufe sind. Überhaupt wird eine Handlung des BERICHTENs meist in der grammatischen Form 'Subst. + Verb + präp. Angabe' beschrieben ('einen Bericht abgeben über etwas'), während das ERZÄHLEN fast immer durch die Form 'Verb + Objekt' dargestellt wird ('etwas erzählen').

Die Frage ist nun, ob die Tätigkeiten des ERZÄHLENs und BERICHTENs stets zur Produktion eines Exemplars der Textsorte 'Erzählung' bzw. 'Bericht' führt. Aus den Beispielen (1) und (1'), die in dieser Art beliebig vermehrbar sind, wird deutlich, daß beide Tätigkeiten in ihrem alltagssprachlichen Verständnis häufig nicht unterschieden werden; wenn hingegen den Textsorten 'Erzählung' und 'Bericht' jeweils konstitutive stilistische Merkmale zugeordnet sind, die sich teilweise gegenseitig ausschließen und zudem deutliche Differenzen in der kommunikativen Funktion beider Textsorten signifizieren, kann ein zwingender Zusammenhang zwischen der Tätigkeit und der jeweils mit ihr begrifflich verwandten Textsorte nicht bestehen.

Die Analyse des ERZÄHLENs und BERICHTENs muß daher unter der Perspektive ihres Handlungscharakters vom traditionellen Verständnis der beiden zugeordneten Textsorten abgekoppelt werden. Beiden Tätigkeiten liegt ein gemeinsames zentrales Globalhandlungsmuster zugrunde, das sich umschreiben läßt mit JEMANDEM EIN 'STÜCK WELT' NÄHERBRINGEN; wenn ich etwas erzähle/berichte, transportiere ich Informationen über einen Realitätsausschnitt in der Annahme, daß dieser für den Hörer/Leser in bestimmter Hinsicht relevant ist.

LABOV/WALETZKY (1967) definieren "Erzählen (...) als e i n e verbale Technik der Erfahrungsrekapitulation (...), im besonderen als die Technik der Konstruktion narrativer Einheiten, die der temporalen Abfolge der entsprechenden Erfahrung entsprechen" (LABOV/WALETZKY 1967: 79, Hervorhebung ebda). Auch VAN DJIK (1974) spricht einer Erzählung im Kontext der Alltagskommunikation die Funktion eines 'Erfahrungsmodells' zu. Darüber hinaus unterscheiden LABOV/WALETZKY in diesem Rahmen zwei Dimensionen des ERZÄHLENs, nämlich die 'referentielle' und die 'evaluative'. Diese Beschreibungselemente treffen für den deutschen Sprachgebrauch ebenso auf das Muster BERICHTEN zu. Dies ist begründet in der handlungstheoretischen Fundierung beider Muster, die im folgenden umrissen werden soll. Dabei wird von einem einheitlichen Muster für beide Tätigkeiten ausgegangen; als Notation wird daher 'ERZÄHLEN/BERICHTEN' verwendet.

3. Die Handlungsmuster

Die Muster, die aus ERZÄHLEN/BERICHTEN erzeugt werden können, decken einen nicht unerheblichen Teil an möglichen sprachlichen Handlungsmitteln ab. Zentrale Muster sind etwa:

BEHAUPTEN
BEGRÜNDEN
BEWEISEN
ARGUMENTIEREN
BESCHREIBEN
BEWERTEN
URTEILEN

Die erzeugten Muster können kombiniert oder auch jeweils selbständig innerhalb des Musters ERZÄHLEN/BERICHTEN auftreten. Die Erzeugungsrelation sieht dann so aus:

$$\text{ERZ/BER} \longrightarrow \left\{ \begin{array}{l} \text{BEH} \\ \text{BEGR} \\ \text{BEWEIS} \\ \text{ARG} \\ \text{BESCHR} \\ \text{BEWERT} \\ \text{URTEIL} \end{array} \right\}$$

Die erzeugten Muster können teilweise ihrerseits mit Erzeugungsregeln einander zugeordnet werden:

$$\text{ERZ/BER} \longrightarrow \left\{ \begin{array}{l} \text{ARG} \longrightarrow \left\{ \begin{array}{l} \text{BEH} \\ \text{BEGR} \\ \text{BEWEIS} \end{array} \right\} \\ \text{BESCHR} \\ \text{BEWERT} \\ \text{URTEIL} \end{array} \right\}$$

Das Handlungsmuster ERZÄHLEN/BERICHTEN kann zudem nicht nur als erzeugendes, sondern ebenso als erzeugtes beschrieben werden. Die erzeugenden Muster leiten sich aus den institutionellen und privaten Handlungsfeldern ab, innerhalb derer nach dem Muster ERZÄHLEN/BERICHTEN gehandelt wird. Solche erzeugenden Muster sind etwa:

(7) jemanden informieren
(8) sich rechtfertigen
(9) etwas bezeugen
(10) etwas bestätigen

(11) jemanden belehren
(12) jemanden unterhalten
(13) angeben, aufschneiden
(14) jemanden bloßstellen

Das Erzeugungsschema hat demnach folgendes Aussehen:

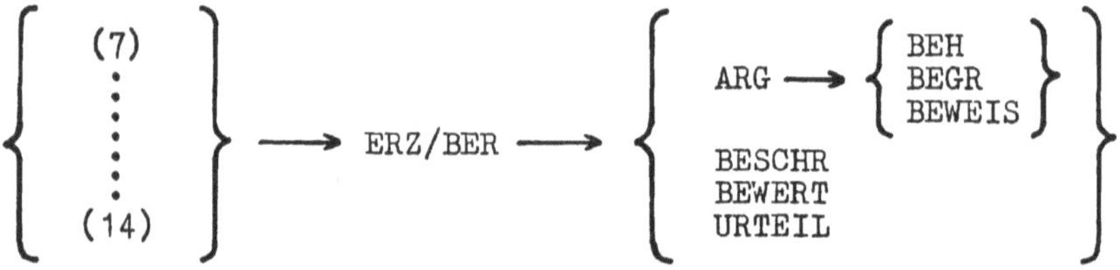

Die Muster (7) - (14) können ebenfalls sowohl kombiniert als auch einzeln das Muster ERZÄHLEN/BERICHTEN erzeugen. Dabei sind allerdings nicht alle Muster untereinander kombinierbar; so schließen sich z.B. (8) und (12) wohl gegenseitig aus, da die kommunikativen Zielrichtungen beider Muster völlig konträr zueinander stehen.

Die Reihe der ERZÄHLEN/BERICHTEN erzeugenden Handlungsmuster kann noch um eine Anzahl von weiteren Beispielen verlängert werden. So könnten etwa auch Muster wie VORWERFEN, AUFFORDERN, BELEIDIGEN, WARNEN, RATGEBEN, EIN PROBLEM BEWUßT MACHEN u.a. in diese Rubrik eingeordnet werden, da solche illokutiven Muster im Einzelfall durchaus auch mit Hilfe von ERZÄHLEN/BERICHTEN realisiert zu werden vermögen (etwa: Jemanden WARNEN, indem man über einen Vorfall erzählt/berichtet, indem man behauptet/begründet/beschreibt/bewertet etc.).

Das Erzeugungsschema verdeutlicht das komplexe Gefüge, in das vom Handlungsaspekt her gesehen das Muster ERZÄHLEN/BERICHTEN eingebettet ist. Diese Komplexität ist u.a. dadurch bedingt, daß für die Dauer einer Handlung nach dem Muster ERZÄHLEN/BERICHTEN keine allgemein gültigen Kriterien genannt werden können, da das Muster sehr offen ist und nur ein Minimum an verbindlichen Regeln für seinen Vollzug aufweist. Die in der rechten Klammer aufgeführten Muster können als Teilhandlungen bzw. als Untermuster des Musters ERZÄHLEN/BERICHTEN aufgefaßt werden.[1] Bei der Realisierung dieses Musters hat der Handelnde aus den möglichen Teilhandlungen jene(s) Muster auszuwählen, mit denen (dem) er seine Handlungsintention(en) am besten verwirklichen zu können glaubt.

Insofern geben die Teilhandlungen der rechten Klammer den Rahmen für die textuelle Gestaltung einer Handlung nach dem Muster ERZÄHLEN/BERICHTEN vor; die Summe der möglichen textuellen Verfahrensweisen bei der Produktion eines Textes nach diesem Muster entspricht der Summe jener Teilhandlungen.[2]

Die in der linken Klammer rubrizierten bzw. rubrizierbaren Muster repräsentieren die Gesamtheit der Intentionen, die eine Handlung nach dem Muster ERZÄHLEN/BERICHTEN auslösen können. Dabei ist denkbar, daß die Erzeugungskette auch unter Ausschluß des mittleren Gliedes funktioniert, was die Entbehrlichkeit des Musters ERZÄHLEN/BERICHTEN für bestimmte Handlungszüge aus dem Reservoir der dort aufgeführten Muster bedeuten würde. So kann die Pfeilrelation, auf ein herausgegriffenes Beispiel abgestellt, auch lauten:

$$\text{SICH RECHTFERTIGEN} \longrightarrow \left\{ \begin{array}{l} \text{ARG} \longrightarrow \left\{ \begin{array}{l} \text{BEH} \\ \text{BEGR} \\ \text{BEWEIS} \end{array} \right\} \\ \text{BESCHR} \\ \text{BEWERT} \\ \text{URTEIL} \end{array} \right\}$$

Zur Klärung der Frage, ob eine Rechtfertigungshandlung nach dem Muster ERZÄHLEN/BERICHTEN vorgenommen wird, gäbe es mehrere Möglichkeiten an Kriterien. Den entscheidenden Aspekt liefern uns LABOV/WALETZKY, die unter 'Erzählung' verstanden wissen wollen "jede beliebige Teilsatzfolge, die zumindest eine temporale Grenze enthält (...)" (LABOV/WALETZKY 1967: 105). Wird also ein Handlungsmuster derart realisiert, daß auf wenigstens zwei Ereignisse oder Handlungen in ihrem temporalen Ablauf referiert wird, so handelt es sich um eine Handlung gemäß dem Muster ERZÄHLEN/BERICHTEN.

4. Zur Struktur des Handlungsablaufs

ERZÄHLEN/BERICHTEN gehört zur Gruppe jener Muster, deren Vollzug auf die Vermittlung eines Ausschnitts von Wirklichkeit gegenüber einem Hörer/Leser abzielt, der in der Regel

(i) keine oder

(ii) eine ohne 'Erinnerungsstütze' von außen nicht mehr aktualisierbare oder

(iii) eine in gewisser Hinsicht nur unvollständige

Kenntnis dieses Wirklichkeitsausschnitts besitzt;[3] dabei kann die Diagnose 'keine'/'nicht mehr aktualisierbare'/'unvollständige' Kenntnis aus der Interessenlage des Sprechers/Schreibers durchaus anders gewichtet sein als aus der Sicht des Rezipienten.

Der Textproduzent muß seine Strategie des ERZÄHLEN/BERICHTENs in Einklang bringen mit der Notwendigkeit, im Hörer/Leser die Rekonstruktion des einschlägigen Wirklichkeitsausschnitts Schritt für Schritt zu erzeugen. Das Ziel ist es, den Rezipienten über den mentalen Nachvollzug von äußeren Zustands-, Geschehens- und Handlungsbeschreibungen für die Aufnahme von daraus durch den Textproduzenten erschlossenen Erfahrungswerten zugänglich zu machen. Die entscheidende Voraussetzung für eine Handlung nach dem Muster ERZÄHLEN/BERICHTEN ist die geistige Verarbeitung des Wahrgenommenen durch den Erzählenden/Berichtenden. Der darin eingeschlossene Vorgang der inneren Abbildung des Wahrgenommenen verläuft auf zwei Ebenen, die wir hier als heuristisch unterscheidbar postulieren wollen, nämlich - analog der von LABOV/WALETZKY in die Diskussion eingeführten Aspekte - auf der referentiellen und evaluativen Ebene. Zwar werden die Vorgänge und Handlungen auf beiden Ebenen weitgehend ineinander verschränkt realisiert; gleichwohl weisen die Verbalisierungen dieser 'inneren Abbilder' zumeist schwerpunktmäßige Präferenzen der einzelnen Verbalisierungsphasen für eine der beiden Dimensionen auf, die insofern für den Vollzug sprachlicher Handlungen nach dem Muster ERZÄHLEN/BERICHTEN strukturbildend wirken.

Die Artikulation von Erfahrungswerten, die sich für den Erzählenden/Berichtenden aus dem erlebten bzw. auch ihm selbst nur sprachlich oder durch elektronische Ton-/Bildträger übermittelten Geschehen[4] ergeben, erfolgt auf der Basis mentaler Evaluationsvorgänge. Der Rezipient muß sich in seiner 'Verstehensstrategie' auf das phasenweise Nacheinander bzw. Nebeneinander und 'Übereinander' von referentiellen und evaluativen Elementen der ERZÄHL-/BERICHTshandlung einstellen. Der Nachvollzug der ex- bzw. impliziten Evaluation erfolgt beim Hörer/Leser auf der Folie seiner Bemühungen um die Rekonstruktion der äußeren Geschehensabläufe, aus der er eigene evaluierende Schlüsse zu ziehen vermag, sowie aufgrund der evaluierenden Vorgaben, die der Sprecher/Schreiber expressis verbis übermittelt.[5]

Die strategischen Grundlagen der Verbalisierung durch den Textproduzenten beruhen auf dessen Annahmen über die Strukturierung des Verstehensprozesses beim Rezipienten und sein Verhältnis zu den Strukturen der mitgeteilten Wirklichkeit. Dies hat zur Folge, daß der Textproduzent nicht immer nur die Evaluation aus dem tatsächlichen Ablauf des eigenen Erlebnisses ableitet, sondern im Einzelfall auch das Verfahren insoweit umkehrt, als er die Auswahl und sprachliche Aufbereitung der Wirklichkeitsausschnitte, auf die er in der ERZÄHL-/BERICHTshandlung referiert, von der erwünschten und im Hörer/Leser als zu erzeugende angestrebten Evaluation abhängig macht. In dieser Weise kann man von einem interdependenten Verhältnis zwischen referentiellen und evaluativen Vorgängen im Bereich der die Textproduktion vorbereitenden und begleitenden mentalen Teilprozesse auf seiten des Sprechers/Schreibers wie auch innerhalb des Verstehensaktes durch den Rezipienten ausgehen.

Das folgende Diagramm soll die bisher diskutierten Zusammenhänge des Ablaufs einer Handlung nach dem Muster ERZÄHLEN/BERICHTEN auf Produzenten- und Rezipientenseite verdeutlichen:

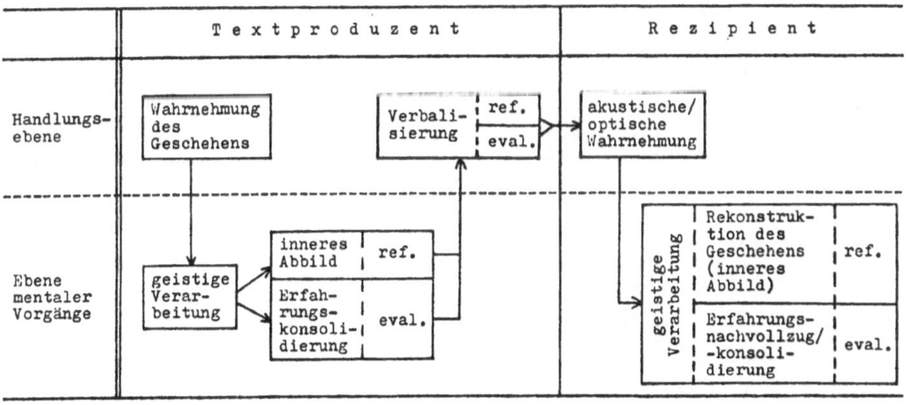

Zum Verständnis des Diagramms ist zu ergänzen, daß die Gesamtheit der mit dem ERZÄHLEN/BERICHTEN verbundenen Operationen prinzipiell differenziert werden kann nach solchen mit eindeutigem Handlungscharakter sowie nach Vorgängen, deren Handlungsqualität nicht eindeutig festlegbar erscheint.[6] Tätigkeiten des erstgenannten Typs werden der Ebene der Handlungen zugeordnet, Vorgänge im Umkreis der 'geistigen Verarbeitung' von Erlebtem werden auf der Ebene mentaler Abläufe abgebildet.

Der Verlauf solcher mentaler Prozesse beim Erzählenden/Berichtenden hängt wesentlich ab von der zeitlichen Distanz, die zwischen der Handlung des ERZÄHLEN/BERICHTENs und dem wiedergegebenen Geschehen liegt. Erfolgt die sprachliche Wiedergabe unter direktem Einfluß des noch nicht völlig abgeschlossenen Geschehens, auf das referiert wird, so weicht dies zwar graduell von der typischen Konstellation im Handlungsfeld des Musters ERZÄHLEN/BERICHTEN ab, liegt aber prinzipiell durchaus noch in dessen Bereich.

Ein Beispiel hierfür ist etwa eine sog. 'Live'-Reportage über ein beliebiges Geschehen im Hörfunk, wie sie REHBEIN (1977: 291 ff) in Gestalt einer Tischtennis-Übertragung diskutiert. Eine exakte Gleichzeitigkeit zwischen dem Geschehen und seiner sprachlich umgesetzten Übermittlung existiert natürlich auch hier nicht, wenngleich der geistige Verarbeitungsprozeß stark verkürzt ist. REHBEIN nimmt daher ein 'vor-organisiertes Wissen' an, das Reporter und Zuhörer über den referierten Gegenstand teilen und das ersterer während der Reportage nur abzurufen und zu aktivieren braucht. Dabei handelt es sich um eine Art standardisierten Vorwissens, "das vor-diskursiv auf einer allgemeineren gesellschaftlichen Ebene organisiert ist" (REHBEIN 1977: 297).

Die damit für den Handlungsvollzug einer 'live' vorgenommenen Sportübertragung umrissenen Bedingungen für ihr Gelingen treffen im weiteren auch auf alle anderen Handlungszüge nach dem Muster ERZÄHLEN/BERICHTEN zu. Der Transfer von Geschehensabläufen in Sprache und ihre Rückerschließung aus der sprachlichen Mitteilung durch den Rezipienten kann nur funktionieren, wenn bestimmte - je nach Referenzobjekt verschiedene - Elemente vor-diskursiven Gemeinschaftswissens zwischen den Kommunikationspartnern existieren. Ihre nähere Analyse könnte wichtige Aufschlüsse nicht nur über Strategien des ERZÄHLEN/BERICHTENs erbringen, sondern auch über die Möglichkeiten und das Ausmaß von Konditionierungen des Verstehensprozesses.[7]

Anmerkungen

1 Die Frage nach ihrem Verhältnis zu der Kategorie 'Basishandlung' bzw. 'Basismuster' kann hier nicht diskutiert werden.

2 Zu den textuellen Handlungsmöglichkeiten, mit denen das Muster ERZÄHLEN/BERICHTEN realisiert wird, gehören zudem Muster wie EINFÜHREN (eines Diskurs-Universums), WEITERFÜHREN, WIEDERHOLEN, EXPANDIEREN u.ä. (vgl. hierzu SANDIG 1978 und HOFFMANN/KLEIN 1979).

3 Daneben gibt es auch die Möglichkeit, daß die Realisierung des Musters erfolgt, obwohl keine der drei genannten Bedingungen vorliegt (etwa bei einem in regelmäßigen Abständen wiederholten 'Sich-Erzählen' von gemeinsamen früheren Erlebnissen unter ehemaligen Kriegs- oder Schulkameraden). Hierbei wäre zu prüfen, ob der Sprecher/Schreiber nicht doch stets von neuem von der impliziten Annahme ausgeht, daß nicht alle Details und Aspekte des gemeinsamen Erlebnisses dem Hörer/Leser präsent sind.

4 Handlungen des Musters ERZÄHLEN/BERICHTEN, die nicht auf selbst Erlebtes referieren, gehören zum Typus des 'Weitererzählens'.

5 Daß der Nachvollzug der Erfahrungswerte, die als zu übermittelnde intendiert sind, auch abhängig ist von einer Reihe von Variablen, die in der Lebensgeschichte, den psycho-sozialen u.a. Determinanten des Rezipienten begründet liegen, sei noch zusätzlich vermerkt. Aus Platzgründen kann hierauf in diesem Rahmen nicht weiter eingegangen werden.

6 Vgl. hierzu das Problem der 'mental acts' (etwa bei GEACH 1957).

7 Für den Bereich des literarischen Verstehens wirft von der literaturwissenschaftlichen Seite her die Rezeptionsästhetik vergleichbare Fragestellungen auf.

Literatur

ANDEREGG, Johannes (1973): Fiktion und Kommunikation. Ein Beitrag zur Theorie der Prosa. Göttingen: Vandenhoeck.

GEACH, Peter Th. (1957): Mental acts. London: Roupledge.

GÜLICH, Elisabeth / RAIBLE, Wolfgang (1977): Linguistische Textmodelle. Grundlagen und Möglichkeiten. München: Fink.

HERINGER, Hans J. (1974): Praktische Semantik. Stuttgart: Klett.

HOFFMANN, Ludger / KLEIN, Klaus-Peter (1979): "Lernziel: Texte verstehen. Handlungstheoretische Grundlagen und didaktische Konsequenzen." GEWEHR, Wolf (ed.) (1979): Sprachdidaktik. Neue Perspektiven und Unterrichtsvorschläge. Düsseldorf: Schwann.

KANZOG, Klaus (1976): Erzählstrategie. Eine Einführung in die Normeinübung des Erzählens. Heidelberg: Quelle & Meyer.

KRUSCHE, Dietrich (1978): Kommunikation im Erzähltext. 1. Analysen: Zur Anwendung wirkästhetischer Theorie. München: Fink.

KUMMER, Ingrid (1975): "Die Formung von Charakterisierungsbegriffen durch Erzählen." EHRICH, Veronika / FINKE, Peter (eds.) (1975): Beiträge zur Grammatik und Pragmatik. Kronberg: Scriptor.

LABOV, William / WALETZKY, Joshua (1967): "Narrative analysis: oral versions of personal experience". HELM, June (ed.) (1967): Essays on the verbal and visual arts. Seattle: University of Washington Press. - Übers.: "Erzählanalyse: Mündliche Versionen persönlicher Erfahrung". IHWE, Jens (ed.) (1973): Literaturwissenschaft und Linguistik. Eine Auswahl Texte zur Theorie der Literaturwissenschaft. II. Frankfurt: Athenäum.

LUDWIG, Otto / WOLFF, Gerhart (1978): "Berichten im Alltag - Berichten in der Schule". Praxis Deutsch 28: 16-24.

REHBEIN, Jochen (1977): Komplexes Handeln. Elemente zur Handlungstheorie der Sprache. Stuttgart: Metzler.

SANDIG, Barbara (1978): Stilistik. Sprachpragmatische Grundlegung der Stilbeschreibung. Berlin: de Gruyter.

STIERLE, Karlheinz (1975): Text als Handlung. Perspektiven einer systematischen Literaturwissenschaft. München: Fink.

VAN DJIK, Teun A. / IHWE, Jens / PETÖFI, János S. / RIESER, Hannes (1972): Zur Bestimmung narrativer Strukturen auf der Grundlage von Textgrammatiken. Hamburg: Buske.

--- (1974): "Action, action-description and narrative". New Literary History 6: 273-294.

WERLICH, Egon (1975): Typologie der Texte. Entwurf eines textlinguistischen Modells zur Grundlegung einer Textgrammatik. Heidelberg: Quelle & Meyer.

DIE METHODE DER SPRECHAKTANALYSE

Kurt Nikolaus

1. Problemstellung

Eine Reihe von Sprachphilosophen und Linguisten - wie z. B. Austin, Searle u. v. a. - haben sich intensiv mit Problemen auseinandergesetzt, die im Zusammenhang mit solchen sprachlichen Handlungen auftauchen wie etwa versprechen, befehlen, warnen, sich entschuldigen, jemandem einen Rat geben etc.; als Ergebnis ihrer Bemühungen haben sie dann sog. Sprechaktanalysen vorgeschlagen, die die genannten Probleme lösen sollten, indem sie angeben, wie man etwas verspricht, wie man befiehlt, warnt, Ratschläge erteilt, sich entschuldigt usw. Leider haben die betr. Theoretiker dabei ein grundlegendes Problem völlig außer acht gelassen, nämlich die Frage, *wie man zu Sprechaktanalysen kommt*. Mit diesem metatheoretischen und methodologischen Problem werde ich mich hier beschäftigen.

Ich möchte zunächst die Relevanz der Fragestellung verdeutlichen. Angenommen, irgendein Autor X schlägt eine sprechaktanalyse Y eines bestimmten Sprechakts Z vor, und er beansprucht dabei, daß die Form seiner Analyse hinreichend allgemein sei, um sie auch auf andere Sprechakte übertragen zu können. Auf diese seine Sprechaktanalyse Y, oder genauer: auf die Annahme ihrer Verallgemeinerbarkeit, mögen einige weitreichende Annahmen über die Natur der Sprache aufbauen - etwa, Sprache sei regelgeleitetes Verhalten o. ä. Um die Gültigkeit dieser Annahmen beurteilen zu können, ist es notwendig, den Anspruch von X, seine Analyse Y von Z sei verallgemeinerbar, kritisch zu überprüfen, und dies wird in vielen Fällen nichts anderes heißen, als daß man einen beliebigen von Z verschiedenen Sprechakt mithilfe derselben Kategorien und Begriffe zu analysieren versucht, die auch X in seiner Sprechaktanalyse Y verwendet hat. (Wir überprüfen also, ob es Fälle gibt, für die der Verallgemeinerbarkeitsanspruch von X bezüglich Y begründet ist.) Der zu überprüfende Anspruch unseres Sprechakttheoretikers X lautet etwa folgendermaßen:

"Zu jedem Sprechakt Z' gibt es eine Sprechaktanalyse Y' von Z' derart, daß Y' (partiell) strukturgleich mit Y ist - was immer dies heißen mag." [1]

X behauptet also, eine Analyse eines bestimmten Sprechakts gegeben zu haben, die gleichzeitig als Paradigma für andere Sprechaktanalysen dienen kann. Wenn es uns gelingt, auch nur einen einzigen Sprechakt Z' zu finden, zu dem keine Sprechaktanalyse Y' von der gewünschten Art existiert, dann muß der Anspruch der Verallgemeinerbarkeit von Y fallengelassen werden. Unabhängig von der Art der paradigmatischen Analyse Y ist es jedoch kaum vorstellbar, daß ein solcher Fall eintritt: Solange die Methode, wie man zu einer Sprechaktanalyse kommt, nicht genau spezifiziert ist, wird man mit einiger Phantasie und etwas Glück stets zu einer Analyse von der gewünschten Art kommen können. Wenn es uns aber wider Erwarten einmal nicht gelingen sollte, eine solche Analyse zu finden, so beweist das nur, daß wir eben eine falsche Methode gewählt haben, und widerlegt nicht zwingend die oben angeführte Behauptung. Das heißt aber nicht anderes, als daß der empirische Gehalt und damit die praktische Anwendbarkeit der vorgeschlagenen Sprechaktanalyse und der darauf aufbauenden Theorie wesentlich reduziert wird.[2] Die Möglichkeit einer Falsifikation des oben formulierten Anspruchs - den meines Wissens alle Sprechakttheoretiker in ähnlicher Weise zumindest implizit erheben - besteht also nur relativ auf eine gegebene Methode der Sprechaktanalyse, die genau spezifiziert, welche Ausgangsdaten wir heranziehen dürfen (oder müssen) und von welchen Methoden und Argumentationsschritten wir Gebrauch machen dürfen (oder müssen). Eine solche Methode ist wesentlich für jede empirische Überprüfung der Sprechakttheorie. Darüber hinaus ist es selbstverständlich notwendig, die Kategorien und Begriffe, die bei Sprechaktanalysen gewöhnlich gebraucht werden, systematisch zu rechtfertigen. Das traditionelle Schema, das Searle vorgeschlagen hat, nämlich: propositional content condition - preparatory conditions - sincerety condition - essential condition, ist oft wiederverwendet und vielleicht noch öfter modifiziert und erweitert worden, vor allem von Wunderlich, aber meines Wissens hat noch niemand die Frage aufgeworfen, warum Sprechaktanalysen denn eigentlich gerade diejenige spezifische Form haben, die sie eben gewöhnlich haben und die uns schon so vertraut geworden ist, daß wir nicht mehr versuchen, sie zu begründen und zu rechtfertigen. Es gibt zwar ausführliche Diskussionen über inhaltliche Probleme bestimmter Sprechaktanalysen, aber diese Diskussionen setzen eine allgemeine Form von Sprechaktanalysen bereits voraus - nämlich das von Searle vorgeschlagene Schema oder eine Modifikation davon -, die nicht hinterfragt wird und deren Adäquatheit offenbar für intuitiv evident gehalten wird. Da dieses Schema niemals prinzipiell in Frage gestellt wurde, hat es unter der Hand de facto einen a-priori-Status angenommen, der ihm keinesfalls zugebilligt werden kann. Auch wenn man nicht glaubt, daß dieses Schema grundsätzlich inadäquat ist, so muß man doch versuchen, es theoretisch oder empirisch zu begründen.

In den bisherigen Bemerkungen sind zwei Aspekte des hier zur Debatte stehenden Problems teilweise vermischt worden, die man auseinanderhalten sollte. Meiner Ansicht nach kann man die Frage "Wie kommt man zu Sprechaktanalysen?" in zwei Teilprobleme aufspalten, nämlich:

1. Von welchen Daten soll man ausgehen, und welche Methoden kann man anwenden, um zu einer Sprechaktanalyse zu kommen? Dies kann als das methodologische Problem der Sprechaktanalyse bezeichnet werden.

2. Welche Form soll eine Sprechaktanalyse annehmen, und welche Kategorien und Begriffe kommen für eine Analyse von Sprechakten in Frage? Dies kann als das metatheoretische Problem der Sprechaktanalyse bezeichnet werden. Wenn bestimmte Kategorien und Begriffe gewählt werden im Gegensatz zu anderen, so ist dies selbstverständlich zu rechtfertigen.

Die beiden genannten Probleme stehen deutlich in engem Zusammenhang miteinander, und beide hängen mit dem weitergehenden Problem zusammen, was Sprechaktanalysen eigentlich analysieren und was der Geltungsbereich dieser Analysen ist. An dieser Stelle möchte ich mich aber auf das methodologische Problem beschränken, da mir dieses als grundlegend für die Beantwortung weiterer Fragen erscheint.

Als Grundlage der Überlegungen zum methodologischen Problem der Sprechaktanalyse werde ich John R. Searles Buch "Speech Acts" von 1969 wählen, da Searles Sprechaktanalysen als typisch gelten können und den meisten anderen Analysen als Vorbild dienten.[3] Obwohl Searles Ausführungen i. allg. recht klar sind, ist die Aufgabe einer Rekonstruktion der sprechakttheoretischen Methode doch äußerst schwierig, denn methodologische Hinweise sind in sprechakttheoretischen Arbeiten sehr selten und zudem meist nicht besonders verläßlich. Als repräsentatives Beispiel werde ich die Analyseprozedur zu explizieren versuchen, die Searle und andere Autoren (nach meiner Interpretation) vermutlich implizit verwendet haben. Ich bin nicht sicher, ob meine Interpretation der sprechakttheoretischen Vorgehensweise korrekt ist; falls nicht, wäre es die Aufgabe der Sprechakttheoretiker, mich zu korrigieren, indem sie die von ihnen tatsächlich benutzten Methoden angeben. Mein Ziel ist hier lediglich, das methodologische Problem der Sprechaktanalyse aufzuwerfen; ich beanspruche nicht, es zu lösen. Die folgenden Überlegungen sollen daher nichts weiter als eine Diskussionsgrundlage schaffen, um das Problem zu verdeutlichen und - vielleicht - e i n e Lösungsmöglichkeit anzudeuten. Das Ziel dieser Arbeit wäre jedenfalls schon erreicht, wenn die folgenden Bemerkungen möglichst bald als überholt gelten könnten, weil sich die Sprechakttheoretiker selber des methodologischen Problems nachdrücklicher angenommen haben, als dies bisher der Fall war.

2. Rekonstruktion der Vorgehensweise bei der Sprechaktanalyse

Da über die Methoden der Erkenntnisgewinnung in der Sprechaktanalyse selten auch nur ein Wort verloren wird, lassen sie sich nur schwer präzise rekonstruieren. Der faktische Weg, der bei der Erstellung von Sprechaktanalysen mehr oder weniger unbewußt eingeschlagen wird, scheint in etwa der folgende zu sein:

Zu untersuchen sei ein beliebiger Sprechakt X (z. B. versprechen, empfehlen, warnen, befehlen o. ä.).

2.1. Ausgangsbasis der Analyse

Unter der Voraussetzung, daß es zu jedem möglichen Sprechakt einen möglichen Satz oder eine Menge von Sätzen (in jeder natürlichen Sprache?) gibt, deren wörtliche Äußerung in einem bestimmten Kontext eine Ausführung des betr. Sprechakts darstellen würde - dies ergibt sich nach SEARLE (1969: 20f.) aus dem Prinzip der Ausdrückbarkeit -, ist als Ausgangsbasis der Analyse eine beliebige, aber hinreichend große menge M von (erfundenen oder beobachteten) Sätzen bzw. Äußerungen (in einer beliebigen Sprache L?) in gegebenen Kontexten anzusehen.

In der Sprechakttheorie ist nicht immer völlig klar, ob jeweils über Äußerungen oder über Sätze geredet werden soll. Searles Ausführungen über Sprechakte als Grundeinheiten sprachlicher Kommunikation (1969: 16) sowie verschiedene andere Gründe legen eher eine Deutung von M als Menge von Äußerungen nahe. Die Notwendigkeit einer Relativierung der Äußerungen auf Kontexte ergibt sich aus der Form konstitutiver Regeln (SEARLE 1969: 34f., vgl. a. 24f.), während eine Relativierung auf natürliche Einzelsprachen in der Sprechakttheorie gewöhnlich unterbleibt.

2.2. Erste Voraussetzung

Das Prinzip der Ausdrückbarkeit ermöglicht es uns nach SEARLE (1969: 20), die Regeln für das Ausführen von Sprechakten mit den Regeln für das Äußern gewisser sprachlicher Elemente gleichzusetzen. Die sprachlichen Elemente, die insbesondere für Sprechakte wichtig sind, sind die sog. illokutionsanzeigenden sprachlichen Ausdrucksmittel (illocutionary force indicating devices). Zu jedem Sprechakt muß es in jeder Sprache mindestens ein solches illokutionsanzeigendes sprachliches Ausdrucksmittel geben: Das Prinzip der Ausdrückbarkeit besagt ja, daß es für jede Bedeutung einen möglichen sprachlichen Ausdruck gibt, der diese Bedeutung genau zum Ausdruck bringt. (Die Relativierung auf natürliche Einzelsprachen fehlt auch hier.) Nun kann aber eine Bedeutung nur dann genau ausgedrückt werden, wenn jeder ihrer Teile genau ausgedrückt werden kann; der Illokutionsmodus ist nach Searle ein wesentlicher Teil der Bedeutung eines Satzes - zumindest legt SEARLEs These, daß "the speech act or acts performed in the utterance of a sentence are in general a function of the meaning of the sentence" (1969: 18) diese Deutung nahe; es muß also zumindest ein sprachliches Ausdrucksmittel geben, mit dessen Hilfe es möglich ist, den Illokutionsmodus einer beliebigen Äußerung genau, d. h. explizit, anzugeben:

> Wherever the illocutionary force of an utterance is not explicit it can always be made explicit. This is an instance of the principle of expressibility, stating that whatever can be meant can be said. (SEARLE 1969: 68)

Sei x die Klasse der illokutionsanzeigenden sprachlichen Ausdrucksmittel in einer Sprache L, deren normale und korrekte Verwendung in illokutionsanzeigender Funktion in einer beliebigen Äußerung y in L anzeigt, daß die betr. sprachliche Äußerung y eine konventionelle Realisierung des Sprechakts X in L ist. (Meist wird so getan, als ob es sich bei x um eine Klasse mit nur einem Element, nämlich einem performativen Verb, handele.) Sei w ein beliebiges Element aus x.

2.3. Erster Analyseschritt

Der erste Analyseschritt der Sprechakttheorie besteht nun darin, daß M in eine "standardisierte" Ausgangsbasis M' überführt wird derart, daß für M' alle die Bedingungen von M gelten, es zu jedem Element aus M ein Äquivalent in M' gibt und umgekehrt und darüber hinaus alle Elemente von M' propositional und illokutionär explizit sind. (Elliptische Redeweisen z. B. werden aufgelöst, indirekte Sprechakte in direkte überführt usw.) Wir erhalten so in M' eine Menge von (teilweise erfundenen oder beobachteten, teilweise konstruierten) Äußerungen in gegebenen Kontexten, die aufgrund unseres intuitiven Vorverständnisses Ausführungen des zu untersuchenden Sprechakts X darstellen und deren propositionaler Gehalt und illokutive Kraft (letztere durch Verwendung von Elementen w aus x) sprachlich explizit gemacht wurde. Die Überführung von M in M' geschieht dabei auf der Basis unserer Kenntnis der normalen und korrekten Verwendung illokutionsanzeigender und anderer Ausdrucksmittel einer gegebenen Sprache in gegebenen Kontexten; an diesem Punkt verläßt sich die sprechakttheoretische Analyse also erstmals auf die sprachliche Kompetenz des Analysators.

2.4. Zweiter Analyseschritt

Aus M' können nun die Bedingungen für den Gebrauch der illokutionsanzeigenden sprachlichen Ausdrucksmittel w aus x in der folgenden Weise abgeleitet werden:

Indem man die Elemente von M' in ihren semantischen und kontextuellen Merkmalen systematisch variiert und feststellt, ob sie bei solcher Variation noch akzeptable sprachliche Äußerungen unter korrekter, normaler und illokutionsanzeigender Verwendung eines Elements w aus x bleiben, kann man die notwendigen Bedingungen für den normalen und korrekten Gebrauch der illokutionsanzeigenden sprachlichen Ausdrucksmittel w aus x (in illokutionsanzeigender Funktion) ermitteln. Alle notwendigen Bedingungen, die man auf diese Weise ermittelt, zusammengenommen, bilden eine hinreichende Bedingung für die Möglichkeit der korrekten und normalen Verwendung eines solchen illokutionsanzeigenden sprachlichen Ausdrucksmittels in illokutionsanzeigender Funktion.

Auf diese Weise kommt man zu einer Aussage folgender Art: "Der Illokutionsmodus einer beliebigen Äußerung y in einer Sprache L, die ein Sprecher S gegenüber einem Hörer H in einem Kontext C macht, kann mithilfe eines illokutionsanzeigenden sprachlichen Ausdrucksmittels w aus x (in illokutionsanzeigender, korrekter und normaler Verwendung) dann und nur dann sprachlich explizit gemacht werden, wenn die folgenden Bedingungen gelten: ...."

Indem man die Bedingungen für den Gebrauch von illocutionary force indicating devices auf diese Weise ermittelt, verläßt man sich ein zweites Mal auf die sprachliche Kompetenz des Analysators. Das Kriterium der Akzeptabilität von Äußerungen in Kontexten ist für diesen entscheidenden Analyseschritt von zentraler Bedeutung. Die verschiedenen Typen von notwendigen Bedingungen für Sprechakte, die Searle vorschlägt (Bedingung des propositionalen Gehalts, Vorbereitungsbedingungen, Aufrichtigkeitsbedingung, für den Sprechakt wesentliche Bedingung) wären dann systematisch aus den verschiedenen Auswirkungen, die eine Verletzung dieser Typen von Bedingungen auf die Akzeptabilität der untersuchten Äußerungen hat, herzuleiten. Daraus, daß die Akzeptabilität in verschiedener Weise und in verschiedenem Umfang verletzt werden kann, ergeben sich dann auch verschiedene Typen des Ausführens von Sprechakten (aufrichtig vs. nicht aufrichtig, defektiv vs. nicht-defektiv), worauf wir hier aber nicht näher eingehen wollen.

Sei B die Konjunktion der notwendigen und in ihrer Gesamtheit hinreichenden Bedingungen für die Möglichkeit der korrekten und normalen Verwendung eines illokutionsanzeigenden sprachlichen Ausdrucksmittels w aus x in illokutionsanzeigender Funktion, wie sie mit dem oben dargestellten, auf dem Kriterium der Akzeptabilität von Äußerungen in Kontexten beruhenden Verfahren ermittelt werden.

2.5. Zweite Voraussetzung

In der Sprechakttheorie wird die Gültigkeit der folgenden Schlußregel als Voraussetzung angenommen:

Wenn der Illokutionsmodus einer beliebigen Äußerung y in einer Sprache L, die ein Sprecher S zu einem Hörer H in einem Kontext C macht, unter genau den Bedingungen B mithilfe eines illokutionsanzeigenden sprachlichen Ausdrucksmittels w aus x (in illokutionsanzeigender, korrekter und normaler Verwendung) sprachlich explizit gemacht werden kann, so sind B die Bedingungen für den Sprechakt (oder genauer: für das Ausführen des Sprechakts) X.

Hierbei ist nicht von Bedingungen für das Ausführen des Sprechakts X in der Sprache L die Rede; zumindest SEARLE nimmt an (1969: 39f.), daß trotz der von Sprache zu Sprache verschiedenen sprachlichen Ausdrucksmittel w aus x die Bedingungen B für alle Sprachen gleich sind.

2.6. Ergebnis der Analyse

Die Bedingungen B sind damit notwendig und in ihrer Gesamtheit hinreichend für die Möglichkeit des Ausführens des Sprechakts X (in jeder beliebigen Sprache L); vielleicht sind sie sogar hinreichend für das tatsächliche Ausführen eines solchen Sprechakts. Daraus lassen sich dann weitere Aussagen ableiten, z. B.:

Unter genau den Bedingungen B zählt eine Äußerung y in einer Sprache L als Realisierung des Sprechakts X.

Wenn ein Sprecher S in einer Sprache L eine Äußerung y macht, die als Realisierung eines Sprechakts X zählen soll, so verpflichtet er sich damit auf die Gültigkeit der Bedingungen B.

Wenn ein Hörer H eine Äußerung y in einer Sprache L als Realisierung eines Sprechakts X akzeptiert, so verpflichtet er sich damit ebenfalls auf die Gültigkeit der Bedingungen B.

3. Kritik der Vorgehensweise der Sprechaktanalyse

In der hier dargestellten sprechakttheoretischen Vorgehensweise sind eine Reihe von Unklarheiten und Problemen enthalten, die in einer intersubjektiv kontrollierten Methode der Erkenntnisgewinnung ausgeräumt werden müßten. (Es sei jedoch nochmals darauf hingewiesen, daß die hier vorgelegte Rekonstruktion der sprechakttheoretischen Methode nur tentativen Charakter hat; die darauf aufbauende Kritik ist daher nicht überzubewerten.)

Zunächst ist nicht klar, ob in der Ausgangsbasis der Analyse von Sätzen-in-Kontexten oder von Äußerungen ausgegangen wird. Weiterhin wird in der Methode der Sprechaktanalyse nicht genügend unterschieden zwischen Bedingungen für den Gebrauch illokutionsanzeigender sprachlicher Ausdrucksmittel, Bedingungen für Sprechakte bzw. für das Ausführen von Sprechakten, Regeln für den Gebrauch illokutionsanzeigender sprachlicher Ausdrucksmittel und Regeln für Sprechakte bzw. deren Ausführung. In der hier dargestellten Rekonstruktion wurde von Regeln überhaupt kein expliziter Gebrauch gemacht; dadurch wurde die wissenschaftstheoretische Problematik des Regelbegriffs umgangen. SEARLE behauptet jedoch (1969: 55, 62), Regeln für den Gebrauch illokutionsanzeigender sprachlicher Ausdrucksmittel aus Bedingungen für Sprechakte abzuleiten; dagegen wurde das sprechakttheoretische Vorgehen hier in der Weise gedeutet, daß vielmehr Bedingungen für Sprechakte aus den Bedingungen für den Gebrauch illokutionsanzeigender sprachlicher Ausdrucksmittel abgeleitet wurden (vgl. 2.5.). Eine weitere Schwierigkeit besteht darin:

Zu einem gegebenen Sprechakt X gibt es in einer gegebenen Sprache mit hoher Wahrscheinlichkeit mehr als nur ein illokutionsanzeigendes sprachliches Ausdrucksmittel, d. h. die Klasse x umfaßt mehr als ein Element. Die Bedingungen für den normalen und korrekten Gebrauch des illokutionsanzeigenden sprachlichen Ausdrucksmittels w_1 aus x (in illokutionsanzeigender Funktion) werden jedoch mit den Bedingungen für den Gebrauch von w_2 aus x nicht völlig übereinstimmen. Dies führt zu Schwierigkeiten bezüglich der eindeutigen Bestimmung der Bedingungen B, die aber gewöhnlich übersehen werden, weil man ein performatives Verb als einzig relevantes Element von x annimmmt. Es ist charakteristisch, daß in der Sprechakttheorie nahezu ausschließlich sprechaktbezeichnende Verben untersucht wurden; dies führt jedoch zu einer Verengung des Gegenstandsbereichs und zu dem Vorwurf, die Sprechakttheorie untersuche nicht Sprechhandlungen, sondern lediglich die Semantik sprechaktbezeichnender Verben. In der Tat folgt aus dem Prinzip der Ausdrückbarkeit bei Searle nur, daß es zu jedem Illokutionsmodus bzw. zu jedem Sprechakt mindestens ein i l l o k u t i o n s a n z e i g e n d e s sprachliches Ausdrucksmittel geben muß:

> ... for any possible speech act there is a possible linguistic element the meaning of which (given the context of utterance) is sufficient to determine that its literal utterance is a performance of precisely that speech act. (SEARLE 1969: 20f.)

Das heißt jedoch weder, daß es genau ein solches sprachliches Element geben müsse, noch, daß es zu jedem Sprechakt auch ein s p r e c h a k t b e z e i c h n e n d e s sprachliches Ausdrucksmittel, z. B. ein performatives Verb, geben muß. (Intonation und Partikel etwa können unter Umständen als illokutionsanzeigende, nicht jedoch als sprechbezeichnende sprachliche Ausdrucksmittel fungieren.) Daraus, daß man bei jeder Äußerung in einer gegebenen Sprache und bei gegebenem Kontext im Prinzip die Art des intendierten Sprechakts sprachlich kennzeichnen kann (Prinzip der Ausdrückbarkeit), folgt eben noch lange nicht, daß es für den betreffenden Sprechakt in der jeweiligen Sprache auch einen Namen gibt: Man kann den Illokutionsmodus einer Äußerung auch anzeigen, ohne ihn zu bezeichnen. Die Sprechakttheorie beschäftigt sich jedoch in der Praxis nur mit solchen Sprechakten, für die es in einer gegebenen Sprache Namen gibt. (Dies sind für verschiedene Sprachen höchstwahrscheinlich verschiedene.) Faktisch liegt der Vorgehensweise der Sprechaktanalyse also nicht die gesamte Klasse x, sondern nur eine echte Teilklasse x' von x zugrunde, die nur diejenigen illokutionsanzeigenden sprachlichen Ausdrucksmittel zu einem Sprechakt X enthält, die gleichzeitig sprechaktbezeichnende sprachliche Ausdrucksmittel sind. Die Teilklasse x' umfaßt in einer jeweiligen Einzelsprache jedoch sehr häufig nur ein Element, so daß sich das oben angeschnittene Problem dann nicht mehr stellt.

Schließlich ist auch nicht hinreichend geklärt, inwiefern die

Sprechakttheorie den Anspruch erheben kann, universale Eigenschaften von Sprechakten aufzudecken, die von einzelsprachlichen Konventionen unabhängig sind. Der Hauptmangel der ganzen Vorgehensweise liegt aber darin, daß sie methodisch im wesentlichen auf der sprachlichen Kompetenz des Sprechakttheoretikers als native speaker aufbaut (vgl. SEARLE 1969: 12f.): Der Begriff der Akzeptabilität von Äußerungen in Kontexten scheint in der Methode der Sprechaktanalyse eine ähnliche Rolle zu spielen wie etwa der Begriff der Grammatikalität von Sätzen in der generativen Grammatik. Während die vorgenannten Unklarheiten und Probleme bei gesteigerter methodischer und theoretischer Sorgfalt im Prinzip durchaus beseitigt werden können - und müssen -, liegt hier ein grundsätzliches methodologisches Problem: Das Verlassen auf "linguistic intuitions" und "tacit knowledge" an wesentlichen Punkten der Sprechaktanalyse entzieht die sprechakttheoretische Methode nämlich einer wirklich stringenten intersubjektiven Kontrolle.

Anmerkungen

1 Unterstrichene Buchstaben stehen für Variablen, nicht unterstrichene für Konstanten.

2 Ähnlich gelagerte Fälle sind in der Wissenschaftstheorie nicht unbekannt; ich verweise etwa auf das Problem der sog. Mehrdeutigkeit statistischer Erklärungen, das zuerst von Hempel aufgeworfen wurde.

3 Die Ausweitung des Sprechaktbegriffs über Sätze hinaus auf längere Diskurseinheiten, wie sie neuerdings etwa von van Dijk diskutiert wird, lasse ich hier völlig außer Betracht.

Literatur

SEARLE, John R. (1969): Speech acts. An essay in the philosophy of language. Cambridge: University Press

HANDLUNGSPRÄSUPPOSITIONEN IN KOMMUNIKATIVEN HANDLUNGEN

Theodossia Pavlidou

1. Einige handlungstheoretische Begriffe; Handlungspräsuppositionen

Menschen führen durch ihr Handeln Veränderungen herbei. Ein Handlungsagent versucht durch die Auswahl und Realisierung eines Handlungsschemas die Handlungssituation so zu beeinflussen, daß er sein Handlungsziel tatsächlich erreicht. Doch Handlungen sind nicht immer erfolgreich; oft entsteht ein Handlungsprodukt, welches dem Handlungsziel nicht entspricht. Beim Vollzug von Handlungen werden Voraussetzungen bzw. Annahmen gemacht, deren Richtigkeit es wahrscheinlicher macht, daß die Handlung erfolgreich ist.[1]

In diesem Zusammenhang vereinbaren wir folgende Redeweise: Erfahrungsinhalte des Handlungsagenten werden *Handlungspräsuppositionen* in einer Handlungssituation genannt, wenn folgendes erfüllt ist: (i) sie stellen einen Teil der Situationstheorie des Agenten dar; (ii) sie gehen in die Gestaltung der Handlung ein, d.h. in die Auswahl des zu realisierenden Handlungsschemas, und somit manifestieren sie sich in den verwendeten Äußerungsoperationen; (iii) ihr Wahrheitswert wird in der Handlungssituation vom Agenten nicht problematisiert.

2. Kommunikative Intention versus Handlungsziel

Wie bei jeder Handlung, so werden auch mit kommunikativen Handlungen Ziele verfolgt; diese gehen üblicherweise über die Verständigung hinaus. Wenn ich jemandem sage, "Hier wird nicht geraucht!", dann begnüge ich mich nicht damit, daß er versteht, was ich meine; vielmehr bin ich daran interessiert, daß er tatsächlich im gegebenen Raum nicht mehr raucht. Eine Äußerung verstanden haben bzw. die kommunikative Intention des Sprechers erschloßen haben bedeutet meistens längst nicht, daß das damit verbundene Handlungsziel gleichzeitig erfüllt wird. Anders gesagt: Das Erkennen der kommunikativen Intention garantiert nicht den Erfolg der Handlung. Andererseits aber ist das Verstehen seitens des Addressaten in den Augen des Agenten notwendige Bedingung für das Erreichen des Handlungsziels. Das Verstehen soll nämlich dem Addressaten den Anstoß geben, der zur Verwirklichung des Ziels führen soll.

3. Handlungspräsuppositionen über den Addresaten; Klassifikation

Im eben erwähnten Sinne sind dann bei kommunikativen Handlungen diejenigen Handlungspräsuppositionen wichtig, die den Partner betreffen; von ihm hängt ja schließlich der Erfolg der Handlung ab. Die Annahmen des Agenten über seinen Partner bzw. des Sprechers über den Addressaten können folgendermaßen unterschieden werden:

A. Annahmen über das Wissen des Partners
- A.1. Kennen von Referenzobjekten, Personen
- A.2. Wissen von Sachverhalten
- A.3. Wissen über Normen, Verhaltensmaximen, Verfahrensweisen etc. innerhalb einer Gruppe
- A.4. Kennen von Sprach-, Kommunikations- und Interaktionsregeln

B. Annahmen über Bedürfnisse, Interessen, u.ä. des Partners

C. Annahmen über Handlungen, Aktivitäten, u.ä. (vergangene oder künftige) des Partners

D. Annahmen über Eigenschaften, Fähigkeiten, u.ä. des Partners

E. Annahmen über Status, Rollen, Funktionen des Partners.

Solche Annahmen über den Partner können verschiedentlich die Auswahl des zu realisierenden Handlungsschemas beeinflussen und sich auf verschiedenen Weisen in den verwendeten Äußerungsoperationen manifestieren: im illokutiven Typ der Sprechhandlung, in der konkreten Formulierung der Illokution (vgl. beispielsweise direkte vs. indirekte Sprechhandlungen), in der Formulierung des propositionalen Gehalts (z.B. in der Verwendung von Abtönungspartikeln), aber auch in paralinguistischen Merkmalen, wie in der Geschwindigkeit der Rede, in der Aufmerksamkeit, mit der etwas gesagt wird, usw.[2]

4. Exkurs über den Präsuppositionsbegriff

In den obigen Klassen von Annahmen sind durchaus solche aufzufinden, die unter anderen Bezeichnungen in der Linguistik bekannt sind, beispielsweise als Präsuppositionen, preparatory conditions, etc. Die meisten Schwierigkeiten scheint der Begriff der Präsupposition zu bereiten. Die Diskussion darüber, was unter "Präsupposition" zu verstehen ist, dauert noch an, wie es aus neueren Arbeiten ersichtlich ist (vgl. z.B. HARDER/KOCK 1976, VERSCHUEREN 1978). Der Präsuppositionsbegriff scheint ein Begriff zu sein, "dessen Unklarheit mit der Anzahl seiner Klärungen zunimmt", wie Grewendorf es formulierte (GREWENDORF 1977: 20).

Aus diesen Gründen ziehe ich es im Moment vor, keine weitere Klärung hinzuzufügen und erst recht nicht nach Verallgemeinerungsmöglichkeiten des Begriffs zu suchen. Stattdessen will ich zunächst auf eine vorhandene Festlegung zurückgreifen:

> Unter "Präsuppositionen" versteht man im Allgemeinen bestimmte "selbstverständliche Voraussetzungen" (FREGE 1892), die ein Sprecher mit einer Äußerung macht; sie werden nicht ausdrücklich behauptet, vielmehr als Annahmen über den vorausgesetzten Hintergrund der Kommunikation mit in die Äußerung eingebracht und müssen vom Hörer auch als solche erschließbar sein. Die Präsuppositionen sind ein Teil der Bedingungen, die erfüllt sein müssen, damit eine Äußerung in einem gegebenen Kontext angemessen sein kann. (FRANCK 1973: 11)

Franck unterscheidet eindeutig zwischen Präsuppositionen und preparatory conditions: erstere hängen "von der speziellen sprachlichen Form der vorgebrachten Äußerung" ab, letztere dagegen wären "unabhängig vom jeweiligen propositionalen Gehalt" (FRANCK 1973: 13f).

Obwohl es oft nicht explizit ausgedrückt wird: Präsuppositionen werden meistens mit Annahmen, die das Wissen des Partners betreffen, assoziiert. Eine der wenigen Ausnahmen hierzu ist Ebert, die explizit auch Annahmen über Interessen, Fähigkeiten, u.ä. des Partners als Präsuppositionen zuläßt (vgl. EBERT 1973: 52). Die Einführung des Begriffs "Handlungspräsupposition"[3] stellt den Versuch dar, auch andere Annahmen (außer Wissensannahmen) und ihre Rolle im Kommunikationsprozeß in die Betrachtung miteinzubeziehen; dabei werden die üblichen Einschränkungen, die an Präsuppositionen gestellt werden, nicht beachtet.

5. Reaktionsmöglichkeiten auf eine Sprechhandlung; Rechtfertigungen

Bekanntlich kann man auf verschiedenen Ebenen auf eine Sprechhandlung reagieren - abgesehen vom üblichen Fortsetzungsmuster, welches mit dieser Sprechhandlung verbunden wird und vom Sprecher erwartet wird (z.B. eine Antwort auf eine Frage, die das angibt, was der Fragende wissen wollte). Nehmen wir das Beispiel:

Hat der hübsche Andreas endlich aufgehört zu rauchen?

In Anlehnung an EBERT 1973 kann man dazu auf der Präsuppositionsebene reagieren, indem man

(a) eine Verständnisfrage stellt, z.B.: "Welcher Andreas?",

(b) eine Bemerkung zu der mitgelieferten Information bringt, wie "Ich wußte gar nicht, daß er raucht.",

(c) mitgelieferte Information bestreitet bzw. einen Einwand erhebt: "Er ist doch gar nicht hübsch!".

Man kann aber auch auf eine Sprechhandlung reagieren, indem man nach dem Sinn der Handlung fragt, einen Grund für die Handlung verlangt oder den Stellenwert der Handlung im Rahmen der gemeinsamen Interaktion in Frage stellt.[4] Auf dieser Ebene geht es weniger um ein Verstehen dessen, was getan bzw. was gesagt worden ist, sondern um ein erklärendes Verstehen im Sinne von Max Weber. In diesem Fall fragen wir nach dem Sinn, welchen der Agent mit seiner Handlung verbunden hat. Um beim Beispiel zu bleiben, könnten solche Reaktionen lauten: "Warum fragst du?", "Wieso ist es jetzt wichtig?", "Was soll das?", "Wie kommst du darauf?", usw.[5], um ihn aufzufordern, die Handlung zu rechtfertigen.

Solche Fragen können natürlich auch kommen, wenn der Sprecher über eigene oder fremde Handlungen berichtet bzw. erzählt, beispielsweise: "Ich habe ihr versprochen, pünktlich da zu sein." Hier könnte ein Grund verlangt werden, sowohl für die Handlung des Versprechens, worüber berichtet wird, als auch für die gerade vollzogene Handlung des Berichtens, d.h.: "Warum hast du es ihr versprochen?" vs. "Warum erzählst du es mir überhaupt?".

6. Rechtfertigung von Sprechhandlungen mit Hilfe von Handlungspräsuppositionen, die sich auf den Addressaten beziehen

Wird der Sprecher mit einer Aufforderung zur Rechtfertigung seiner Handlung konfrontiert, so hat er auch mehrere Möglichkeiten, dieser Aufforderung nachzukommen. Er kann das Handlungsziel nennen, welches er mit der Handlung verfolgt hat. Oder er kann ein Bedürfnis angeben, welches zu der Handlung geführt hat. Er kann aber auch eine Handlungspräsupposition angeben, die in die Gestaltung der Handlung eingegangen ist.[6]

Wir werden nun anhand von Beispielen untersuchen, inwiefern die verschiedenen Klassen von Handlungspräsuppositionen die Funktion des Rechtfertigens der zugehörigen Sprechhandlung übernehmen können.[7] Dabei sind folgende Fragen relevant: Kann der Agent jede Art von Handlungspräsupposition als Grund für seine Sprechhandlung angeben? Wie können diejenigen Handlungspräsuppositionen, mit denen gerechtfertigt werden kann, charakterisiert werden?

Es wird bei den Beispielen ein Schema S_A S_B S'_A verwendet, welches eine Folge von Sprechhandlungen zwischen den Kommunikationspartnern A und B darstellt; dabei ist

S_A eine beliebige Sprechhandlung von A,
S_B eine Aufforderung von B an A, S_A zu rechtfertigen,
S'_A ein Versuch von A, S_A mit Hilfe einer Handlungspräsupposition von S_A, HPräs(S_A) zu rechtfertigen.

Es wird zunächst nicht unterschieden, ob HPräs(S_A) zutrifft oder nicht. Allerdings sollte an dieser Stelle darauf hingewiesen werden, daß falls HPräs(S_A) zutrifft, dann weist der Addressat eine darauf sich stützende Rechtfertigung höchst wahrscheinlich zurück: Der Addressat weiß ja bereits den angegebenen Grund; er möchte wohl etwas darüber hinaus wissen.

Betrachten wir die einzelnen Klassen von Handlungspräsuppositionen, die in 3. angegeben wurden:

ad A.1.: Die Annahmen, die in dieser Klasse enthalten sind, gelten typischerweise als Präsuppositionen:

(1) S_A: Ich habe Werner gesehen.
HPräs(S_A): B kennt Werner.
S_B: Na und? Warum erzählst du mir das?
* S'_A: Ich dachte, du kennst Werner.

Unabhängig davon, ob die Handlungspräsupposition von A zutrifft oder nicht, kann A sie nicht zur Rechtfertigung von S_A gebrauchen und zwar unabhängig vom illokutiven Typ von S_A.

ad A.2.: Ähnliches gilt auch für Annahmen über das Wissen von Sachverhalten, zumindest insofern als sie als Präsuppositionen im gängigen Sinne aufgefaßt werden können:

(2) S_A: War Werner's Frau sehr aufgeregt?
HPräs(S_A): B weiß, daß Werner verheiratet ist.
S_B: Warum fragst du?
* S'_A: Wußtest du denn nicht, daß Werner verheiratet ist?

Auch hier kann die Handlungspräsupposition nicht zur Rechtfertigung benutzt werden.

Wenn dagegen eine solche Handlungspräsupposition in die Sprechhandlung als preparatory condition eingeht, dann kann sie eher als Grund für die Sprechhandlung akzeptiert werden:

(3) S_A: Ich habe Werner getroffen.
HPräs(S_A): B weiß nicht, das ich Werner getroffen habe.

S_B: Na und?
S_A': Ich dachte, du hast es nicht gewußt.

In diesem Fall kann S_A' als Rechtfertigung gelten, im Zusammenhang mit einem angenommenen Interesse, eben daß B etwas wissen möchte, was hier nicht angesprochen wird.

Darüber hinaus gibt es Wissensannahmen, mit denen doch gerechtfertigt werden kann, wie folgendes Beispiel (in Anlehnung an HARTMANN 1977: 105) zeigt: A bietet B einen Kaffee an; B antwortet: "Nein, danke; ich darf ja keinen Kaffee mehr trinken." Der angegebene Grund für die Ablehnung stellt eine Handlungspräsupposition über das Wissen von A dar, die mit dem Abtönungspartikel <u>ja</u> markiert wird.

ad A.3/A.4.: Es scheint, daß mit Annahmen, die die Bekanntheit von Normen, Verfahrensweisen, u.ä. innerhalb einer Gruppe unterstellen, wohl gerechtfertigt werden kann:

(4) Während einer Abstimmung in der Mitgliedervollversammlung einer Organisation:
S_A: Ich habe einen Antrag zur Geschäftsordnung.
HPräs(S_A): Es ist allgemein bekannt, daß die Geschäftsordnung für Vollversammlungen einen G.O.-Antrag während einer Abstimmung zuläßt.
S_B: Was soll das? Das stört doch die Abstimmung!
S_A': Kennen Sie denn unsere G.O. nicht?

Dasselbe gilt auch für pauschale Unterstellungen, z.B. daß der Partner die Sprache kennt, jedoch nur dann, wenn sie nicht zutreffen:

(5) In einem Ausländerviertel fragt ein Deutscher:
S_A: Wissen Sie vielleicht, wann der nächste Bus kommt?
HPräs(S_A): B kann Deutsch.
S_B: Nix verstehen!
S_A': Ach, entschuldigen Sie, ich dachte, Sie sprechen Deutsch.

Das eben angegebene Beispiel stellt eine ziemlich paradoxe Situation dar: Wenn B tatsächlich kein Deutsch kann, dann kann er auch nicht S_A' verstehen. Daß es sich um eine Rechtfertigung handelt, könnte er aus Merkmalen wie Intonation, Gestik u.ä. eventuell erschließen. Sollte B die Sprache kennen, so wäre eine solche Rechtfertigung seitens A, wie im Beispiel (5), völlig absurd. Es sei denn HPräs(S_A) bezieht sich z.B. auf eine einzelne Wortbedeutung und wird benutzt um einen Vorwurf zurückzuweisen:

(6) S_A: Und in dieser Plethora der Sinneseindrücke ...
HPräs(S_A): B kennt das Wort "Plethora".
S_B: Du verwendest ja schon wieder ein Fremdwort!
S_A': Ach, entschuldige, ich dachte, du wüßtest, was "Plethora" heißt.

Eine Zwischenstellung in diesem Spektrum erhalten Annahmen, die das Vermögen des Partners betreffen, z.B. Sätze in einem bestimmten Kontext zu interpretieren, eine primäre Illokution aus einer sekundären zu erschließen (i.S.v. Searle) oder Gesprächsimplikaturen zu folgern (i.S.v. Grice), und die Unterstellung der Bekanntheit der damit verbundenen Sprach- bzw. Kommunikationsregeln. Betrachten wir zunächst den Fall von indirekten Sprechhandlungen:

(7) S_A: Möchtest du, daß ich gehe?
HPräs(S_A): B weiß, daß mit der Frage ein Vorschlag gemeint ist.
S_B: Warum fragst du?

Hier kann unmöglich die unterstellte "Regel" zur Rechtfertigung benutzt werden. Wohl aber etwas, was diese Unterstellung weiterhin aufrechterhält:

S'_A: War nur ein Vorschlag.

In den Fällen, wo der Sprecher etwas implikatiert, unterstellend, daß der Addressat in der Lage ist die Implikatur zu erschließen, kann genausowenig die entsprechende Handlungspräsupposition zur Rechtfertigung dienen. Nehmen wir Grice's Beispiel:

(8) B fragt A, wie es C mit seiner neuen Stelle geht.
S_A: Ziemlich gut, denke ich; er mag seine Kollegen und im Knast war er noch nicht.
HPräs(S_A): B kennt die nötigen Schritte, Maximen, usw. um die Implikatur, C ist nicht besonders ehrlich, abzuleiten.
S_B: Was soll das mit dem Knast heißen?

A kann kaum HPräs(S_A) benutzen, um B einen Grund zu geben; er könnte höchstens eine Handlungspräsupposition hier verwenden, die sich auf das Wissen von B bezieht, über den Charakter, Vergangenheit, etc. von C:

S'_A: Ich dachte, du wußtest, daß C schon mal gefeuert wurde, weil er in dunkeln Geschäften verwickelt war.

Die hier enthaltene Annahme scheint als Evidenz für die implikatierte Behauptung - daß C nicht besonders ehrlich sei - zu fungieren.

ad B./C./D./E.: Für Handlungspräsuppositionen dieser Art gilt folgendes:

- Handlungspräsuppositionen, die mit Präsuppositionen identifiziert werden können, taugen nicht als Grund für die zugehörige Handlung:

(9) S_A: Warum hast du das Auto in die Garage gestellt?

HPräs(S_A): B hat das Auto in die Garage gestellt.
S_B: Wieso?
* S_A': Ich dachte, du hättest das Auto in die Garage gestellt.

- Handlungspräsuppositionen, die als preparatory conditions in die Sprechhandlung eingehen, können zur Rechtfertigung verwendet werden:

(10) S_A: Ich verspreche dir, dich mal nach Athen mitzunehmen.
HPräs(S_A): B will unbedingt nach Athen.
S_B: Wie kommst du denn darauf?
S_A': Willst du denn nicht mit?

- Sonnst scheinen solche Handlungspräsuppositionen im Rechtfertigungsprozeß durchaus einsetzbar zu sein; zumindest insofern als sie für den Handlungserfolg relevant sind, beispielsweise:

(11) S_A: Die Straßen sind verstopft.
HPräs(S_A): B fährt gleich los.
S_B: Na und?
S_A': Wolltest du nicht gleich fahren?

(12) S_A: Kannst du das Buch da oben hinstellen?
HPräs(S_A): B ist größer als ich.
S_B: Wieso fragst du mich?
S_A': Du bist ja größer als ich.

Wenn dagegen die Handlungspräsupposition mit dem Handlungsziel nicht zusammenhängt bzw. nur indirekt relevant ist, dann kann damit nicht gerechtfertigt werden:

(13) S_A: Tut mir leid, daß ich das Auto nicht repariert habe.
HPräs(S_A): B kann nicht das Auto selbst reparieren.
S_B: Wieso?
* S_A': Weil du das nicht kannst.

7. Schlußbemerkungen

Abschließend können wir folgendes festhalten: Zum einen gibt es offensichtlich eine Hierarchie der Handlungspräsuppositionen, nämlich: von ganz a l l g e m e i n e n, z.B. solchen die gültig für eine ganze soziale Gruppe in jeder Situation wären (z.B. 1.4.), zu ganz s p e z i f i s c h e n, d.h. solchen die nur für einen Partner in der gegebenen Situation gültig sind. Innerhalb dieser Hierarchie scheinen addressatenspezifische Handlungspräsuppositionen für die Rechtfertigung der zugehörigen Handlung eher verwendbar zu sein als allgemeine.[8] Zum anderen scheint es, daß nur solche Handlungspräsuppositionen als Grund angegeben werden können, die relevant für den Erfolg der Handlung, d.h. die wichtig für die Verwirklichung des Handlungsziels sind; relativ unproblematisch sind dagegen solche, die nur die kommunikative Intention, und somit das Verstehen der Äußerung betreffen (wie beispielsweise Präsuppositionen).

Diese beiden Aspekte, nämlich die Stellung einer Handlungspräsupposition innerhalb der Hierarchie und ihre Relevanz in Bezug auf das Ziel, scheinen also zu bestimmen, ob die Handlungspräsupposition zur Rechtfertigung der Handlung verwendet werden kann oder nicht.

Es gibt noch Unklarheiten in diesem Zusammenhang, und vor allem sehr viele Fragen, die noch zu untersuchen wären, z.B.: Wie hängen Handlungspräsuppositionen, mit denen gerechtfertigt werden kann, mit den preparatory conditions zusammen? Bleiben solche Handlungspräsuppositionen invariabel, wenn wir von einer direkten zu einer indirekten Sprechhandlung übergehen? usw. All dies wäre relevant für eine umfassendere Untersuchung von praktischer Argumentation, insbesondere unter dem Gesichtspunkt: W e r gibt w a s f ü r Gründe für seine Handlungen?

Anmerkungen

1 Zur Erläuterung der in diesem Abschnitt angedeuteten Begriffe vgl. PAVLIDOU 1978.

2 Zum Beispiel reden wir langsamer und deutlicher mit Leuten, von denen wir annehmen, sie können die verwendete Sprache nicht so gut.

3 Das Wort "Handlungspräsupposition" habe ich von EHLICH/REHBEIN 1972 entnommen. Dort wird der Terminus "Handlungspräsupposition" als handlungstheoretische Explikation des Begriffs der Erwartung verwendet; mit einer Handlungspräsupposition wird demnach ein Vorgriff auf die Handlungen des Anderen unternommen (EHLICH/REHBEIN 1972: 109).

4 Weitere Möglichkeiten auf Handlungen zu reagieren, worauf hier nicht eingegangen wird, wären das Bestreiten der Zweckmäßigkeit der Handlung in Hinblick auf das verfolgte Ziel oder das Ablehnen der Handlung aufgrund von Wertsystemen.

5 Die Untersuchung dessen, wie man nach einem Grund für eine Handlung fragen kann bzw. was man an einer Handlung (auf der Ebene des subjektiven Sinns) anfechten kann, muß einer anderen Arbeit vorenthalten bleiben.

6 Der Agent kann sogar eine Handlungspräsupposition benutzen, um selber eine Rechtfertigung für die Erwiderung des Addressaten zu verlangen, z.B.:
Mach die Tür zu!
Mach die doch selber zu!
Gehst du denn nicht sowieso in die Küche?

Die letzte Sprachhandlung enthält eine Handlungspräsupposition, die in die erste Sprechhandlung eingegangen ist, nämlich: der Addressat geht in die Küche.

7 Die folgenden Beispiele und Diskussion beschränkt sich auf Sprechhandlungen, die Überlegungen lassen sich aber genauso auf nichtsprachlichen Handlungen übertragen.

8 Präsuppositionen sind zwar addressatenspezifisch, können aber leicht aufgrund der Kenntnis der Sprache erschlossen werden.

Literatur

EBERT, Karen (1973): "Präsuppositionen im Sprechakt". Ten CATE, Abraham P./JORDENS, Peter (eds.) (1973): Linguistische Perspektiven. Referate des VII. Linguistischen Kolloquiums, Nijmegen 1972. Tübingen: Niemeyer.

EHLICH, Konrad/REHBEIN, Jochen (1972): "Erwarten". WUNDERLICH, Dieter (ed.) (1972): Linguistische Pragmatik. Frankfurt a.M.: Athenäum: 99-114.

FRANCK, Dorothea (1973): "Zur Problematik der Präsuppositionsdiskussion". PETÖFI, Janos S./FRANCK, Dorothea (eds.) (1973): Präsuppositionen in Philosophie und Linguistik. Frankfurt a. M.: Athenäum: 11-41.

GREWENDORF, Günter (1977): "Präsuppositionen bei disjunktiven Fragen". Linguistische Berichte 52: 13-31.

HARDER, Peter/KOCK, Christian (1976): The theory of presupposition failure. Copenhagen: Akademisk Forlag.

HARTMANN, Dietrich (1977): "Aussagesätze, Behauptungshandlungen und die kommunikative Funktion der Satzpartikeln <u>ja</u>, <u>nämlich</u> und <u>einfach</u>". WEYDT, Harald (ed.) (1977): Aspekte der Modalpartikeln. Studien zur deutschen Abtönung. Tübingen: Niemeyer: 101-114.

PAVLIDOU, Theodossia (1978): Wahrheit-Handlung-Argumentation. Beeinflussen kommunikative Faktoren die Wahrheitsfindung? Hamburg: Buske.

VERSCHUEREN, Jef (1978): "Reflections on presupposition failure: a contribution to an integrated theory of pragmatics". Journal of Pragmatics 2: 107-154.

EIN BEITRAG ZUR THEORIE DER INDIREKTEN SPRECHAKTE

Werner Sökeland

0. Bei den Klassikern der Sprechakttheorie taucht das Konzept der Indirektheit noch nicht auf. AUSTIN (1962) und SEARLE (1969) erläutern ihre Theorie an völlig direkten, expliziten Äußerungen. In der Rezeption der Sprechakttheorie durch MAAS/WUNDERLICH (1972) und andere zeigte sich jedoch bald, daß in der Kommunikationspraxis viele Äußerungen vorkommen, deren kommunikative Funktion in einem problematischen Verhältnis zu ihrer sprachlichen Struktur steht. Seitdem gehört Indirektheit zu den meistdiskutierten Konzepten der linguistischen Pragmatik.[1] Allerdings ist eine kohärente Theorie bisher nicht erarbeitet worden. Auch in diesem Beitrag kann nur der Ansatz einer Theorie vorgeschlagen werden.

1. Eine vorläufige Bestimmung der Begriffe "direkter" und "indirekter Sprechakt" ist im Gesagten schon enthalten: Während bei direkten Sprechakten die kommunikative Funktion der sprachlichen Struktur entspricht, sind indirekte Sprechakte Äußerungen, bei denen sprachliche Struktur und kommunikative Funktion in einem problematischen Verhältnis zueinander stehen. Zur sprachlichen Struktur zähle ich neben phonologischen und syntaktischen Merkmalen auch die Bedeutung, verstanden im Sinne der strukturalen Semantik. Der Terminus "kommunikative Funktion" wird im folgenden gleichbedeutend mit "illokutionäre Rolle" verwendet.

1.1. Ein Spannungsverhältnis zwischen Struktur und Funktion empfindet man beispielsweise an Fragesätzen, mit denen ein anderer Sprechakt als der des Fragens ausgeführt wird:

(1) Wer kannte ihn nicht, diesen charmanten Plauderer?

oder bei Fragehandlungen, die nicht mit Fragesätzen ausgeführt werden:

(2) Ich suche den Bahnhof.

Äußerung (1) hat die Struktur eines Fragesatzes, ist aber als Behauptungsakt intendiert. Explizit lautet er:

(1a) Fast alle kannten diesen charmanten Plauderer.

Äußerung (2) erscheint von der Struktur her wie eine Aussage, stellt aber eine Fragehandlung dar, die folgendermaßen zu paraphrasieren ist:

(2a) Wo ist der Bahnhof?

1.2. Problemlos ist die vorläufige Definition der Indirektheit allerdings nicht. Denn die Rede von einem "Spannungsverhältnis" oder einem "Widerspruch" zwischen z.B. Fragesatz-Struktur und Behauptungs-Funktion ist zugespitzt und metaphorisch. Widersprüche entstehen nur zwischen Entitäten, die in ihren grundlegenden Merkmalen gleich sind. Es gibt daher keinen objektiven Widerspruch zwischen einer Struktur und einer Funktion, sondern allenfalls zwischen der kommunikativen Funktion F_1, die man einer bestimmten sprachlichen Struktur "normalerweise" oder "in wörtlicher Interpretation" zuschreibt, und der Funktion F_2, die die Äußerung dieser sprachlichen Struktur im jeweiligen Einzelfall hat.

1.3. Damit stellen sich drei Fragen, die eine Theorie der indirekten Sprechakte beantworten muß, will sie die Brauchbarkeit des Indirektheit-Begriffs nachweisen und ihn operationalisierbar machen.

a) Worin äußert sich der Widerspruch zwischen der mit einer Sprachstruktur "normalerweise" verbundenen und der bei der aktuellen Äußerung intendierten Funktion?

b) Was ist die "normale" oder "wörtliche" Interpretation einer Sprachstruktur?

c) Was kann als über-intuitives Kriterium für das Vorliegen eines Spannungsverhältnisses zwischen zwei Illokutionen dienen?

2. Zur ersten Frage. Nach der präzisierten Begriffsbestimmung sind in jedem indirekten Sprechakt zwei gegensätzliche Illokutionen angelegt. Für jede von ihnen muß es Indikatoren geben. Eine Gruppe von Indikatoren (ich nenne sie "Basisindikatoren") deutet auf die im indirekten Sprechakt häufig nicht realisierte Illokution hin. So wird man bei

(3) Ich brauch' mal einen Schraubenzieher

die Aussagesatz-Struktur als Basisindikator einer Mitteilungs-Illokution auffassen. Diese Illokution, die ich als Basisillokution bezeichne, ist aber nicht voll realisiert. Denn der Sprecher kann nicht erwarten, daß der Hörer ein Interesse daran hat, den geäußerten propositionalen Inhalt zu erfahren. Das aber wäre Voraussetzung für einen geglückten Mitteilungs-Akt. Tatsächlich stellt (3) eine Aufforderungshandlung dar. Explizit lautet sie:

(3a) Gib' mir (bitte) einen Schraubenzieher.

Die tatsächliche illokutionäre Rolle der Äußerung erkennt der Hörer mit Hilfe weiterer Indikatoren, die ich "Sekundärindikatoren" nenne. Bei (3) zeigen z.B. die Partikel <u>mal</u> und der Kontext als Sekundärindikatoren die Aufforderungs-Funktion an.

Ein zweites Beispiel. Die Äußerung

(4) Wer arbeitet schon gern bei dieser Hitze?

stellt trotz Fragesatz-Struktur keine Fragehandlung dar, sondern eine Behauptung:

(4a) Niemand arbeitet gern bei dieser Hitze.

Dies wird durch die Intonation, die Partikeln <u>denn schon</u> und den Kontext offenbar.

2.1. Typische Basisindikatoren sind die explizit performative Formel und die Satzmuster, d.h. Grundstrukturen von Sätzen, in die noch keine Lexeme aus offenen Paradigmen eingesetzt sind. Typische Sekundärindikatoren sind Modalpartikeln, Intonation und Akzent sowie der Kontext. Sie arbeiten auf der Grundlage der Basisindikatoren und können zu ihnen in Gegensatz treten. Häufig harmonieren zwar einige der Sekundärindikatoren (z.B. die Intonation) mit den Basisindikatoren, andere aber (etwa die Partikeln) widersprechen den Basisindikatoren u n d den mit ihnen verträglichen Sekundärindikatoren.

2.2. Sekundärindikatoren sind, wie dargestellt, sprachlicher und kontextueller Natur. Somit lassen sich die indirekten Sprechakte in zwei Gruppen unterteilen.

2.2.1. Äußerungen, bei denen sprachliche Sekundärindikatoren zu den Basisindikatoren (und eventuell mit diesen harmonierenden Sekundärindikatoren) in Gegensatz treten. In

(5) Du wirst deinen Führerschein machen

beispielsweise deutet das Satzmuster (Nomen + futurisches Prädikat) als Basisindikator auf einen Sprechakt des Vorhersagens hin. Intonation und Akzent können als Sekundärindikatoren jedoch anzeigen, daß tatsächlich ein Befehl

(5a) Du wírst deinen Führerschein machen!

oder eine Frage vollzogen wird:

(5b) Du wirst deinen Führerschein machen?

2.2.2. Bei indirekten Sprechakten der zweiten Gruppe harmonieren alle sprachlichen Sekundärindikatoren mit den Basisindikatoren; aus dem Kontext ergibt sich jedoch, daß eine wörtliche Interpretation der sprachlichen Illokutionsindikatoren nicht angezeigt ist.[2] So mag im Kontext einer Familie die Äußerung

(6) Es ist gleich acht

den Angesprochenen dazu auffordern, die Nachrichten im Fernsehen anzuschalten, ohne daß irgendwelche sprachlichen Indikatoren dafür vorlägen.

2.3. In der Pragmatik wird gemeinhin dem Situationskontext großes Gewicht beigemessen. Die Untersuchung indirekter Sprechakte bringt hier eine gewisse Korrektur. Denn zur eindeutigen Interpretation eines indirekten Sprechakts der ersten Gruppe, also einer Äußerung mit widersprüchlichen sprachlichen Illokutionsindikatoren, ist der Kontext wenigstens tendenziell nicht erforderlich. Der Grund: Zwischen sprachlichen Illokutionsindikatoren bestehen stabile Dominanzverhältnisse. Die Indikatoren, die oben scheinbar willkürlich als sekundäre bezeichnet wurden, sind über die sogenannten Basisindikatoren dominant. Das heißt: Intonation, Akzent und Modalpartikeln können in Gegensatz zu einer performativen Formel oder einem Satzmuster treten und dadurch deren wörtliche Interpretation unterdrücken. Daher ist bei widersprüchlichen sprachlichen Indikatoren aus der Äußerung kontextfrei ersichtlich, welche Illokution intendiert ist.[3]

Zur Disambiguierung unentbehrlich ist der Kontext nur, wenn die sprachlichen Illokutionsindikatoren einer Äußerung homogen sind, d.h., wenn alle sprachlichen Sekundärindikatoren die illokutionäre Interpretation der Basisindikatoren bestätigen. Die Äußerung

(7) Wie spät ist es?

kann - auch ohne Veränderung der Intonation - mal als direkte Frage nach der Uhrzeit, mal als indirekte Aufforderung

(7a) Geh' jetzt ins Bett!

o.ä. gemeint sein. Welche Deutung zutrifft, hängt vom Kontext ab. Er ist hier der dominante Sekundärindikator.

3. Wenn unter Berücksichtigung aller Indikatoren, der sprachlichen wie der kontextuellen, zusammen die Illokution einer Äußerung stets hinreichend genau und eindeutig bestimmt werden kann - dann mag man mit MEYER-HERMANN (1976) fragen, ob das Konzept der Indirektheit überhaupt sinnvoll und notwendig ist.

M.E. ist es zumindest sehr nützlich. Und das aus einer Reihe von Gründen.

3.1. Ohne dies Konzept wären performative Formeln und Satzmuster als z.T. hochgradig polysem anzusehen. Der Fragesatz etwa hätte nicht nur die kategoriale Bedeutung des Fragens, sondern auch die des Drohens, Behauptens, Warnens, Anbietens usw. - da man mit Fragesätzen all diese Sprechakte vollziehen kann. Damit aber würde der Bedeutungsbegriff überstrapaziert.

3.2. Bestimmte Lexeme und Syntagmen könnten semantisch nicht analysiert werden, weil sie das Gegenteil ihrer selbst bedeuten würden. Das zeigen vielzitierte Beispiele wie

(8) Ich verspreche dir, daß ich dir das heimzahlen werde.

Das Lexem versprechen müßte im Lexikon u.a. den Bedeutungseintrag 'androhen' erhalten, weil es in Äußerungen wie (8) in diesem Sinne eingesetzt wird. Versprechen und androhen aber besitzen polar entgegengesetzte Bedeutungskomponenten.

3.3. Der Verzicht auf den Begriff der Indirektheit vermindert auch die Leistungsfähigkeit der linguistischen Theorie. Dieser Begriff ist ja keineswegs Selbstzweck, sondern hilft, erklärungskräftige allgemeine Aussagen zu formulieren. Daraus bezieht ein solches theoretisches Konzept allererst seine Daseinsberechtigung. Der Begriff der Indirektheit nun hat sich bereits insofern bewährt, als festgestellt wurde: Mit indirekten Sprechakten sind bestimmte kommunikative Ziele verbunden, die mit ihren direkten Äquivalenten nicht gleichermaßen erreicht werden können. FRANCK (1975) nennt die Aufgabe, kommunikativen Kontakt sozial akzeptabel, unverbindlicher, unverfänglicher und höflicher zu gestalten. Indirekte Sprechakte sind also oft tentativer als direkte. Auch wenn damit nicht alle Aufgaben indirekter Sprechakte genannt sind, zeigt sich doch schon der Nutzen des Konzepts Indirektheit: Ein theoretischer Begriff, der über eine bestimmte Klasse von Merkmalen (nämlich über strukturelle) definiert wurde, läßt sich verwenden, um allgemeine Aussagen über eine völlig andere Klasse von Phänomenen (nämlich über kommunikative Strategien) zu machen.

Die genannten kommunikativen Aufgaben werden sowohl von indirekten Sprechakten mit widersprüchlichen sprachlichen Illokutionsindikatoren erfüllt als auch von jenen, bei denen Kontextindikatoren zu den homogenen sprachlichen Indikatoren in Gegensatz treten.[4] Daher empfiehlt es sich, beide unter dem Oberbegriff "indirekte Sprechakte" zu belassen. Selbstverständlich ist das nicht, da - wie gezeigt - bei den indirekten Sprechakten des ersten Typs die sprachliche Form als Ganzes auf Grund der Dominanzverhältnisse unzweideutig die tatsächliche illokutionäre Rolle signalisiert. Diese Äußerungen deshalb als direkte Sprechakte einzustufen, würde die Theorie jedoch verkomplizieren, ohne daß damit ein Erkenntnisgewinn verbunden wäre.

4. Nach dieser Grundsatzentscheidung drängt sich um so stärker die Frage auf, was die "wörtliche" Interpretation von Basisindikatoren ist und wie diese abgesichert werden kann. Warum ordnet man der Struktur des Fragesatzes die Frageillokution zu? Die Rechtfertigung liegt in der naiven Intuition des native speaker. Der native speaker schreibt der Fragesatz-Struktur als solcher die Fragefunktion als Normalfunktion zu. Am einfachsten ermittelt man die wörtliche Interpretation von Satzmustern anhand von Nonsense-Sätzen. Zweifellos signalisiert

(9) Pratschen die Lobos den Glumpf ins Salf?

für sich genommen eine Fragefunktion. Basisindikatoren wie die Fragesatz-Struktur sind also Grundformen von Äußerungen mit einer Standard-Illokution. Man kann ihnen zwar sprachliche oder kontextuelle Indikatoren hinzufügen und dadurch ihre illokutionäre Interpretation ändern. Man kann sie aber nicht weiter reduzieren und so eine von der Basis-Illokution abweichende Interpretation herbeiführen.

4.1. In diesem Zusammenhang ist der in der strukturellen Linguistik geprägte Begriff der Markierung hilfreich. Die Fragefunktion stellt quasi die unmarkierte Interpretation eines Fragesatzes dar. Erhält der Fragesatz dadurch, daß ihm ihm widersprechende Sekundärindikatoren hinzugefügt werden (z.B. eine für Fragen atypische Intonation, bestimmte Partikeln oder besondere Kontextinformationen), eine andere Interpretation, so ist diese markiert.

4.2. Ein bemerkenswertes sprachliches Indiz stützt die hier bezogene theoretische Position. In der nicht-wörtlichen Redewiedergabe können, wie schon MAAS/WUNDERLICH (1972) bemerken, indirekte Sprechakte auf zwei Weisen referiert werden: mit dem performativen Verb, das für die intendierte Illokution steht, oder mit dem Verb, das die nicht realisierte Basisillokution bezeichnet.
Ein Beispiel:

(10) Hast du nicht mit dem Streit angefangen?

(10a) Er fragte sie, ob sie nicht mit dem Streit angefangen habe.

(10b) Er warf ihr vor, mit dem Streit angefangen zu haben.

Diese doppelte Redewiedergabe ist bei direkten Sprechakten nicht möglich.

Die Tatsache, daß manche Sprechakte mit zwei performativen Verben referiert werden können, andere nur mit einem, läßt sich ohne das Konzept der Indirektheit nicht adäquat erklären.

5. Wenn indirekte Sprechakte Äußerungen sind, die durch zwei verschiedene performative Verben referiert werden können, wäre folgende Äußerung allerdings ebenfalls ein indirekter Sprechakt:

(11) Geh' bitte mal aus dem Weg.

Denn (11) kann sowohl mit auffordern als auch mit bitten wiedergegeben werden:

(11a) Er bat ihn, aus dem Weg zu gehen.

(11b) Er forderte ihn auf, aus dem Weg zu gehen.

Anders als bei den Wiedergabe-Alternativen von (10) aber widersprechen sich hier die beiden performativen Verben bzw. die durch sie ausgedrückten Illokutionen nicht. Was kann also - und damit komme ich zur dritten Frage - als Kriterium dafür dienen, daß zwei Illokutionen in einem Gegensatz zueinander stehen?

5.1. M.E. wäre hierfür eine ausgearbeitete Klassifikaton illokutionärer Typen hilfreich.[5] In ihr würden zwar z.B. Auffordern und Mitteilen (Äußerung 3) zu unterschiedlichen Illokutionsklassen gehören, nicht aber Auffordern und Bitten (Beispiel 11). Auffordern steht vielmehr hierarchisch über Bitten - jede Bitte ist eine Aufforderung, aber nicht jede Aufforderung ist eine Bitte.

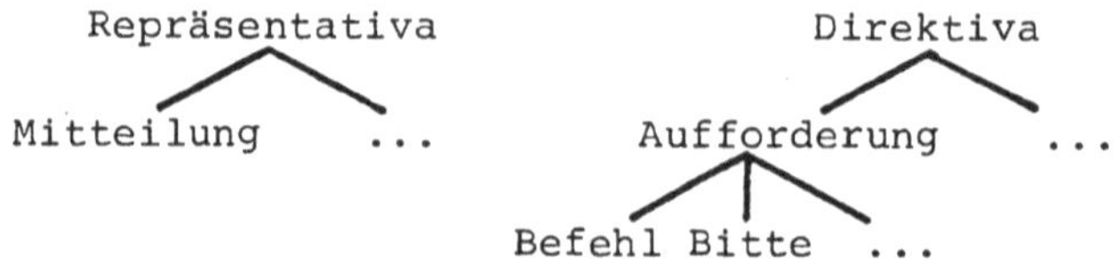

5.2. Eine solche Klassifikation, in der auch die jeweiligen Indikatoren einer Klasse aufgeführt sind, würde eine operationalisierbare und theoretisch einwandfreie Definition der indirekten Sprechakte ermöglichen: Indirekte Sprechakte wären Äußerungen, die Basisindikatoren für eine Illokutionsklasse besitzen, zugleich jedoch für eine andere Klasse sekundäre sprachliche und/oder kontextuelle Indikatoren, die über die Basisindikatoren dominant sind.

Eine rhetorische Frage wie (1) z.B. ist ein indirekter Sprechakt, weil sie einen Basisindikator aus der Klasse der Interrogativa oder, wie WUNDERLICH (1976) sagt, der erotetischen Sprechakte enthält, aber zugleich Sekundärindikatoren aus der Klasse der Repräsentativa aufweist.

5.3. Bei einer ausgefeilten Klassifikation sind auch Indirektheitsbeziehungen zwischen Illokutionstypen innerhalb einer Klasse denkbar. Bedingung ist, daß die beiden Typen in keiner hierarchischen Beziehung zueinander stehen. Ein als Bitte ausgesprochener Befehl wäre ein solcher indirekter Sprechakt.

6. Wenn indirekte Sprechakte Äußerungen mit Indikatoren zweier verschiedener Illokutionsklassen oder -typen sind, liegt der Schluß nahe, daß mit ihnen zwei illokutionäre Akte zugleich ausgeführt werden. SEARLE (1975) ist ein Hauptvertreter dieser These. Sie ist jedoch als globale Aussage nicht haltbar, da sie eine inakzeptable Folgerung nach sich zöge.

Würde mit der Äußerung von (1) nicht nur ein repräsentativer Sprechakt mit der Paraphrase (1a) vollzogen, sondern auch ein Frageakt, dann müßte der Sprecher einander ausschließenden Gelingensbedingungen genügen. Denn eine konstitutive Bedingung der Fragehandlung lautet, daß der Sprecher die erfragte Information nicht besitzen darf, während jemand, der eine Behauptung aufstellt, gerade über eine Kenntnis sicher verfügen muß. Beides kann nicht zugleich der Fall sein. Ähnliches gilt für in Frageform vollzogene Aufforderungen:

(12) Schaltest du mal das Radio an?

(12a) Schalte das Radio an!

oder für Drohungen in Form von Versprechen (Beispiel 8).

Deshalb sollte man daran festhalten, daß wenigstens bei einem Teil der indirekten Sprechakte nur die durch die Sekundärindikatoren angezeigte Illokution realisiert ist, während die in den dominierten Basisindikatoren angelegte Rolle nicht zum Tragen kommt.

Anmerkungen

1 Die wichtigsten Beiträge sind: EHRICH/SAILE (1972), FRANCK (1975), MAAS/WUNDERLICH (1972), MEYER-HERMANN (1976), SEARLE (1975).

2 Ein direkter Sprechakt liegt folgerichtig dann vor, wenn weder unter den verschiedenen sprachlichen Indikatoren noch zwischen sprachlichen und kontextuellen Indikatoren ein Widerspruch besteht.

3 Das wurde am Rande schon von MAAS/WUNDERLICH (1972: 126ff.) und WUNDERLICH (1972: 18f.) vermerkt, ohne daß diese Spur systematisch weiterverfolgt worden wäre.

4 Das gilt unabhängig davon, ob die Indirektheit einer Äußerung konventionell ist oder nicht. Deshalb wäre es nicht sinnvoll, mit HENNIG/HUTH (1975) Äußerungen wie Hast du mal Feuer? nicht zu den indirekten Sprechakten zu zählen.

5 Einen ersten Ansatz bietet SEARLE (1973). Inzwischen hat Searle weitere Klassifikationsvorschläge ausgearbeitet.

Literatur

AUSTIN, John L. (1962): How to do things with words. Oxford: University Press. - Übers.: Zur Theorie der Sprechakte. Stuttgart: Reclam, 1972.

EHRICH, Veronika / SAILE, Günter (1972): "Über nicht-direkte Sprechakte". WUNDERLICH (ed.) (1972b): 255-287.

FRANCK, Dorothea (1975): "Zur Analyse indirekter Sprechakte". EHRICH, Veronika / FINKE, Peter (eds.) (1975): Beiträge zur Grammatik und Pragmatik. Kronberg: Scriptor: 219-231.

HENNIG, Jörg / HUTH, Lutz (1975): Kommunikation als Problem der Linguistik. Eine Einführung. Göttingen: Vandenhoeck.

MAAS, Utz / WUNDERLICH, Dieter (1972): Pragmatik und sprachliches Handeln. Mit einer Kritik am Funkkolleg "Sprache". Frankfurt/Main: Athenäum.

MEYER-HERMANN, Reinhard (1976): "Direkter und indirekter Sprechakt". Deutsche Sprache 4: 1-19.

SEARLE, John R. (1975): "Indirect Speech Acts". COLE, Peter / MORGAN, Jerry L. (eds.) (1975): Syntax and Semantics, Bd. 3: Speech Acts. New York etc.: Academic Press, 1975: 59-82.

--- (1973): "Linguistik und Sprachphilosophie". BARTSCH, Renate / VENNEMANN, Theo (eds.) (1973): Linguistik und Nachbarwissenschaften. Kronberg: Scriptor: 113-125.

--- (1969): Speech Acts. Cambridge: University Press. - Übers.: Sprechakte. Ein sprachphilosophischer Essay. Frankfurt/Main: Suhrkamp, 1971.

WUNDERLICH, Dieter (1976): Studien zur Sprechakttheorie. Frankfurt/Main: Suhrkamp.

--- (1972a): "Zur Konventionalität von Sprechakten". WUNDERLICH (ed.) (1972b): 11-58.

--- (ed.) (1972b): Linguistische Pragmatik. Frankfurt/Main: Athenäum.

THE ARGUMENTATIVE STATUS OF MODALITY

Viggo Sørensen

Anybody familiar with HALLIDAY (1970) or BRINKMANN (1971) will recall the idea of two modal systems, parallel and somehow related. First, epistemic modality or Informationsmodus, see (1) and (2). Next, modulation or Realisierungsmodus, see (3) and (4).

(1) You must be very careless.

(2) Du kannst Migräne haben.

(3) You must be very careful.

(4) Du kannst ruhig weiterschlafen.

With this paper I want to characterize in pragmatic terms the modal types mentioned, and to venture in the end a pragmatic definition of such 'classical' modality as a whole.[1]

Starting on safe ground, we know epistemic modality to signal in broad outline source and strength of evidence in case a statement does not rely (entirely) on speaker's own observation or knowledge. For instance, the modals of (1) and (2) reflect traits of conclusion and mere guess, resp, on part of the speaker. His evidence for concluding or guessing has been left unsaid or presupposed here, to be made explicit just on demand, say this way:[2]

(5) (You must be very careless.) We seem to find your books left all over the school.

(6) (Du kannst Migräne haben.) Die Lisa hatte das, und das gab genau so ein Flimmern vor den Augen.

The argumentative situation is that of figure 1, where I distinguish between communication level and subject level, following HABERMAS (1971), and between illocutive, perlocutive, and propositional aspects of speech action, following SEARLE (1969). In stating V the speaker hopes to bring about the hearer's conviction (or just consideration) Z by supporting his claim (or just fancy) X with a hint to some presupposed evidence Y of source or strength M.

In the model of pragmatic hierarchy by SØRENSEN (1976 / 1978) Y and X show up as contents of an auxiliary act and its head, resp, see figure 2. M can be represented as a specification of

Figure 1

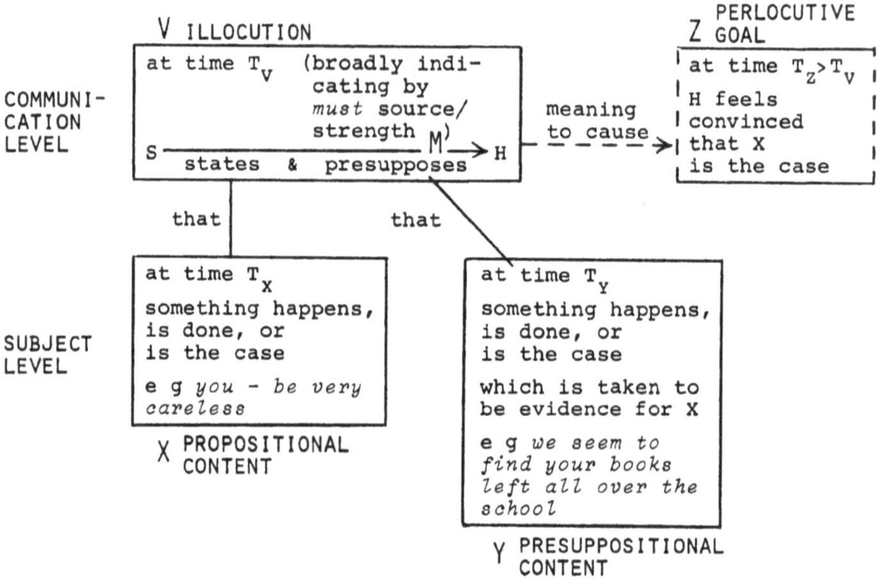

Figure 2

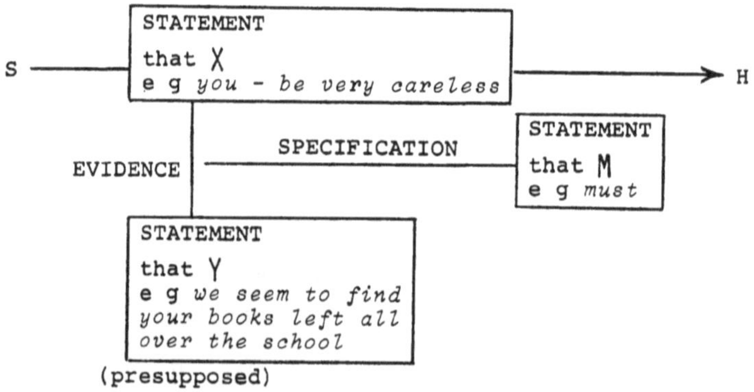

the very relation between X and Y. By further translation into the argument model of TOULMIN (1958), see figure 3, the tacit warrant W underlying Y's supporting X proves to be some cognitive 'law' (whether a stereotype or based upon logic, experience and the like).[3]

Figure 3 [4]

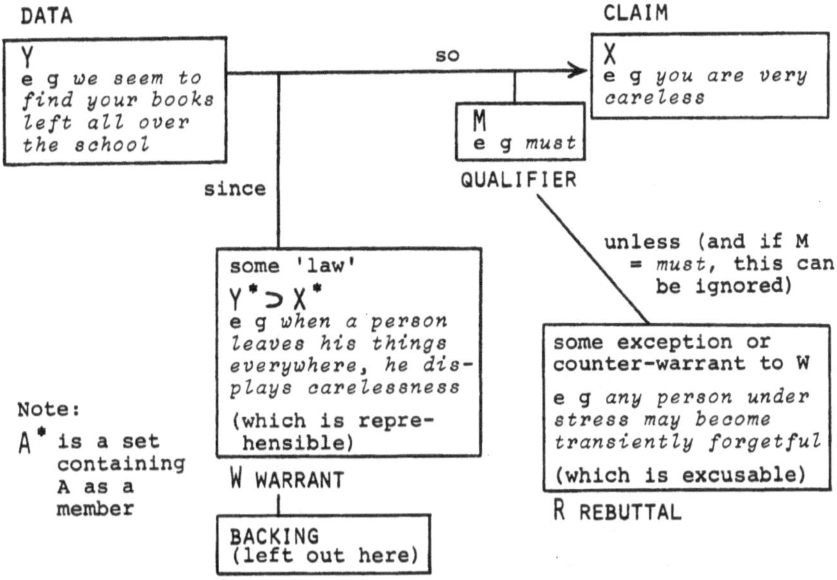

Instead of attacking directly the modality of (3) and (4), I shall make a strategic detour. Let me start by considering (7) and (8), near equivalents of (3) and (4), resp, in a way to be specified later on. As opposed to constative utterances like (1) and (2) we are here confronted with performative ones, in the sense of AUSTIN (1962).

(7) Be very careful !

(8) Schlaf ruhig weiter !

As for the action theoretical structure of such performative utterances, see figure 4 (and compare with figure 1). The perlocutive goal Z is now some action on part of the hearer or the speaker (or both).[5] And, owing to the anticipatory use, the propositional content X allows no support of evidence, but at most some justification Y'.

The over-all picture is, however, very much the same in both figures, and from a text hierarchal angle justification simply parallels evidence as a main sort of auxiliary action, see figure 5 (and compare with figure 2). In accordance with this WUNDERLICH (1974) faces the possibility that we extend the scope of Toulmin's

Figure 4

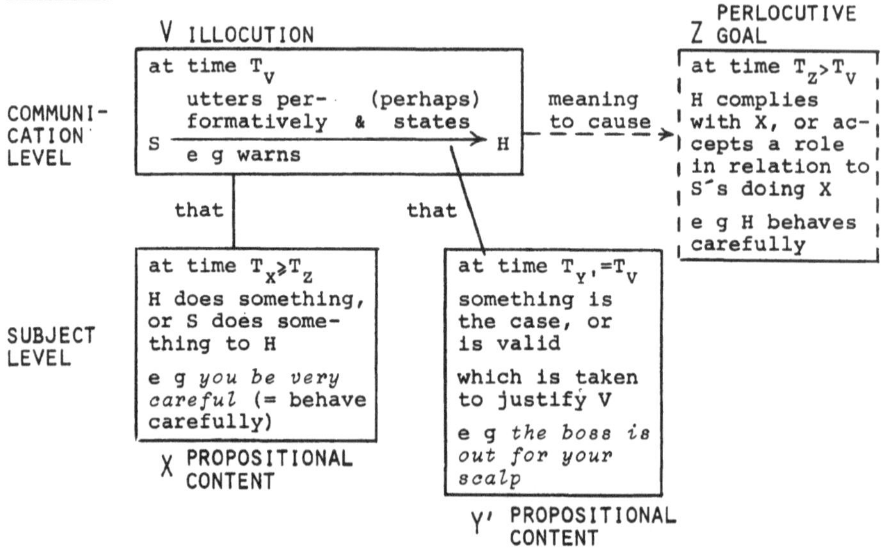

Figure 5

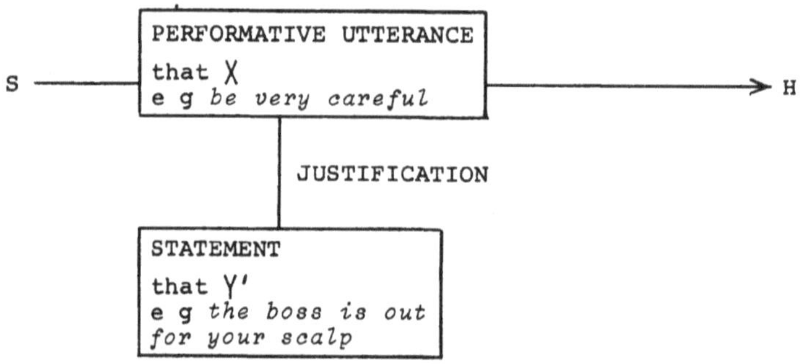

argument model so as to cover justification, too. I have done so by figure 6.[6] You will notice that the warrant here is some 'law' of proper and expedient behaviour, the communicative aspects of which are captured in the preparatory speech act rules of SEARLE (1969).[7]

Strangely enough, justification cannot be signalled (as presupposed) in direct performative utterances the same way as can evidence in a constative frame. We have no simple counterpart of epistemic modality at our disposal, and must if necessary make our justification explicit, as shown in (9) and (10).

Figure 6

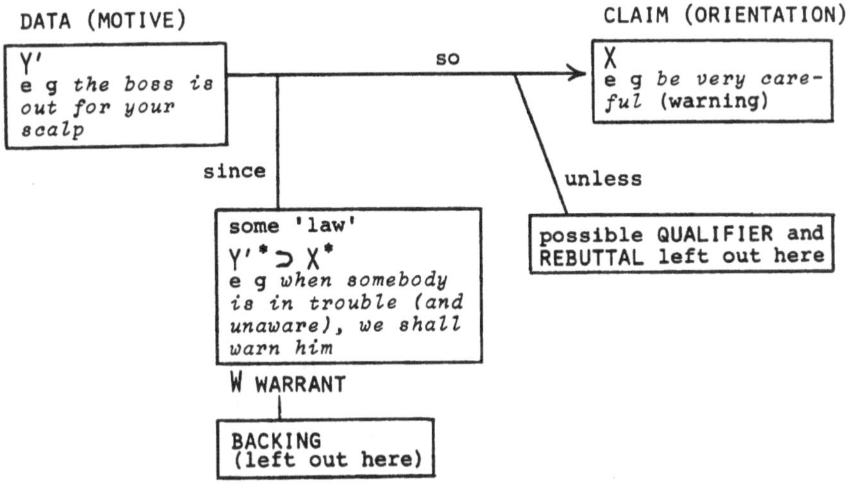

(9) Be very careful ! The boss is out for your scalp.

(10) Schlaf ruhig weiter ! Du hast wohl kein Interesse daran.

But then, what about the modals of (3) and (4) ?

Most likely these utterances will be used performatively, as warning and reassurance, resp, and no doubt their modals reflect the difference of illocutive force. Nevertheless, such illocutive function does not seem the essence of the modality in question. Just introduce a third person subject and a verb in preterite tense, as shown by (11) and (12), and back we are in the constative.[8]

(11) He must (or: had to) be very careful.

(12) Sie könnte ruhig weiterschlafen.

The second part of my detour shall lead through this subspecies of constative utterances.

What is achieved by the modal in (11) and (12) can be seen from figure 7. The speaker signals in broad outline something or somebody A motivating (or enforcing) on subject level the activity B explicitly referred to. Thus, by (11) he is understood to be under pressure from some social norm or from his own conscience. Whereas in (12) she has got some corresponding license.

Events, states, and anything in the past fall completely out-

Figure 7

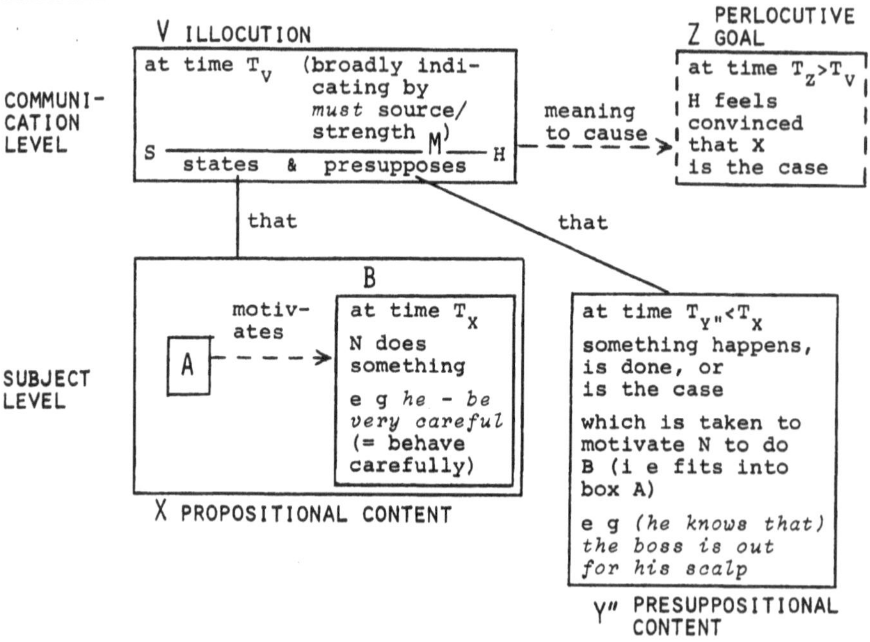

side the scope of this, I should say: motivational modality, so that utterances like the following have to be interpreted along other lines:[9]

(13) He had to be a fanatic.

(14) Jeden Augenblick könnte das Regen anfangen.

On the other hand, any propositional content even if including a motivational modality may become modalized epistemically, this giving birth to bimodal constructions:

(15) He must have had to be very careful.

(16) Sie muss ruhig haben weiterschlafen können.

The textual structure produced by motivational modality can be seen from figure 8. You will observe that a signal of motive (or pressure) on subject level falls under the heading of specification, a third main type of auxiliary action. Nothing prevents a further translation into Toulmin's argument model, see figure 9. The warrant here proves to be some ethic, moral, or other social 'law'.

And now, at length time has come to consider such utterances as (3) and (4), or for that sake: (17) and (18).

Figure 8

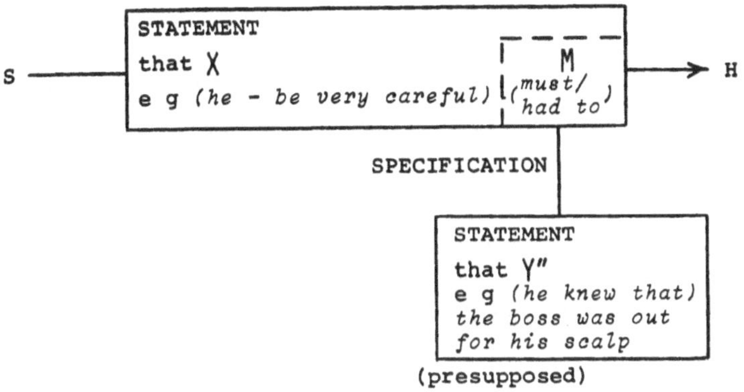

Figure 9

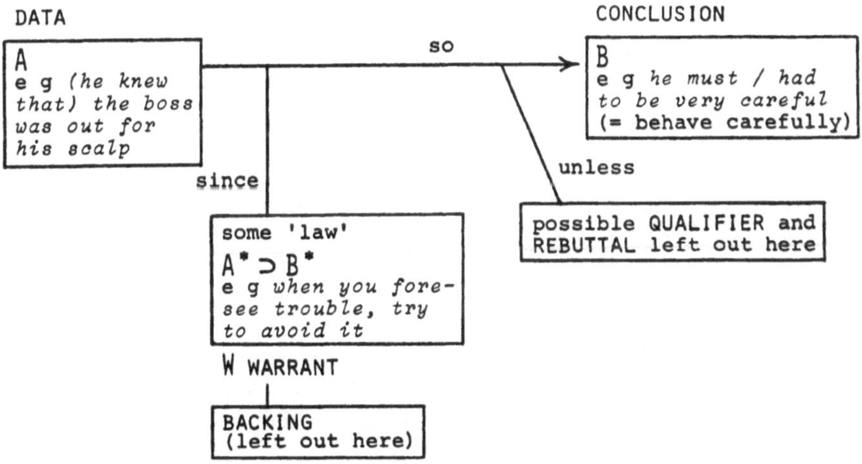

(17) I must leave now.

(18) Ich soll dir einen neuen Ball kaufen.

Taken in a literal sense they form a subclass of the motivationally modalized constatives just discussed. The speaker refers to some future action on part of himself or the hearer, and hints at a motive or a pressure operating at utterance time. But as you know, any statement can be sensibly questioned as to its 'true' function, and this question will become the more urgent the more odd appears a 'pure' informative interpretation. With the utterances at issue it seems obvious we should look for some such non-informative point.

The matter proves to be one of indirection, along the lines of GRICE (1975) and SEARLE (1975). Most probably the speaker, by referring to an obligation or a license of the hearer's or his own, wants the hearer to carry out such steps of inference which I have spelled out in some detail by figure 10. For the lack of space I shall just summarize here. The hearer is to identify the activity on communication level with that one hinted at on subject level, and so to copy the 'true' perlocutive goal Z' out from the propositional part B.[10]

Interpreted by way of such level transfer (3) will normally be understood as a warning, (4) as a piece of reassurance, (17) as an excuse, and (18) as a promise.[11]

In fact it is no easy matter to draw a line between direct and indirect speech action. However derived appear from a formal point of view the utterances under discussion, in practice they represent the ordinary way of warning and reassuring, excusing and promising. When combined with proper subject and tense the modals seem perfectly conventional illocutive devices, far more precise than the mood in (7) and (8), and colloquial as opposed to possible explicit performative formulae:[12]

(19) Excuse me for leaving now.

(20) Ich verspreche, dir einen neuen Ball zu kaufen.

Whether a question of indirection or not, the operation of modality on communication level implies such new aspects that we shall anyway be justified in recognizing here a third type of modality. Just look at figure 10 once more.

What is modally signalled by way of level transfer proves to be interest, intention, obligation, and the like on part of the interlocutors. But that is exactly the sort of things referred to in speech act conditions, as constitutive features of illocution. So that, in the light of figure 4-6, justification c a n be modally signalled after all, not by epistemic, but by this specific modal type, for which I suggest the name of illocutive modality.[13]

And now let me conclude. The unity of 'classical' modality, re-

Figure 10

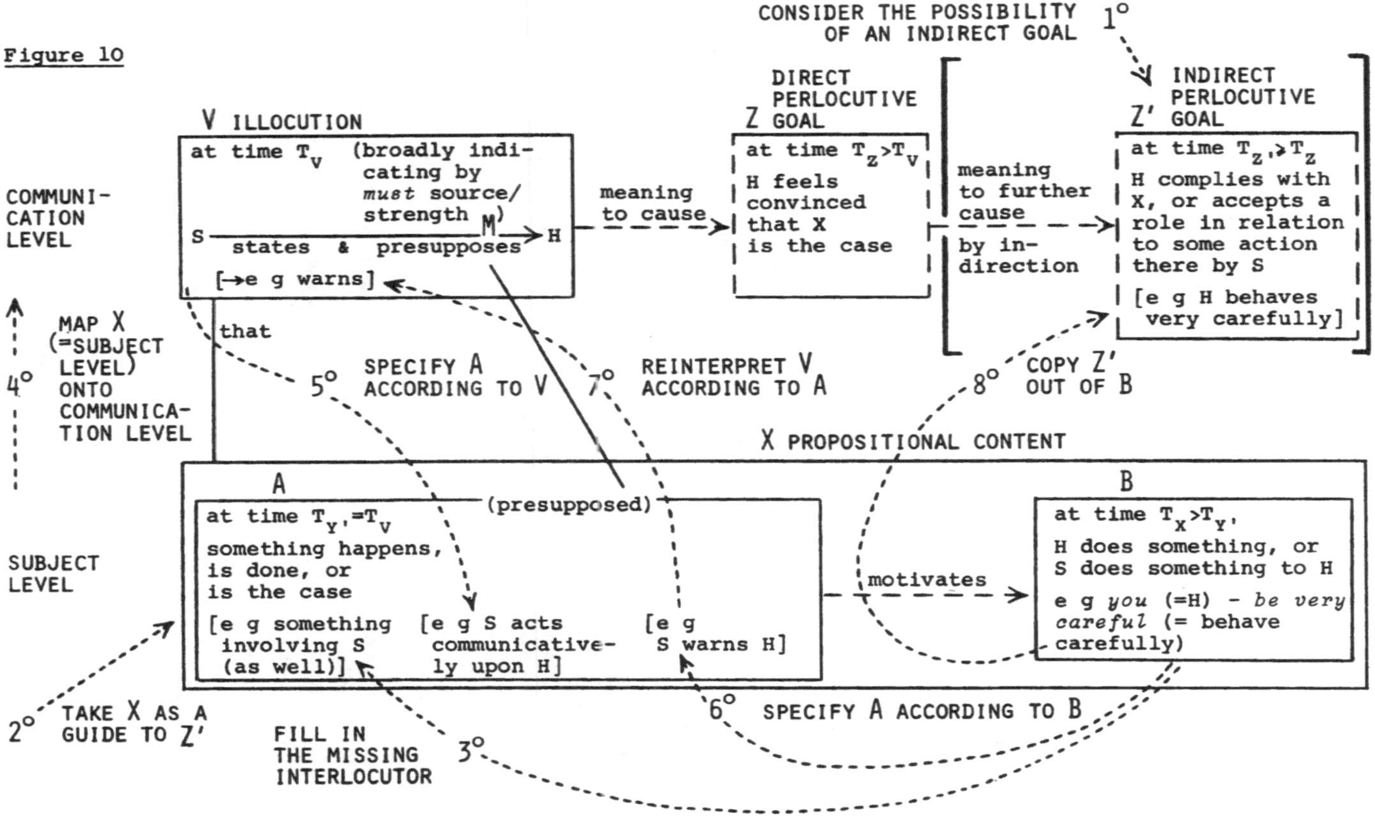

Notes: [ ] mark features created in the process of indirect interpretation.

Y" (cf figure 7) has been left out for the lack of place.

flected lexically, semantically, and syntactically in HALLIDAY (1970) and BRINKMANN (1971), should ultimately be localized on a pragmatic level.[14] The common feature of epistemic, motivational, and illocutive modality is the argumentative status of dummy explanation. That is to say, each type broadly signals (as presupposed) one main sort of explanation, if we take this term in the wide sense of LANG (1976) and SØRENSEN (1976).

With a definition of modality along these lines I dare hope that the modal field so richly expressed[15] shall turn into a more clear-cut and coherent one, easier for us to handle.

Notes

1 The term modality is often used in a liberal sense so as to include e g the entire field of signalled speaker's attitude. Cf BUBLITZ (1978). I shall ignore this terminological expansion.

2 The parts of (5) and (6) placed between brackets may constitute prior moves, now being 'further explained' by the rest because of an intervening display of hearer's perplexity or scepticism. Alternatively (5) and (6) may be uttered as they stand if the speaker anticipates and tries to prevent such hearer's reaction. Cf SØRENSEN (1976).

3 A stereotype will often include some evaluative feature crucial for its efficiency, such as those at the bottom of box W and R of fig 3. Cf HUTH (1975). If so the evaluative feature is liable to affect the illocutive function of the claim X, making (5) a reproach, etc. Cf the notion of 'true' or indirect function below.

4 The warrant W of an argument is normally kept tacit (and has to be reconstructed by the text analyst), but in rare cases it may replace the data D as explicit part of the argumentation. Cf TOULMIN (1958).
The figure implies the 'topical interpretation' of a rebuttal R as counter-warrant. Cf HUTH (1975).

5 In (7) and (8) the verbs <u>be</u> and <u>weiterschlafen</u> do refer to actions (meaning <u>behave</u> and <u>sich weiterschlafen legen</u>, resp).

6 Out from the complicated nature of the relation DATA : CLAIM I have focussed the motivational aspect of the former, whereas SCHWITALLA (1976) has focussed the goal aspect (positively or negatively) implied, cf Z of fig 4. For this reason the orientation of time in fig 6 (reflected by the arrow) is opposite to that of Schwitalla's corresponding figures.

7 So, in the argument of fig 6 the justification works (mainly) by satisfying the rule of hearer's interest. You might justify via speaker's interest or relevance as well, or even combine

such ways, e g "Be very careful! I love you. And it is such a tricky job."

8 The double function as performative and non-performative device could be ascribed to any verb (or predicate). Just that the two usages are never felt semantically so wide apart as with modals, maybe except for verbs of judging, cf FILLMORE (1969).

9 Most likely both (13) and (14) are taken to be epistemically modalized. The preterite tense will then be explained as an instance of consecutio temporum trickered by some omitted inquit. As for (13) a motivational interpretation might be admitted provided we understand be in the sense of note 5. The utterance will then be about a player's obligation or the like.

10 Of course the single steps (1°-8°) of the said figure are pure logical constructs without any ontological implications in themselves. What cannot be denied is the operation of such all-over search strategies in interpretation.

11 In a proper setting (3) and (4) might be otherwise meant and understood. E g consider (3) as an order, to be justified this way: "(Be very careful!) I don't want any complaints from our costumers." And if we allow for ironic readings, (4) might actually be a rebuff meaning: "Don't meddle in my business".

12 The difficulties in making a clear-cut distinction between direct and indirect usage seem to reflect the role of indirection in the history of language. According to SCHLIEBEN-LANGE (1976) the body of speech act types shows considerable diachronical change (extensionally as well as intentionally). And according to GRAYSHON (1977) the strategy of indirection is a main device for expressing new illocutional forces.

13 Cf the volitive modality of BUBLITZ (1978).

14 EHLICH/REHBEIN (1972) can be taken to agree. Their subdivision of Realisierungsmodus according to its operation on Ebene der Sachverhalte or Ebene der Verbalisierungen matches exactly my distinction of motivational and illocutive modality.

15 My selection of modal verbs and no other modal devices throughout the examples of this paper has the one and only purpose to facilitate the exposition.

Literature

AUSTIN, John L. (1962): How to do things with words. London: Oxford University Press.

BRINKMANN, Hennig (1971): Die Deutsche Sprache. Düsseldorf, Schwann.

BUBLITZ, Wolfram (1978): Ausdrucksweisen der Sprechereinstellung im Deutschen und Englischen. Tübingen, Niemeyer.

COLE, Peter / MORGAN, Jerry L. (eds) (1975): Syntax and semantics. III: Speech acts. New York, Academic Press.

EHLICH, Konrad / REHBEIN, Jochen (1972): "Einige Interrelationen von Modalverben". WUNDERLICH, Dieter (ed) (1972): Linguistische Pragmatik. Frankfurt a.M.: Athenäum.

FILLMORE, Charles J. (1969): "Verbs of judging". Papers in Linguistics 1.

GRAYSHON, Matthew C. (1977): Towards a social grammar of language. The Hague etc: Mouton.

GRICE, H.P. (1975): "Logic and conversation". COLE / MORGAN (eds).

HABERMAS, Jürgen (1971): "Vorbereitende Bemerkungen zu einer Theorie der kommunikativen Kompetenz". HABERMAS, Jürgen / LUHMANN, Niklas (1971): Theorie der Gesellschaft oder Sozialtechnologie. Frankfurt a.M.: Suhrkamp.

HALLIDAY, M.A.K. (1970): "Functional diversity in language as seen from a consideration of modality and mood in English". Foundations of Language 6.

HUTH, Lutz (1975): "Argumentationstheorie und Textanalyse". Der Deutschunterricht 27.6.

LANG, Ewald (1976): "Erklärungstexte". DANEŠ, František / VIEHWEGER, Dieter (eds) (1976): Probleme der Textgrammatik. Berlin: Akademie-Verlag.

SCHLIEBEN-LANGE, Brigitte (1976): "Für eine historische Analyse von Sprechakten. WEBER, Heinrich / WEYDT, Harald (eds) (1976): Sprachtheorie und Pragmatik. Tübingen, Niemeyer.

SCHWITALLA, Johannes (1976): "Zur Einführung in die Argumentationstheorie: Begründungen durch Daten und Begründungen durch Handlungsziele in der Alltagsargumentation". Der Deutschunterricht 28.4.

SEARLE, John R. (1969): Speech acts. London: Cambridge University Press.

--- (1975): "Indirect speech acts". COLE / MORGAN (eds).

SØRENSEN, Viggo (1976): "Textuality - a pragmatic approach". KARLSSON, Fred (ed) (1976): Papers from the 3rd Scandinavian Conference of Linguistics. Turku: Academy of Finland.

--- (1978): "Dialogue: negotiation of texts". GREGERSEN, Kirsten (ed) (1978): Papers from the 4th Scandinavian Conference of Linguistics. Odense: Odense University Press.

TOULMIN, Stephen (1958): The uses of argument. Cambridge: Cambridge University Press.

WUNDERLICH, Dieter (1974): Grundlagen der Linguistik. Reinbek: Rowohlt.

WITTGENSTEIN AND LINGUISTIC PRAGMATICS,

Some remarks on Wittgenstein's "meaning-and-use" dictum and its influence.

Guido Thys

0. Introduction.

Unlike all other contributions in the section pragmatics, this paper does not deal with an actual topic in the field of pragmatical research, but rather with one of its (claimed) foundations.
I will investigate to what extent it is justifiable to call Wittgenstein a predecessor and his "Philosophical Investigations" (henceforth: PI), and more specifically his "meaning-and-use" dictum, an early form of linguistic pragmatics.
Much depends of course on one's interpretation of Wittgenstein's writings on the meaning of linguistic entities and his conception of philosophy in general. In the first chapter I will therefore briefly discuss two quotations which are considered to be representable for what is called his "use-theory". My interpretation will substantially weaken Wittgenstein's predecessorship, as will be shown in chapter two.
The present paper is an adaptation of the last chapter of a larger work in which I thoroughly analyzed the value of the dictum, as well as Miss Anscombe's leading translation of the "Philosophische Untersuchungen". (THYS 1978).

1. On the interpretation of Wittgenstein's "meaning-and-use" dictum.

The claim that Wittgenstein was the first philosopher to come up with a semantic theory which attributed the meaning of a linguistic entity to the way it is actually used, are mostly based upon (one of) the following quotations:

(1) Don't look for the meaning, look for the use.

(2a) For a <u>large</u> class of cases -though not for all- in which we employ the word "meaning", it can be defined thus: the meaning of a word is its use in the language.
And the meaning of a name is sometimes explained by pointing to its bearer. (ANSCOMBE 1976: §43)

(2b) Man kann für eine <u>grosse</u> Klasse van Fällen der Benützung des Wortes "Bedeutung" -wenn auch nicht für <u>alle</u> Fälle seiner Benützung- dieses Wort so erklären: Die Bedeutung eines Wortes

ist sein Gebrauch in der Sprache.
Und die Bedeutung eines Wortes erklärt man manchmal dadurch, dass man auf seinen Träger zeigt.

(1) is quoted without reference because it is not to be found in any of the publications of Wittgenstein's work; he is only reported to have said this on a few occasions. It has moreover rather the form of a slogan and it is mostly quoted as such. We can therefore not use it to explain what the relation is between meaning and use according to Wittgenstein. The only passage where he did write something substantial about this relation is the notorious PI § 43. Since most studies of Wittgenstein are done in English, most quotations are taken then from the translation rather than from the original text. The most current translation of the "Philosophische Untersuchungen" is the one by Anscombe. A comparison I made of over 1500 instances in the text where "Bedeuten", "Sinn", "Gebrauch" -and their equivalents- occured on the one hand and their translation on the other proved that Anscombe obscured certain useful distinctions of the German text and actually committed a few severe mistakes. I will restrict myself here to the relevant results of this study by showing that the translation helps to create certain false impressions about particular aspects of Wittgenstein's work, including the "meaning-and-use" dictum in PI §43. Next I will argue that it is not only misleading but also contrary to Wittgenstein's own methodology to consider his writings about use on their own, without the context of the other paragraphs, of his conception of philosophy and indeed also to a certain extent semantic theories of his time.

1.1. The implications of Anscombe's translation on the interpretation of the dictum.

Apart from a few unfortunate and mistaken translations, "to mean" corresponds to "bedeuten", "meinen" and "heissen", "sense" to "Sinn" and "use" to "Verwendung", "Gebrauch" and "Anwendung".
The distinction between "Sinn" and "Bedeutung" is not the fregeian since Wittgenstein uses the latter for small, isolated linguistic entities (such as words) and the former for combinations of them (such as sentences). Moreover, it is possible to establish a difference between the three German equivalents of "meaning" quoted above: as far as these verbs are used to denote things or activities that have to do with linguistics one can attribute to them the following values. The subject of "meinen" is always human and the verb denotes the intentional act of meaning something by an utterance. "Heissen" is used to give paraphrases (Und "denken" hiesse hier etwas wie:

zu sich selber reden. PI §32) or to denote "to imply, to entail". "Bedeuten" stands for the property which can be attributed to isolated linguistic entities, viz. that they mean something. This distinction is lost in the translation.
It is evenso possible that there is an analoguous distinction between the German terms for "use" although I did not find any evidence for that at first sight. Richardson however does suggest one which is certainly worth considering (RICHARDSON 1976: 48f.).
It is clear that the translation impoverishes the text in this respect. This has also got some implications on PI §43. This paragraph features indeed only the term "Bedeutung", which means that the "meaning" in the English text stands for less aspects than one is lead to believe. Consequently, PI §43 is not a formulation of a "general rule of meaning" as it is sometimes called. And this is crucial in the decision whether Wittgenstein did or did not propose a semantic theory at all.

And there is also the fact that in the very same paragraph 43 the German "erklaeren" gets two different translations: "define" and "explain". I fail to see Anscombe's motivation for diversifying here, specially since the English terms are so different in meaning. "Erklaeren"-"explain" is regular and the last sentence of the paragraph is thus in conformity with Wittgenstein's opinion that the meaning of a name is indeed not to be identified with the referent (cf. also PI §§ 1f.) but that the latter one can only help to explain. The phrase "it can be defined thus" in the first part creates the impression that what follows is indeed a definition of the meaning of a word and thus a semantic theory, whereas the German text only claimes there to be an explanatory relationship between meaning and use. In other words: the use of a word explains its meaning but both terms are not intersubstitutable since there is something more to the meaning than just the use, much in the same way as there is something more to a name than just its bearer. And this will indeed prove to fit in with Wittgenstein's other writings.

1.2. A coherent interpretation.

Armed with this knowledge about the translation one can have a second look at PI § 43.
For a start there is a widespread misunderstanding to be eliminated. The "large class of cases" does not refer to certain categories of words to which the dictum applies, as e.g. HINTIKKA (1973: 54) supposes. A careful

reading shows that it applies to <u>cases in which we apply the word "meaning".</u> In other words, the explanation of the term "meaning", in its restricted sense of "Bedeutung", only concerns certain instances in which we use it. What are these instances?

First of all it are the occasions on which we are talking about the meaning of <u>a</u> word, as PI §43 clearly says. It forms indeed part of Wittgenstein's "ordinary language philosophy" that he is not interested anymore in abstract entities but in the properties of concrete linguistic phenomena (cf "Philosophical Grammar §19). It is only in these cases, or as Wittgenstein would say: in these language-games that the use can explain the meaning. We are here at the core of Wittgenstein's later philosophy. Indeed, when one reads the PI it becomes clear that his primary aim was to <u>dissolve</u> rather than to solve philosophical problems (as he tried in the "Tractatus") by showing that they emerge from treating words in isolation, rather than looking at them in their "natural environment", i.e. in their everyday use in language. In order to reconstruct such environments he invented language-games which are actual parts of our language or artefacts but which both illustrate how certain words are used. "Meaning" is one of these words which are investigated (cf. the title "Philosophical <u>Investigations</u>") with this method. So PI §43 should be read as follows: in a large class of language-games in which we use the phrase "the meaning of a word", this meaning can be be explained by its use in the language.

In this way Wittgenstein wants to refute a number of existing semantic theories whichhe finds erroneous: his own picture theory (cf. "Philosophical Grammar" § 19), nominalism and atomism (cf PI §§ 383 and 5 resp.), the fregeian theory that meaning is a question of reference alone (cf. PI §1f.), the theory that meaning is a mental process (cf."Blue Book" p.65 / "Philosophical Grammar" §23), in short:

> The mistake we are liable to make could be expressed thus: We are looking for the use of a sign, but we look for it as though it were an object <u>co-existing</u> with the sign. (One of the reasons for this mistake is again that we are looking for a "thing corresponding to a substantive"). The sign (the sentence) gets its significance from the system of signs to which it belongs. Roughly: understanding a sentence means understanding a language. ("Blue Book" p. 5)

and similarly PI §§ 190, 563 and XI and "Philosophical Grammar" §§ 10 and 44.

But in the end, what are then the cases to which PI § 43 applies? When we have a close look at all paragraphs in which Wittgenstein says significant things about the relationship between meaning and use <u>in practice</u>, we see that we can, as HUNTER (1971: 382f.) does, substitute "means" by "has the same use as". In other words: PI §43 only means that the meaning of a word is not something independant of it, that what two synonyms have in common, their meaning, is not some third entity. But how this works in practice is not of Wittgenstein's concern.

Neither does he investigate why two words have the same meaning. This why-question has to do with another aspect of the meaning of a word. And Wittgenstein knows that there is indeed such an aspect: it is the one he refers to when he talks about "grasping the meaning in a flash" (PI §138) This is also the reason why knowing how to use a word is a necessary though not sufficient condition for knowing its meaning. The why-question forms part of another language-game that does not belong to the cases referred to in PI §43. More precisely, it is the aspect which is related to "meinen". But, of course, since the translation of the "Philosophische Untersuchungen" does not make any distinction between both (except in PI §XI on one occasion), they become synonymous and the reader is not only confused by contradictions, but is also led to believe that Wittgenstein's dictum is some sort of a general theory of meaning, whereas it only deals with one aspect of meaning (which is confirmed by "Philosophical Grammar §23).

Indeed, as a consequence of all this, we see that there is no longer any doubt that Wittgenstein did not propose such a theory. And this is again entirely consequent with one of his principles: on several occasions (e.g. PI §§ 109 and 128) he states that a philosopher should not propose theories or definitions, but make clear the role, service, use, function, employment, etc. of a word, and then certainly of what he calls philosophical "odd-jobs words" ("Blue Book p.44).

2. Is Earl Wittgenstein the father of Princess Linguistic Pragmatics? [1]

It is clear that there is indeed a relation between Wittgenstein's dictum and linguistic pragmatics, which after all also deals deals with the way in which linguistic entities are used. Furthermore, many references to Wittgenstein can be found in the most diverse works on linguistic pragmatics. What kind of relationship we have here is less easy to point out: it depends on one's interpretation of Wittgenstein's works on the one hand and on one's ideas about pragmatics on the other.
It is not the aim of this chapter to define and evaluate goals, methods and scope of linguistic pragmatics; the reader will find this in literature. Note that I always refer to linguistic pragmatics since I don't want to include the study of other human activities than speech.

I will restrict myself here to indicating some of the consequences of the interpretation of the dictum that was set up in the previous chapter on what can be its relation -and to a lesser extent that of language-games and of forms-of-life- to linguistic pragmatics.

2.1. Wittgenstein's writing do not belong to linguistic pragmatics.

A first thing to point out is that it is nonsense to say that part of what Wittgenstein was doing actually <u>was</u> pragmatics as several authors do (e.g. ZABEEH 1971: 357f. / BARBOSA 1975: 51) or call language-games "speech-acts" (as BRAUNROTH 1975: 111). If terminology were retro-active, i.e. if terms from new theories would be simply applicable to aspects of older ones, all humanist sciences would become a giganteske hodge-podge.

2.2. Wittgenstein was not alone.

Wittgenstein was not the only one to work in the field of "use-theories" in his time. We therefore must be careful when we study his influence. A theory which came into being independantly from his and which played a crucial part in the development of linguistic pragmatics is Austin's. Because of several similarities between both, one could think that the latter has been influenced by the first. FURBERG (1963: 50f.) concludes, after a detailed comparison, that, even though there are important similarities in both works, the dissimilarities outweigh them. And similarities are rather coincidental anyhow since there cannot be any influence: by the time Austin was developing his theory Wittgenstein's ideas were only known in a small circle of friends and students to which Austin did not belong.
And, more generally, influence on British Analytical Philosophy is much overestimated: Ordinary Language Philosophy as such developed fairly independantly.

2.3. The consequences of the interpretation.

Some authors claim that Wittgenstein influenced linguistic pragmatics by stating that meaning is use in the language or in the language-game (SCHLIEBEN-LANGE 1975:30f. / VAN DER AUWERA 1978: 40). So do BRAUNROTH (e.a.)(1975: 111f.), and here we can even read that Wittgenstein in this way designed a "pragmatical semantics" of which the speech-act theory would be an elaboration, a view which is shared by COOPER (1973: 38f.) and KOPPERSCHMIDT (1973: 75). BATES (1976: 8) emphazises the importance to linguistic pragmatics of Wittgenstein's evolution from the picture theory to saying that meaning is use.

Let us, just for a moment, suppose that Wittgenstein has indeed written a use-theory and state that his language-games are in fact speech-acts. This would clearly be wrong since the list of language-games Wittgenstein gives in PI §23 contains several items which go far beyond the scope of speech-acts, such as"telling a story" which are characterised by PITCHER (1969: 230f.) as expressing speech-activities.
Furthermore, there are several dissimilarities to be found between this supposed use-theory and studies in linguistic pragmatics. For instance, performance utterances, illocutionary forces, etc. presuppose language-games because they need a broader context which is expressed by the latter: we cannot make a promise if we do not know what the custom of promising is (cf. HINTIKKA 1977: 236f.). Wittgenstein does not provide a means either to account for two oppositions that would occur against such a kind of semantic theory, namely Grice's account of intention (which corresponds more or less to Wittgenstein's "meinen") and Austin's distinction between illocutionary and perlocutionary forces (cf. BOGEN 1972: 200f.). And he would seem to overlook that a definition of meaning in terms of activity does not imply necessarily that meaning is use (cf.HARDWICK 1971: 38).

But in view of the interpretation summarized in the previous chapter we see that all this has no sense. For a start, the dictum cannot apply to the discipline within linguistic pragmatics that studies utterances in connection with intention and belief since Wittgenstein's term for this kind of meaning is "meinen", whereas PI §43 clearly is about "bedeuten". And much more important, the dictum is not a theory. The reason herefore is that what Wittgenstein says about the meaning of a word only concerns one aspect of the meaning, viz. the fact that in some cases "meaning" can be substituted by "having the same use as". And this is clearly not one of the language-games linguistic pragmatics is engaged in.
For rightful use-theories it is however possible to be linked up with speech-acts and other theories, as shown in COOPER 1973 , ALSTON 1968, CRESSWELL 1973 and SCHWAYDER 1970.

2.4. Rules, conventions and forms-of-life.

Another link which is sometimes made concerns the conception of "convention" and "rule". Especially in connection with PI §199 where rules are called customs, habits, this indeed very much looks like "conventions" in linguistic pragmatics (cf. VAN DER AUWERA 1978: 41 / BRAUNROTH 1975: 111f.).

The general line we can draw in all pragmatic theories about these concepts does however not correspond to Wittgenstein's conception of them.
Indeed, in the end we see that for Wittgenstein everything seems to culminate in his "form-of-life" in spite of its only being mentioned five times in his entire work (PI §§ 19, 43, 241, I, XI). He doesn't explain the concept and therefore interpretations are as numerous as people interpreting it. It is again one of those Wittgensteinian concepts which probably don't serve any other purpose than stimulate the reader to think about the topic himself. The form-of-life has to do with the determination of language by the way we live, i.e. by the conventions of a community.
There seems to be empirical evidence for this claim: HARDWICK (1971: 53) discusses the work of Br. Malinowski and quotes:

> Language is essentially rooted in the reality of the culture, the tribal life and customs of a people and...it cannot be explained without constant reference to these broader contexts of verbal utterance.

Words refer to objects, situations, etc. which are important to all people in a speech community that has adopted such a form-of-life (HARDWICK 1971: 68). A form-of-life indicates a common acting of people; following rules, communicate, etc. (and thus all language-games) are "Gebraueche, Institutionen" embedded in it (cf. PI §199).
In other words:

> Speaking a language -i.e. speaking and understanding it- is engaging in certain modes of behavior that exhibit a variety of abilities or skills. It is to engage in what Wittgenstein calls "forms-of-life". (PITCHER 1969: 243)

However, what is most relevant in this study is the fact that for Wittgenstein the use of words in language, in language-games, is largely determined by the conventions of a community. The conventional praxis of speaking a language is not reducable to rules, but to forms-of-life. It is clear that conventions here do not have the meaning they usually have in linguistics or, for that matter, in linguistic pragmatics.

3. Conclusions

What then, is the relevance of Wittgenstein's writings to linguistic pragmatics?
When one wants to look at the history of thoughts that underlie linguistic pragmatics, one inevitably comes across the PI. I fully agree with SEARLE (1977: 14; the same idea is in a way also expressed by BATES 1976: 8) when

he emphasizes that Wittgenstein's main merits are on the one hand stating, against existing theories, that semantics should not be concerned with meaning as a mental entity, and, on the other, indicating a new philosophical method. For Wittgenstein speaking is engaging in behaviour. Linguistic pragmatics corresponds to this as the study of meaning as an act.
But we cannot say that Wittgenstein's use-theory had any influence since this "use" is not the kind of use linguistic pragmatics is concerned with and his theory is not even a theory.
In these circumstances we can hardly call (as BRAUNROTH 1975: 111) Wittgenstein the "father of linguistic pragmatics " and the PI "its Bible".

Note

1 Adapted from a phrase by KASHER (1977: 106)

Literature

1. works by Wittgenstein

Philosophische Untersuchungen . Frankfurt: Suhrkamp, 3rd ed. 1975
The Brown and Blue Books . Oxford: Blackwell, 2nd ed. 1965
Philosophical Grammar . Oxford: Blackwell, 2nd ed. 1974

2. other works

ALSTON, W.P. (1968): "Meaning and Use". PARKINSON (1968)
ANSCOMBE, G. (1953): Wittgenstein - Philosophical Investigations - translated by... . Oxford: Blackwell, 3rd ed. 1976
BARBOZA, B. (1975): "Sur la notion de signification dans les Investigations Philosophiques de Wittgenstein". Leuven: unpublished.
BATES, E. (1976): Language and context . New York etc.: Academic Press.
BOGEN, J. (1972): Wittgenstein's Philosophy of Language . London: Routledge and Kegan.
BRAUNROTH, M. (e.a.) (1975): Ansaetze und Aufgaben der linguistischen Pragmatik . Muenchen: Athenaeum Fischer Taschenbuch Verlag.
COOPER, D.E. (1973): Philosophy and the nature of language . London: Longman.
CRESSWELL, M. (1973): Logics and Languages. London: Methuen.

FURBERG, M. (1963): Saying and Meaning . Oxford: Blackwell.
HARDWICK, C.S. (1971): Language learning in Wittgenstein's later philosophy . The Hague: Mouton.
HINTIKKA, J. (1973): Logic, Language-games and Information . Oxford: The Clarendon Press.
--- (1977): "Language-games". Dialectica: vol 31, fasc. 3-4.
HUNTER, J.F.M. (1971):"Wittgenstein on Meaning and Use".KLEMKE (ed.) 374-393.
KASHER, A. (1977): "What is the theory of use?". Journal of Pragmatics 1.
KLEMKE, E.D. (ed.) (1971): Essays on Wittgenstein . Urbana etc.: University of Illinois Press.
KOPPERSCHMIDT, J. (1973): Allgemeine Rhetorik . Stuttgart etc.: Kohlhammer.
PARKINSON, G. (ed.) (1968): The Theory of Meaning . Oxford: OUP.
PARRET, H.(1977): "De sprekende mens". Brussels: Belgian Radio and Television typescript.
PITCHER, G. (1969): The philosophy of Wittgenstein . Englewood Cliffs: Prentice Hall.
RICHARDSON, J. (1976): The Grammar of Justification . Sussex UP.
SCHLIEBEN-LANGE, B. (1975): Linguistische Pragmatik . Stuttgart etc.: Kohlhammer.
SCHWAYDER, D. (1968): "Uses of language and uses of words". PARKINSON (1968).
SEARLE, J. (1977): Interview with H. Parret. PARRET (1977).
THYS, G. (1978): "An interpretation of Wittgenstein's "meaning-and-use" dictum with some remarks on its relation to linguistic pragmatics". Universitaire Instelling Antwerpen: unpublished.
VAN DER AUWERA, J. (1978): Inleiding tot de linguïstische pragmatiek . Leuven: Acco.
ZABEEH, (1971): "On language-games and forms-of-life". KLEMKE (ed.) 328-373.

DIRECT SPEECH ACTS[1]

Johan Van der Auwera

1. Introduction

In this paper I want to analyse the following thesis: for a speech act ("SA") to be a direct SA ("DSA"), it is necessary and sufficient that there is a formal characteristic of the SA explicitly indicating the illocutionary type.[2] The determination of the direct illocutionary force would be context-independent.[3] I will confront this thesis with the claim that (1) and (2), in contrast with the indirect SA ("ISA") of (3), exemplify direct orders or, at least, direct directives.[4]

(1) Leave the room.

(2) I order you to leave the room.

(3) I wonder if you would mind leaving the room.[5]

Many investigators subscribe to both the general thesis on DSAs and the analysis of imperatives and performatives. Unfortunately, these workers are mostly only clearing the ground for a study of ISAs. I claim that the groundwork is not clear at all.[6]

2. The imperative

Notice, first of all, that the structure of (1) does not indicate the type of the directive. (1) could be an order, a request, a piece of advice, a suggestion, etc.. Let me call these illocutionary forces "subtypes". They are all subtypes of the directive, which will be called the "type". So, one might think that the imperative pattern as such only indicates a SA type and not a SA subtype. However, if one looks at a few other imperative sentences, this is no longer self-evident.

(4) Come again.

(5) Come in.

(6) Have some cookies.

(7) Have a good time.

(8) Come again and I will break your neck.

One may think of (4) and (5) as expressives. The first can function as a farewell and the second as a welcome. (6) could be an offer and (7) a wish.

Whatever SA type they would belong to in the Searlian classification, they are not directives. The illocutionary force of the imperative in (8), ANSCOMBRE (1977: 22, 24) calls a "supposition". This is not covered by Searle's classification, but one can safely say that it is not a directive.

Does this force me to revise the view that the imperative structure only indicates directiveness? I do not think it does. This decision implies two problems. First, how to deal with the troublemakers in (4) to (8)? Secondly, if I hold that the imperative only indicates the directive SA type, what then is the relation between the imperative mood and the directive SA subtypes?

As for the problem of the non-directive readings of (4) to (8), I believe that if the concept of an ISA has any value, these non-directively interpreted SAs are ISAs. That is, I believe that one can say goodby, welcome, offer, wish and suppose t h r o u g h directing in a way that is similar enough to the way one can request through asserting or questioning. To fully vindicate this claim, I would have to come up with a detailed account of ISAs that accomodates the traditional requests through asserting and questioning as well as the SAs in (4) to (8). But a study of ISAs is not the point of this paper. So my intuition will have to suffice.

As for the second problem, I hold that the imperative pattern by itself does not indicate a subtype, but only the content common to a set of subtypes, viz. order, demand, request, plea, advice, suggestion, permission, exhortation and maybe others. I will here assume that this subtype set is the set characterized by the type. In other words, an imperative would indicate what is common to a l l directives.[7]

On the alternative analysis[8], the imperative form does signal a subtype, viz. that of the order. It is not being denied then that imperatives very often function as directives other than orders, but these would be ISAs. I find it unconvincing, both that the order would be a directive that carries the literal meaning of the imperative formula and that the relation between an order - the would-be literal SA - and (say) a request, as both are associated with a particular imperative, are related in a way that is similar to the relation between a primary and a secondary illocution of an ISA. On top of that, I am happy with my intuition that the imperative would indicate a type and not a subtype.

From the fact that an imperative only signals a type and not a subtype, one may want to conclude that corresponding to the imperative structure there is only a direct SA type and there are no direct SA subtypes. In a Searlian framework this conclusion is not very elegant, because there would be a discrepancy between the concepts of directness and indirectness as applied to SAs. SEARLE (1975a) speaks about indirect requests as well as about indirect directives. So indirectness matters for both SA types and SA subtypes. Directness, on the other hand, would only be relevant for SA types.

This conclusion is unnecessary, though. It does not follow from the fact that the imperative only indicates a SA type that a SA subtype cannot in a relevant way be considered direct. Here is a viable suggestion: a SA subtype is direct iff the type that subsumes the subtype is unambiguously identified by the linguistic structure of the utterance, i.e. iff the type is direct. The directness of a subtype is thus defined in terms of the directness of a type. So, to get back to example (1), it is a direct directive and, as the case may be, a direct order, request, suggestion, advice, etc.. Which particular SA subtype is involved, depends on context. In this conception, the fact that an utterance is vague as to the subtype does not diminish its directness. In other words, the context-dependence of a SA subtype is not in contradiction with its directness. This proposal captures the following facts: (i) an imperative signals a SA type; (ii) the various SA subtypes that can be expressed by the imperative do not seem to be different with respect to the directness dimension; (iii) the (say) requestive reading of something like (3) and that of something like (1) d o seem to differ in directness, to the extent that one could defend calling the interpretation of (3) indirect and that of (1) direct.

The fact that it is possible to define the directness of a SA subtype and thereby to avoid the unelegant discrepancy in the application of directness and indirectness to SAs, does not imply that this solution is itself ideal in all respects. Remember that I started off from the thesis that directness is a context-independent notion. To some extent, this thesis has to be revised now, for in my analysis of the imperative the identification of a direct SA subtype is context-dependent. This means that for SA subtypes I have so far given up the link between context-independence and directness. The directness of a subtype would no longer relate to a context-independent identification of that subtype, but only to that of the appropriate type. This was done in order to be able to say that both SA types and subtypes can be direct and indirect. Alternatively,

I could have held on to a stipulation that a SA, whether type or subtype, is direct when it can be identified without recourse to context. (1) would no longer be called a direct request, order, etc.. Notice that it would not necessarily become an indirect request, etc., since one may still believe that it is very different from the traditional ISAs. It would at least be a "non-direct" request, order, etc..[9, 10]

This account of the necessity and sufficiency of SA type directness for a SA subtype, subsumed under the type, to be direct, allows for an utterance to contain overt signals for the SA subtype. These signs would be sufficient for directness, but not necessary. Thus both (1) and (9) can be direct orders.

(9) I hereby order you to leave the room.

(1) is direct because the necessary and sufficient condition for subtype directness, viz. the type directness, obtains. (9) is a direct order because of a sufficient and unnecessary condition, the fact that the utterance forces a peculiar subtype reading. This sufficiency entails that the necessary and sufficient condition, that of the context-independent determination of the type, is also fulfilled. Observe that I have chosen a performative to exemplify this case. This is the structure that will now be examined in some detail.

3. The performative

The explicit performative pattern of (2) is contrasted with the ISA of (3).

(2) I order you to leave the room.

(3) I wonder if you would mind leaving the room.

At first sight, (2) seems to leave no doubt on its subtype. It is an order and n o t a request, a piece of advice or a suggestion. Similarly, (10), (11) and (12) seem to clearly indicate their subtype.

(10) I request you to leave the room.

(11) I advise you to leave the room.

(12) I suggest that you should leave the room.

Does this mean that a SA employing a performative formula not only indicates type but also subtype?

The answer is negative. The problem with performative utterances is, as AUSTIN (1975: 64) already noticed, that these utterances can often be interpreted in a constative or representative sense. So (2), (10), (11) and (12) could all be interpreted as statements, e.g. when they occur as answers to the question in (13).

(13) What is your normal reaction when I make too much noise?

This vagueness is not present in all utterances that exhibit a performative pattern, since utterances may have other illocutionary indicators. "Hereby", e.g., secures a performative reading. But the performative pattern itself is vague and it is important to see that the vagueness is one of SA type - representative versus directive in cases (2), (10), (11) and (12) - and not of SA subtype, as was the case for the imperative structure. So, for performatives, even type determination can be context-dependent. If one would like to call (2), (10), (11) and (12) direct directives, it looks as if one has to give up the context-independence of SA type directness.

Should we not hesitate to call them direct? This is a partially terminological issue, I fear. Let me compare (1), (2) and (3) again.

(1) Leave the room.

(2) I order you to leave the room.

(3) I wonder if you would mind leaving the room.

(3) is indirect and (1) is direct, though I indicated that it would not be hard to defend calling the latter non-direct, though not necessarily indirect. (2) and the business of the ambiguity between a constative and a performative reading is fairly different from both (1) and (3). So whether the directive reading of (2) were called direct, non-direct or indirect, it would need a special subcategory anyway. I will first take a Searlian point of view and make the directive reading of (2) a direct one. I will investigate an alternative afterwards.

A different but related question is whether the constative and the performative readings have the same status. One might argue that (2) is either constative or performative. Alternatively, one might claim that one interpretation can only be acquired t h r o u g h the other in some sense. Again, I will first adopt a Searlian position and say that they have the same status. I will soon question this hypothesis.

So much for the directness of type. The directness of subtype has to be considered, too. Can we call (2) a direct order? This is easy. For a subtype to de direct, the type has to be direct. We have just described what directness of type for a performative means. So the subtype can also be called direct. Moreover, a performative has overt subtype signals.

Even though an imperative and a performative have now been given a uniform treatment in the sense that both of them are considered to carry a direct type as well as a direct subtype, this uniformity covers up some large differences. The direct type of a performative is context-dependent, that of the imperative is not. The direct subtype of the imperative is context-dependent. The direct subtype of a performative is context-dependent to the extent that the type depends on context, but once the latter is recognized, the utterance indicates the subtype.

It is quite disconcerting that the attempt to define the directness of (1) and (2), whether of type or subtype, in terms of linguistic correlates has had mainly negative results. It could not be maintained that for any utterance x, type and subtype are explicitly indicated. (1) is a counter-example. But one cannot even hold on to the weaker thesis that any utterance x at least indicates its type, since now the performative is the counter-example. The latter conclusion can perhaps be obviated, though. It has been argued persuasively and on independent grounds[11] that a performative utterance is basically assertoric and that the performative reading is a derived one. I will not repeat the arguments here. Let me accept it, even if only for the sake of the argument, and see what if can offer in this context. Now, as to the type of derivation, I would not necessarily go as far as ANSCOMBRE (1977: 37-38) and call it the derivation of an ISA. I do not have to go into the details of performative utterances here. It suffices that the utterance could be called non-direct.

In such an account the type of (2) is obviously no longer context-independent. (2) is simply a direct representative. Thus it ceases to be a counter-example to the claim that English SAs indicate their type.

3. Conclusions

SA directness is often defined in terms of linguistic correlates. However, the thesis that every SA has a direct t y p e as well as a direct s u b t y p e which are encoded in the structure of the utterance is argued to be false. This means that some SA subtypes which are said to be direct, on closer scrutiny, come out as non-direct. The weaker thesis that every SA has a direct t y p e which is indicated by formal linguistic means can only be maintained if one gives up a Searlian account of the performative, that is, if a performative reading is considered to be a derived one. If one then does not want to lose the ability to say that every SA has a direct type as well as a direct subtype, subtype directness has to be redefined. It would no longer be dependent on linguistic correlates for the subtype but on linguistic indicators for the type subsuming the subtype.

Notes

1 An earlier version of this paper was read at the Department of Linguistics of the University of Göteborg (September 1978). It is part of a larger enterprise reported on in VAN DER AUWERA (1978 / 1979). Special thanks are due to Jens Allwood. This work was supported through a research fellowship of the Belgian National Science Fund.

2 In VAN DER AUWERA (1978: § 2) it is argued that SEARLE (1975a) would subscribe to this. It matches definitions given by EHRICH/SAILE (1972: 256), MEYER-HERMANN (1976) and BROWN/LEVINSON (1978:137).

3 "Context-independence" is a technical notion designed to capture the conventional relation between a peculiar sign explicitly indicating a particular meaning. Clearly, these conventions only exist in the c o n t e x t of specific human societies (cp. SEARLE 1978).

4 For the purpose of this paper, I will use Searle's typology. See SEARLE (1975b / 1976a).

5 The examples are those of SEARLE (1975a: 64). This strategy of contrasting ISAs with grammatical patterns like the imperative and with the performative is also applied by FORMAN (1974: 162-163), SADOCK (1974:74), GARVEY (1975: 45-46), CLARK/LUCY (1975: 56), DILLER (1977: 9) and SÖKELAND (1979), but n o t, interestingly, by AUSTIN (1975: 32-33). He only contrasts ISA-like things with performatives. The third possibility, viz. a contrast between ISAs and sentence patterns, but not with performatives, can be found in DORE (1977: 235), LYONS (1977: 784-785), DILLER (1978: 1) and GEUKENS (1978).

6 Compare with MEYER-HERMANN (1976).

7 The following authors do not seem to believe that any particular subtype is the literal meaning carried by the imperative: SADOCK (1974: 139ff, 148 ff / 1977: 70ff), AUSTIN (1975: 32-33, 73-74, 76-77, 101-102), SEARLE (1976b: 1119), ALLWOOD (1977: 57), LYONS (1977: 745ff) and GEUKENS (1978: 262, 272).

8 Such account is implied by ANSCOMBRE (1977: 24), DILLER (1978: 1) and probably BROWN/LEVINSON (1978: 104, 113, 137, 196, 234).

9 CAFFI (1978: 188-189) introduces a notion of "implicit" SAs, including both ISAs and SAs which, to use my terminology, seem to literally express a SA type of which the refinement to subtype is dependent on context. Using this term, an imperative would be an implicit SA that is not indirect. It is interesting to see that Caffi was inspired by AUSTIN (1975) (cp. note 5). So there is at least one person that has had attention for this non-Searlian idea of Austin's. GEUKENS (1978: 262) agrees with Caffi, be it that SAs that would be "implicit", do not get a special label.

10 See EHRICH/SAILE (1972) for "non-direct" SAs.

11 See CORNULIER (1975), ALLWOOD (1976: 207-208, 211 / 1977: 55, 65), ANSCOMBRE (1977: 64ff) and HARDER (1978: 195).

ALLWOOD, Jens (1976): Linguistic Communication as Action and Cooperation. Göteborg: University, Linguistics Department. (Gothenburg Monographs in Theoretical Linguistics 2).

--- (1977): "A critical look at speech act theory". DAHL, Osten (ed.) (1977): Logic, pragmatics and grammar. Göteborg: University, Linguistics Department: 53-69.

ANSCOMBRE, Jean-Claude (1977): "La problématique de l'illocutoire dérivé". Langage et Société 2: 17-41.

AUSTIN, J.L. (1975): How to do things with words. ed. by URMSON, J.O./ SBISA, M.. Cambridge, Massachusetss: Harvard University Press.

BROWN, Penelope / LEVINSON, Stephen (1978): "Universals in language usage: Politeness Phenomena". GOODY, Ester (ed.) (1978): Questions and politeness. Cambridge: Cambridge University Press: 56-324.

CAFFI, Claudia (1978): "Atti linguistici indiretti". CONTE, Maria-Elisabeth / RAMAT, Anna-Giacalone / RAMAT, Paolo (eds.) (1978): Sprache im Kontext. Tübingen: Niemeyer: 187-198.

CLARK, Herbert / LUCY, Peter (1975): "Understanding What Is Meant from What Is Said". Journal of Verbal Learning and Verbal Behavior 14: 56-72.

CORNULIER, Benoît de (1975): "La notion d'auto-interprétation". Etudes de Linguistique Appliquée 19: 52-82.

DILLER, Anne-Marie (1977): "Le conditionel marqueur de dérivation illocutoire" Semantikos 2: 1-17.

--- (1978): "La forme en -rais dans les actes indirectes de langage". Forthcoming in Actes de la Session de Linguistique de Bourg-Saint-Maurice. Paris: Conseil Scientifique de la Sorbonne Nouvelle.

DORE, John (1977): "Children's Illocutionary Acts. FREEDLE, Roy O. (Ed.) (1977): Discourse Production and Comprehension. Norwood, New Jersey: Ablex Publishing Corporation: 227-244.

EHRICH, Veronika / SAILE, Günter (1972): "Uber nicht-direkte Sprechakte". WUNDERLICH, Dieter (1972) (Ed.) Linguistische Pragmatik. Frankfurt am Main: Athenäum Verlag: 255-287.

FORMAN, Donald (1974): "The speaker knows best principle: Or why some complicated facts about indirect speech acts are really obvious facts about questions and declaratives" Papers from the tenth regional meeting of the Chicago Linguistic Society: 162-177.

GARVEY, Catherine (1975): "Requests and responses in children's speech". Journal of Child Language 2: 41-63.

GEUKENS, Steven K.J. (1978): "The distinction between direct and indirect speech acts: towards a surface approach". Journal of Pragmatics 2: 261-276.

HARDER, Peter (1978): "Language in action. Some arguments against the concept "illocutionary"". GREGERSEN, K. (Ed.) (1978): Papers from the Fourth Scandinavian Conference of Linguistics. Odense: Odense University Press: 193-197.

LYONS, John (1977): Semantics: 2. Cambridge: Cambridge University Press.

MEYER-HERMANN, Reinhard (1976): "Direkter und indirekter Sprechakte". Deutsche Sprache 4: 1-19.

SADOCK, Jerrold M. (1974): Toward a Linguistic Theory of Speech Acts. New York: Academic Press.

--- (1977): "Aspects of Linguistic Pragmatics". ROGERS, Andy / WALL, Bob / MURPHY, John P. (eds.) (1977): Proceedings of the Texas Conference on Performatives, Presuppositions, and Implicatures. Arlington, Virginia: Center for Applied Linguistics: 67-78.

SEARLE, John R. (1975a): "Indirect Speech Acts". COLE, Peter / MORGAN, Jerry L. (eds.) (1975): Syntax and Semantics 3: Speech Acts. New York: Academic Press: 59-82.

--- (1975b): "A taxonomy of illocutionary acts". GUNDERSON, Keith (ed.) (1975): Language, Mind, and Knowledge. Minneapolis: University of Minnesota Press: 344-369.

--- (1976a): "A classification of illocutionary acts". Language in Society 5: 319-332.

--- (1976b): "The rules of the language game". Times Literary Supplement 3,887: 1118-1120.

--- (1978): "Literal Meaning". Erkenntnis 13: 207-224.

SÖKELAND, Werner (1979): "Ein Beitrag zur Theorie der indirekten Sprechakte" VAN DE VELDE, Marc / VANDEWEGHE, Willy (eds.) (1979): Bedeutung, Sprechakte und Texte. Akten des 13. Linguistischen Kolloquiums, Gent 1978, Bd. 2. Tübingen: Niemeyer.

VAN DER AUWERA, Johan (1978): Indirect Speech Acts Revisited. Antwerp: University, Linguistics Group. (Antwerp Papers in Linguistics 16).

--- (1979): "On the irrefutable meaning of basic speech acts". Journal of Pragmatics.

A METHOD FOR THE EVALUATION OF RECIPIENT RESPONSE TO METAPHORICAL PROPOSITIONS

Jean-Pierre van Noppen

The aim of this paper is 1. to outline in short the objectives pursued within the frame of our research in general; 2. to outline the method devised to provide a relatively satisfactory answer to our questions, and 3. to highlight a number of problems raised by our approach, i.e. to draw attention to a number of possible pitfalls in the use of the method.

Our general preoccupation is with theographic (i.e. descriptive theological) statements in popular religious prose. In an earlier paper (VAN NOPPEN 1979a, 1979b) we have attempted to describe the mode of meaning obtaining in these propositions in terms of metaphor logic. This model of description was found to be more adequate than, say, the more traditional logics of literalism, fundamentalism, revelation, and analogy.

The procedure outlined here was conceived as the third prong of a project endeavouring to elucidate a number of communication problems that had been observed in the realm of popular theology. Previous phases in the study involved lexical scrutiny of a number of texts, and a content analysis of the correspondence between the author of the text and his audience (VAN NOPPEN 1976). From these enquiries it became clear that the obstacle to effective exchange between the professional theologians and their lay public at large was less a problem of word meaning than one of prepositional sense and logic (VAN NOPPEN 1978). The asymmetry in status between author and reader, it was observed, frequently led to a situation in which the sense of the propositions was determined almost exclusively from the specialist's point of view, with an apparent disregard for the layman's more limited insight into the precise doctrinal implications of the expressions. A perusal of the author's complaints on the one hand, and of the readers' reactions on the other, indicated that the connotations associated with the propositions by the readers were not invariably coextensive with the shades of meaning intended or used by the writer. It would be an error, therefore, to consider "meaning" as a solid and indivisible entity, as an inherent property of expressions: one must maintain a distinction between the content of a proposition as a type, and its interpretation, i.e. the various ways in which the expression can be understood and responded to in its various potential communication contexts, and study the value of the terms in their contexts to both the "sender" and the "receiver".

In the present case example, the aim pursued is to indicate the nature and the intensity of the recipients' response to spatial terms framed within theographic messages; in other words, to measure a number of propositions' effectiveness in conveying theographic sense from the author towards the addressee. This procedure assumes a communication model like, say, SHANNON and WEAVER's (1949:7). Our preoccupation would, in Weaver's terms, be with the B-level of communication : "how precisely do the transmitted symbols convey the desired meaning ?" The objective is not to measure the quantity of factual information conveyed by the utterances, since as a consequence of the subject matter (i.e. a putative mode of supranatural reality) and of the mode of logic prevailing within the particular register, the sense of the utterances relies on intension rather than on extension. Nor is it our goal to examine the recipient's process of comprehension, in which case we might have resorted to some natural language understanding system. The method relevant to our purpose should enable us to chart the public's ability to recognize the potential metaphoric, <u>in casu</u> theographic, value of the focus within its context and to interpret it "correctly", that is, optimally with regard to the meaning originally intended.

The rules of transference (LEECH 1969:89), which explain how metaphorical sense can be made out of apparently contradictory or sortally deviant propositions, and approaches which consider metaphorical propositions as ambiguous because of the polysemy of the focus (KOOIJ 1971:117-8) bring us a step closer to our aim inasmuch as they seek to determine the features which constitute the (typical or arbitrary) basis for (actual or potential) transferred use, but do not allow one to assess the connotative value of an image unless it be based on relatively predictable and/or typical features (AARTS/CALBERT 1976: 240-3). The demand at this stage, then, is for a method that enables us to deal with connotative aspects of meaning and that offers the possibility to measure this meaning in objective, quantifiable terms. Following the approach advocated, but not worked out, by JEFFNER (1972:71) we sought to apply to our specific objectives a technique inspired by OSGOOD's semantic differential (1971:20-26, 273-5). The choice proved to be a fruitful one, inasmuch as the approach offers a "combination of controlled association and scaling procedures (...) designed to measure the meanings (states of language users) of signs (units of messages)". In Osgood's technique, the meaning associated with a sign is judged against a series of scales representing a continuum between polar pairs of adjectives which verbally determine various dimensions of experience. Early research on the development of the method drew attention to the kinship between synesthetic association of various forms of experience on the one hand, and the process of metaphoric assimilation on the other, when the associations were given a linguistic form. This observation led to the proposal to use the semantic differential as an index of certain aspects, especially connotative aspects, of linguistic meaning.

Basing ourselves on these experimental and theoretical premisses, we deemed it must be possible to adapt this method to our purposes in order to measure and quantify the connotative meanings associated with spatial foci in theographic frames against a series of scales between "contrary" adjectives, substantives, and phrases representing the various theological categories involved, and thus to assess the "theographic cash-value" of particular terms and categories without having to go through the laborious intermediate stage of defining the transferred or transferrable components. Before this adaptation could be achieved, however, three problems remained to be solved :

1. It was, obviously, impossible to require the future respondents to rate the theographic value associated with every single proposition in the Corpus of texts, or even with one occurrence of every type. For this reason, a selection was effected among the various theographic propositions so as to confront the respondents with a manageable sample of expressions reflecting the atmosphere and the issues of the kind of popular theology under study, as well as the categories typical of the age's theography -- thus to give us as safe and representative a basis as possible for tentative generalizations at a later stage. As a matter of choice, and for various reasons developed in another paper (VAN NOPPEN 1974), it was decided to focus on theographic expressions based on the crucial category of spatial imagery, i.e. of relation, localization, orientation, polarity, etc.

2. Our option to assess the respondents' reactions to whole expressions rather than to individual word symbols raised problems of its own, since one might justifiably argue that the subjects' attention could be diverted from the spatial focus to another phase of the proposition. However, on the basis of our earlier

observation that the main source of misunderstanding resided in the sense of theographic statements rather than in the meaning of individual terms, it was deemed that the results obtained by response to spatial foci embedded within their original theographic contexts would yield a more adequate picture of the public's recognition of their theographic value than the elicitation of connotations associated with single lexical items. Care was taken, however, to present the spatial foci within frames as coherent as possible (at least according to the authors of the statements) in order to obtain responses to one theographic category at a time.

3. We disposed of no absolute standard of orthodoxy with reference to which the public's reactions could be differentiated and charted. For this reason, the respondents were asked first to delineate their personal notion of "God" by associating the term with the given pairs of polarities, and only afterwards to chart their reactions to the various theographic propositions. In this manner, the degree of accuracy or distortion resulting from particular modes of theography could be measured for each individual respondent in terms of the deviation with regard to his own conception of the divine.

These considerations eventually led us to the elaboration of a series of semantic association grids, the warp of which was constituted by characteristic spatial-theographic propositions, the woof by ten continua between theological opposites, divided into seven columns to facilitate the quantification of the intensity of the various responses, and the calculation of the deviations observed. For practical purposes, the equi-conceptual poles were grouped on one side of the grid. (A sample grid is reproduced below)

Alongside the completion of one of these grids for every statement submitted to them, the respondents were given the opportunity to formulate freely their own paraphrase of the propositions, or to comment on the responses given in the grid Appropriate spaces for this purpose were provided in the questionnaires. The combination of forced-choice scales and open-ended questions was deemed to offer several advantages : it allowed us to elicit responses from subjects not accustomed to think or speak of God in the categories defined by the pairs of opposites, and it partly resolved the ambiguity of responses in the central "zero" column : subjects failing to chart their degree of association with either of the opposites or failing to indicate any response at all were thus offered a chance to comment on their reasons for doing so.

Having thus fixed our objectives, decided upon a method, and elaborated a questionnaire, we must now briefly comment upon our options with regard to the practical organization of the enquiry.

The questionnaire was administered to a sample of native English speakers from a variety of social, religious, and educational backgrounds. No systematic attempt was made at representativity or stratification of the sample according to the canons of sociological research. This rather arbitrary choice of respondents was primarily conditioned by constraints imposed by the time and facilities available, although a number of other arguments may be invoked as well :

1. Our aim was not to infer from the sample any general conclusions about the public or the readers of popular theology as a whole, but to study various responses to a particular mode of language use.

2. Since the propositions used were borrowed from a Corpus of "popular" literature, all adult native speakers of English could, in principle, be considered as potential addressees, and therefore qualify as informants. We are conscious, however, that the compromise constituted by the casual composition of the sample seriously limits the possibility of generalizing about any correlations between the variables in the Corpus of informants (age, education, religious affiliation,...) and possible differences that might be observed in their responses. As in the use of a Corpus, the validity of the conclusions is limited to the sample used; yet, although no claims at exhaustiveness or representativity of the observed responses can be made, the information yielded by a body of informants, albeit limited, provides more objective and reliable indications than can be gained from introspective methods of investigation.

3. Although precedents do not per se provide a justification, it may be pointed out that whereas sociolinguistic enquiries which seek to correlate linguistic data with specific social variables give evidence of at least some degree of rigour in the construction of their samples, many linguistically oriented studies, say in acceptability, use, and attitude, make little or no attempt to draw on a systematically composed body of informants (Cf. for example the majority of contributions to GREENBAUM (ed.) 1977.)

It is interesting to note that a fair number of respondents spontaneously appended notes and comments to the questionnaires, either to describe the difficulties they had encountered in responding to the propositions, or to voice their dissatisfaction with particular formulations or aspects in the conception of the enquiry. It appeared from these reactions that the test (twenty statements to be checked against ten continua each) required a great deal of patience, concentration, and time from the volunteers. The test, we are conscious, was a long one, and manifestations of satiation are to be observed throughout the second half of the questionnaire. Some other respondents' reactions revealed that, in spite of the objective formulated in the instructions, the test was occasionally experienced as an inquiry into the respondents' belief or commitment rather than into their response to the statements. More understandable are the comments of those who deem that the statements cannot be properly understood outside a wider context -an inevitable consequence of our sampling compromise- and of those, mostly philosophically or theologically trained people, who recognize the statements' potential to be interpreted on more than one level, and complain about their ambiguity. Finally, some criticism was addressed to the various polarities. The polar oppositions were felt to form too compelling a corset, although the possibility of free paraphrase alleviated this situation for those able to verbalise their impressions. Other respondents experienced some difficulty in relating the scales to the statements, apparently because they failed to ask the intermediate question, as per instructions, "what does this tell me about God ?".

The responses obtained were transferred onto punched cards, and subsequently computed and processed mechanically. For this purpose, the various divisions on the scales were numbered from 1 to 7, from left to right. For each statement, the average response on each of the scales was calculated for the total population as well as for each particular class constituting it. This procedure allowed us to compare the effect of the categories under study on the subjects' conception of God as described by means of the same scale, and thus to gain at least some indications about what has been called the theographic "cash-value" of the various modes of theography.

It is impossible, within the bounds of a short paper, to give an intimation of the results obtained. The reader will find an analysis of the results for a few of the statements in a forthcoming article (VAN NOPPEN 1979c). Since the aim of this paper is mainly methodological, we must satisfy ourselves with the reproduction of the grid reflecting the responses to one of the statements.

First, an analysis was made of the responses to the term "God". The average of the responses between 1 and 7 on each scale was calculated up to the first decimal, and represented by a line on the grid. Next, an item-per-item comparison between the responses obtained for the term "God" and for each of the statements was carried out. The average responses to the statements were represented graphically by dots on the scales, in such a manner that the degree of deviation could be visualised by means of the distance and the direction of the gap between the line and the dot.

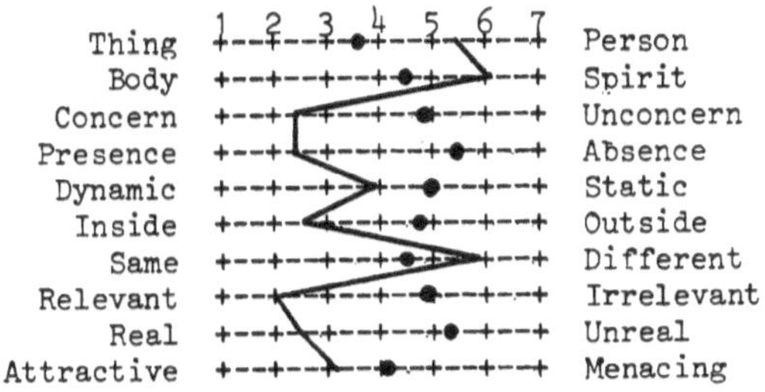

Statement 19 : "There is no being out there at all. The skies are empty."

Average response to:	"God"	Stat. 19	Difference
Thing/Person	5.5	3.6	- 1.9
Body/Spirit	6.0	4.5	- 1.5
Concern/Unconcern	2.4	4.9	+ 2.5
Presence/Absence	2.4	5.4	+ 3.0
Dynamic/Static	3.8	5.0	+ 1.2
Inside/Outside	2.6	4.8	+ 2.2
Same/Different	5.8	4.5	- 1.3
Relevant/Irrelevant	2.0	4.9	+ 2.9
Real/Unreal	2.4	5.3	+ 2.9
Attractive/Menacing	3.1	4.1	+ 1.0

The analysis of these data shows that the responses in the grids concord with the informal results obtained in the correspondence and in the open paraphrases. We may infer, therefore, that despite the limitations and reserved which we have deemed proper to impose on our observations, valuable indications and insights into the readers' attitudes may be gained. Future experience must prove the degree of reliability of the method, and notably whether the application of

Osgoodian techniques to whole propositions elicits the same degree of consensus from the informants as the semantic association with single terms. In the converse case, a new experiment should attempt to operate with pairs or groups of sentences featuring minimal differences. If, however, the method proves as reliable and practically workable as the present state of our study allows us to hope, its potential applications in the whole realm of verbal communication are manifold.

BIBLIOGRAPHY

AARTS, J.M./CALBERT, J.P. (1976) : Metaphorical and Non-Metaphorical Extensions of Meaning. Löwen und Sprachtiger, Proceedings of the VIIIth Conference on Linguistics : 240-3.

GREENBAUM, S. (1977) : Acceptability in Language. The Hague : Mouton.

JEFFNER, A. (1972) : The Study of Religious Language. London : SCM Press.

KOOIJ, J.G. (1971) : Ambiguity in Natural Language. The Hague : North Holland.

LEECH, G. (1969) : Towards a Semantic Description of English. London : Longmans.

OSGOOD, C./SUCI, G./TANNENBAUM, P. (1971) : The Measurement of Meaning. Urbana : University of Illinois Press.

SHANNON, C.E./WEAVER, W. (1949) : The Mathematical Theory of Communication. Urbana : University of Illinois Press.

VAN NOPPEN, J.-P. (1974) Spatiality in Religious Discourse. Revue des Langues Vivantes XL n° 1 : 7-24.

--- (1976) : Alter Wein in neuen Schläuchen ? Linguistica Biblica 37 : 1-9.

--- (1978) : Communication et Vulgarisation : Le problème posé par une situation de discours asymétrique, étudié dans la théologie populaire britannique. Réseaux 34-35 (forthcoming).

--- (1979a) : Théograpie, métaphore et erreurs d'interprétation. Revue des Langues Vivantes (forthcoming).

--- (1979b) : Theographie, Metaphertheorie, und Interpretationsfehler. Linguistica Biblica (forthcoming).

--- (1979c) : A Method for the Evaluation of Recipient Response. Technical Notes for Bible Translators (forthcoming).

ENTERING LINGUISTIC ACTION THROUGH THE BACK DOOR

Jef Verschueren

(The form in which this paper was presented can by no means be regarded as final. It was merely a manifesto. Therefore I confine myself to publishing an abstract. A thoroughly revised version of the complete text will appear sooner or later as the introductory chapter to a longer work in preparation to be called "What people say they do with words".)

ABSTRACT

In order to gain insight into any form of social behavior a researcher can either operate with a preconceived framework of abstract and general concepts (the Front Door Method) or he can attempt to penetrate into the conceptual space that the participants in the action associate with it (the Back Door Method). In practice, neither of these methods exists in its pure form. Thus, the framework of abstract and general concepts with which the object of investigation can be approached is (consciously or unconsciously) based on the Front Door researcher's previous observations of concrete and particular phenomena and, hence, absolute universality can never be reached. On the other hand, the Back Door researcher will probably never succeed completely in giving up his habits of generalization; more importantly, an actor's conceptualization of his own actions can only be studied indurectly, for instance by scrutinizing the ways in which he talks about his actions or, on a higher level of generality, the ways in which his language allows him to talk about them. No doubt, some universals of behavior exist; therefore the Front Door Method is necessary. But since abstract and general concepts can never capture the full meaning of social phenomena which are themselves concrete and particular, and since there is a fundamental unity between the actor's concepts and the practices they serve to conceptualize, also the Back Door Method should be applied. A combination of the two methods may yield an optimal understanding of social behavior. Social studies have mainly used the Fornt Door Method, though some pleas for the alternative have been made (e.g. P. Winch, 1958, "The Idea of a Social Science and Its Relation to Philosophy", London: Routledge and Kegan Paul).

In this paper the above argument was further developed. Moreover, it was argued that linguistic action, which is no doubt a form of social behavior, has almost exclusively been studied by means of the Front Door Method (as represented, e.g. in Austin's and Searle's speech act theory and most of the linguistic studies based on it). Further, a cross-linguistic and cross-cultural study of the words and expressions which speakers of different languages have at their disposal to talk about linguistic actions was proposed as a first attempt to sneak in through the back door. Considerable time was devoted to the problems of how to describe the meaning of those words and expressions and how to avoid the Whorfian error of completely identifying words with concepts. By undertaking this type of research, a significant and totally new contribution to the study of linguistic action can be made. Briefly, whereas theorists of linguistic action have been trying to explain how people do things with words, I propose to undertake a study of what people say they do with words. Both studies are necessary to obtain an optimal understanding of the nature of linguistic action.

3. TEXTLINGUISTIK

REPRÄSENTATION VON TEXTSINN ALS WELTKOMPLEX

Claudia Biasci

Ich werde hier einen kurzen objektsprachlichen Text im Rahmen der Textstruktur-Weltstruktur Theorie (TeSWeST) analysieren, mit dem Ziel die Sinnstruktur des Textes zu rekonstruieren.

1. Einige Grundbegriffe der TeSWeST

Die TeSWeST operiert mit drei Grundelementen: (i) dem objektsprachlichen Text (der linearen Manifestation des Textes (TLim)), (ii) der Repräsentation der Sinnstruktur des Textes (der intensional-semantischen Textrepräsentation (TInR)), (iii) der Repräsentation der dem Text durch die Sinnstruktur zuordbaren extralinguistischen Korrelate (der welt-semantischen Repräsentation (WSeR), wobei der Ausdruck 'Welt' mit einem komplexen Sachverhalt synonym ist).

Die TeSWeST besteht aus drei Hauptkomponenten: der text-grammatischen Komponente, der weltsemantischen Komponente, der Lexikon-Komponente.

Ich werde mich hier ausschließlich mit der textgrammatischen Komponente und besonders mit der Sinnrepräsentation (TInR) eines Textes beschäftigen.

TInR ist gleich zu $\langle T^{\diamond}$, TwD$\rangle$ d.h. die Sinnrepräsentation eines objektsprachlichen Textes (O-Text) ist gleich zu dem Text ($T^{\diamond}$; der Struktur der Propositionen, aus denen der Text besteht) und der Beschreibung der Textwelten (TwD; die Welten haben in der TeSWeST eine derjenigen der Quantoren in der Prädikatlogischensprache ähnliche Funktion).

$T^{\diamond}$ besteht aus der Verbindung von Sinnrepräsentation von Textsätzen.

Die Sinnrepräsentation eines Textsatzes wiederum besteht aus drei Teilen (Propositionstypen):

i) der Beschreibung eines Sachverhaltes (deskriptiven Proposition, Bsp.: Keiner der Vorübergehenden beachtet den Soldaten,

ii) der Einstellung des Sprechers gegenüber diesem Sachverhalt (weltkonstitutiven Proposition), Bsp.: Ich sehe, daß keiner der Vorübergehenden den Soldaten beachtet,

iii) der performativen Modalität des Sprechers (performativen Proposition), Bsp.: Ich teile euch mit, daß ich sehe, daß keiner der Vorübergehenden den Soldaten beachtet.

Diese drei Teile können in Form einer Konfiguration von Propositionen dargestellt werden:

$$P^{Pc} :: P^{Ww} :: \mathcal{P}^{Dr}$$

Das Zeichen '::' markiert die Einbettung, P^{Pc} und P^{Ww} stehen für Propositionen, $\mathcal{P}^{Dr}$ für eine einzige Proposition oder eine Propositionenkonfiguration. Jede Proposition besteht aus einem Teil, der die lokale und temporale Informationen enthältet, und aus dem sogenannten Propositionskern.

P^{Pc} ist eine in ihrer modal-performativen Funktion gebrauchte kommunikative Proposition. Dieses bedeutet, daß der Zeitpunkt, der Ort und die kommunizierende Person als JETZT & HIER & ICH in der Proposition repräsentiert werden sollen.

P^{Ww} ist eine in ihrer weltkonstitutiven Funktion gebrauchte weltkonstitutive Proposition. Wie die performative, verlangt auch sie die Repräsentation des Zeitpunktes, Ortes und der kommunizierenden Person als JETZT & HIER & ICH.

$\mathcal{P}^{Dr}$ ist eine in deskriptiver Funktion gebrauchte Proposition oder Propositionenkonfiguration.

Es ist hier wichtig zu beachten, daß nur die Beschreibung des Sachverhaltes in den meisten Fällen explizit ist. Die anderen Informationen müssen der Beschreibung des Sachverhaltes vom Interpreten zugeordnet werden.

Die kanonische Sprache. Die Sprache der Sinnrepräsentation ist eine wohldefinierte (kanonische) Sprache. Auf die Prinzipien, auf die sie sich stützt, werde ich hier nicht eingehen.[1] Ich möchte nur betonen, daß die syntaktische Normierung in dieser Arbeit privilegiert wird. Was die semantische Normierung angeht, muß ich sagen, die semantische Einheiten, mit denen ich operiere, sind sehr nahe der natürlicher Sprache.

2. Die Sinnrepräsentation eines Textes. Ein Beispiel

2.1. Der O-Text

Der objektsprachliche Text, den ich analysieren werde, stammt aus La Storia von Elsa Morante. Zuerst gebe ich die originale Version auf Italienisch und dann meine deutsche Übersetzung an. Es gibt zwar eine Übersetzung des Romans, sie ist aber für mein Ziel unbrauchbar.[2]

> Erano circa le due del dopopranzo, e a quell'ora, come d'uso, poca gente circolava per le strade. Nessuno dei passanti poi, guardava il soldato, perché i Tedeschi, pure se camerati degli Italiani nella corrente guerra mondiale, non erano popolari in certe periferie proletarie. (MORANTE 1974: 15)

> Es war ungefähr zwei Uhr des Nachmitags, und in jener Zeit, wie üblich, waren wenige Leute auf den Straßen. Außerdem schaute keiner der Vorübergehenden den Soldaten an, weil die Deutschen, obwohl sie im laufenden Weltkrieg Kamerad der Italiener waren, nicht populär in gewissen proletarischen Vororten waren.

Einfachheitshalber habe ich einige Umformungen vorgenommen und einiges außer Betracht gelassen. Ich gebe hier die deutsche Version des von mir analysierten objektsprachlichen Textes an:

(OT_1) Wenige Leute waren auf den Straßen, weil es ungefähr zwei Uhr war. Außerdem schauten die Vorübergehenden den Soldaten nicht an, weil die Leute (= die Vorübergehenden) in gewissen proletarischen Vororten die deutschen (= den Soldaten) nicht gern hatten.

2.2. Die Propositionenstruktur ($T^{\diamond}$)

In diesem Abschnitt gebe ich die Repräsentation des ersten Elementes des Tupel $\langle T^{\diamond}, TwD \rangle$ an.

In (R1) wird die innere Struktur der Propositionen angegeben, die in Fig.1 anhand einer Abbildung dargestellt wird. Was (R1) betrifft, muß man dazu sagen, daß sie vereinfacht und nicht formal vollständig ist. Es fehlt zuerst die Beschreibung einer Kommunikationssituation, die die Parameter für die Interpretation der in (R1) anwesenden ICH, JETZT und HIER Elemente angibt. Es fehlen ferner die Argumentrollenindikatoren, die Topikalisierungsoperation wurde nicht durchgeführt und deshalb ist die Negation in einfacher Art (vgl. (20b) und (29b)) betrachtet. Die Namen- und Individuenindizes sind, um die Lektüre zu vereinfachen, direkt in (R1) angegeben und stehen aus Einfachheitsgründen nicht zwischen den '§' Zeichen (solche Zeichen markieren diejenigen

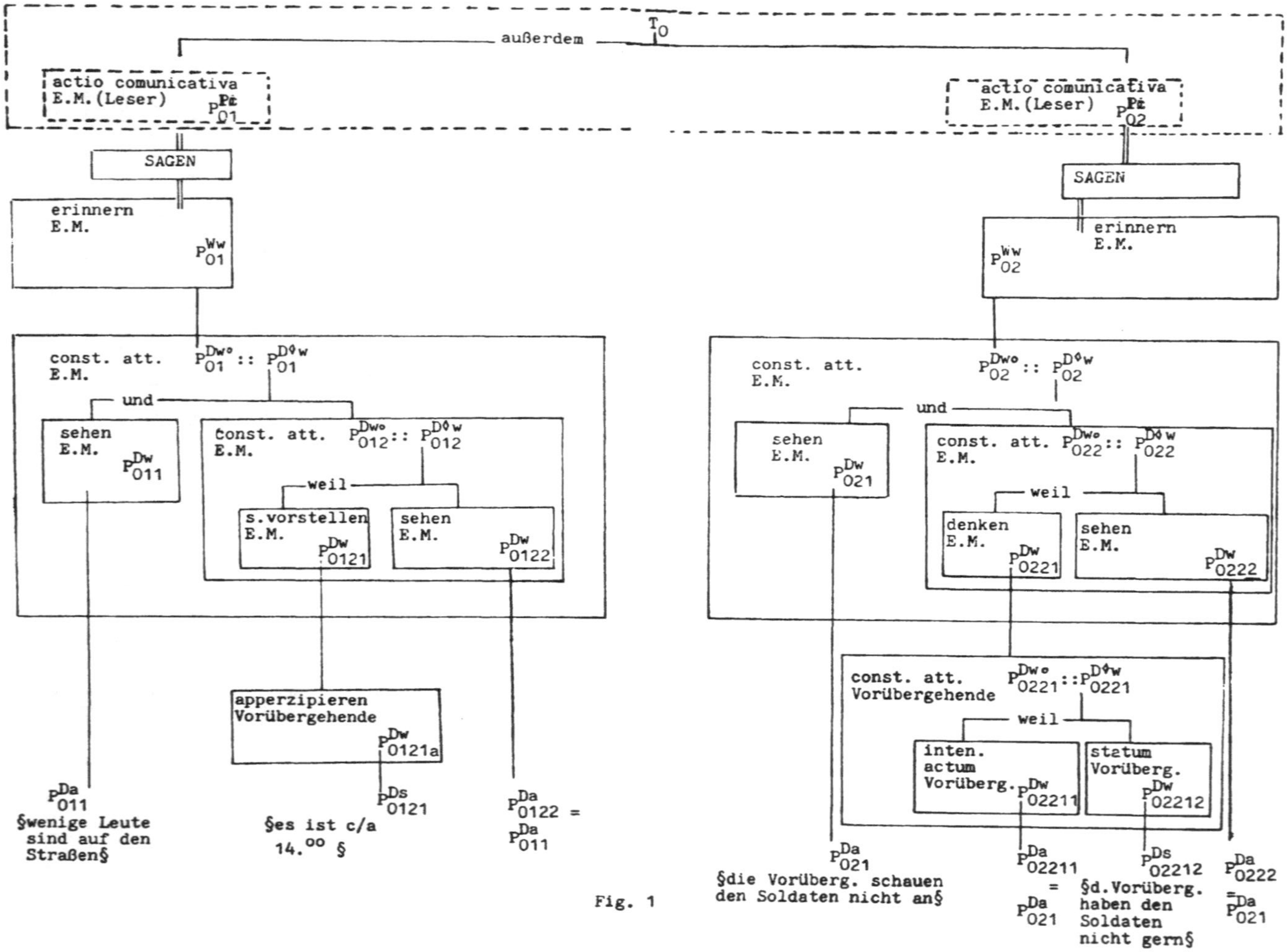

T_0
außerdem
actio comunicativa E.M.(Leser) P_{01}^{Pt}
actio comunicativa E.M.(Leser) P_{02}^{Pt}
SAGEN
SAGEN
erinnern E.M. P_{01}^{Ww}
erinnern E.M. P_{02}^{Ww}
const. att. E.M. $P_{01}^{Dw\circ} :: P_{01}^{D\phi w}$
und
sehen E.M. P_{011}^{Dw}
const. att. E.M. $P_{012}^{Dw\circ} :: P_{012}^{D\phi w}$
weil
s.vorstellen E.M. P_{0121}^{Dw}
sehen E.M. P_{0122}^{Dw}
apperzipieren Vorübergehende P_{0121a}^{Dw}
P_{011}^{Da}
§wenige Leute sind auf den Straßen§
P_{0121}^{Ds}
§es ist c/a 14.^{oo} §
P_{0122}^{Da} = P_{011}^{Da}
const. att. E.M. $P_{02}^{Dw\circ} :: P_{02}^{D\phi w}$
und
sehen E.M. P_{021}^{Dw}
const. att. E.M. $P_{022}^{Dw\circ} :: P_{022}^{D\phi w}$
weil
denken E.M. P_{0221}^{Dw}
sehen E.M. P_{0222}^{Dw}
const. att. Vorübergehende $P_{0221}^{Dw\circ} :: P_{0221}^{D\phi w}$
weil
inten. actum Vorüberg. P_{02211}^{Dw}
statum Vorüberg. P_{02212}^{Dw}
P_{021}^{Da}
§die Vorüberg. schauen den Soldaten nicht an§
P_{02211}^{Da} = P_{021}^{Da}
P_{02212}^{Ds}
§d.Vorüberg. haben den Soldaten nicht gern§
P_{0222}^{Da} = P_{021}^{Da}

Fig. 1

Einheiten, die nicht wohlgeformt sind).

(R1) (1)	T_0	:=:	$P_{01}^{\mathbb{P}\mathfrak{c}}$ außerdem $P_{02}^{\mathbb{P}\mathfrak{c}}$	
(2a)	$P_{01}^{\mathbb{P}\mathfrak{c}}$	:=:	dFi	JETZT/=t$\mathfrak{c}$01/,HIER/=l$\mathfrak{c}$01/,$p_{01}^{\mathfrak{c}}$
(2b)	$p_{01}^{\mathfrak{c}}$	:=:	sagen	ICH/=E.M./,EUCH/=Leser/,P_{01}^{Ww}
(3a)	P_{01}^{w}	:=:	dFi	JETZT/=tw01/,HIER/=lw01/,p_{01}^{w}
(3b)	p_{01}^{w}	:=:	erinnern	ICH/=E.M./, $P_{01}^{Dw\bullet}$
(4a)	$P_{01}^{Dw\bullet}$	:=:	dFi	tDw•01, lDw•01, $p_{01}^{Dw\bullet}$
(4b)	$p_{01}^{Dw\bullet}$	:=:	const. att.	E.M., $P_{01}^{D\diamond w}$
(5)	$P_{01}^{D\diamond w}$	:=:	P_{011}^{Dw} und $P_{012}^{Dw\bullet}$	
(6a)	P_{011}^{Dw}	:=:	dFi	tDw011, lDw011, p_{011}^{Dw}
(6b)	p_{011}^{Dw}	:=:	sehen	ich/=E.M./, P_{011}^{Da}
(7a)	P_{011}^{Da}	:=:	dFi	tDa011,lDa011/=auf den Straßen/, p_{011}^{Da}
(7b)	p_{011}^{Da}	:=:	sein	wenige Leute

(8a)	$P_{012}^{Dw\bullet}$	:=:	dFi	tDw•012, lDw•012, $p_{012}^{Dw\bullet}$
(8b)	$p_{012}^{Dw\bullet}$	:=:	const. att.	E.M., $P_{012}^{D\diamond w}$
(9)	$P_{012}^{D\diamond w}$	:=:	P_{0121}^{Dw} weil P_{0122}^{Dw}	
(10a)	P_{0121}^{Dw}	:=:	dFi	tDw0121, lDw0121, p_{0121}^{Dw}
(10b)	p_{0121}^{Dw}	:=:	s. vorstellen	ich/=E.M./, P_{0121a}^{Dw}
(11a)	P_{0121a}^{Dw}	:=:	dFi	tDw0121a, lDw0121a, p_{0121a}^{Dw}
(11b)	p_{0121a}^{Dw}	:=:	apperzipieren	Vorübergehende, P_{0121}^{Ds}
(12a)	P_{0121}^{Ds}	:=:	dFi	tDs0121, lDs0121, p_{0121}^{Ds}
(12b)	p_{0121}^{Ds}	:=:	identisch	tDs0121 und c/a 14.^{00}

(13a)	P_{0122}^{Dw}	:=:	dFi	tDw0122, lDw0122, p_{0122}^{Dw}
(13b)	p_{0122}^{Dw}	:=:	sehen	ich/=E.M./, P_{0122}^{Da}
(14)	P_{0122}^{Da}	=	P_{011}^{Da}	

==

(15a)	$P^{P¢}_{02}$	:=:	dFi	JETZT/=t¢02/,HIER/=l¢02/,$p^{¢}_{02}$
(15b)	$p^{¢}_{02}$	:=:	sagen	ICH/=E.M./,EUCH/=Leser/,P^{Ww}_{02}
(16a)	P^{w}_{02}	:=:	dFi	JETZT/=tw02/,HIER/=lw02/,p^{w}_{02}
(16b)	p^{w}_{02}	:=:	erinnern	ICH/=E.M./, $P^{Dw\circ}_{02}$
(17a)	$P^{Dw\circ}_{02}$	:=:	dFi	tDw∘02, lDw∘02, $p^{Dw\circ}_{02}$
(17b)	$p^{Dw\circ}_{02}$	:=:	const. att.	E.M., $P^{D\Diamond w}_{02}$
(18)	$P^{D\Diamond w}_{02}$	:=:	P^{Dw}_{021} und $P^{Dw\circ}_{022}$	
(19a)	P^{Dw}_{021}	:=:	dFi	tDw021, lDw021, p^{Dw}_{021}
(19b)	p^{Dw}_{021}	:=:	sehen	ich/=E.M./, P^{Da}_{021}
(20a)	P^{Da}_{021}	:=:	dFi	tDa021, lDa021, p^{Da}_{021}
(20b)	p^{Da}_{021}	:=:	¬anschauen	die Vorübergehenden,den Soldaten

---

(21a)	$P^{Dw\bullet}_{022}$	:=:	dFi	tDw•022, lDw•022, $p^{Dw\bullet}_{022}$
(21b)	$p^{Dw\circ}_{022}$	:=:	const. att.	E.M., $P^{D\Diamond w}_{022}$
(22)	$P^{D\Diamond w}_{022}$	:=:	P^{Dw}_{0221} weil P^{Dw}_{0222}	
(23a)	P^{Dw}_{0221}	:=:	dFi	tDw0221, lDw0221, p^{Dw}_{0221}
(23b)	p^{Dw}_{0221}	:=:	denken	ich/=E.M./, $P^{Dw\bullet}_{0221}$
(24a)	$P^{Dw\bullet}_{0221}$	:=:	dFi	tDw•0221, lDw∘0221, $p^{Dw\circ}_{0221}$
(24b)	$p^{Dw\circ}_{0221}$	:=:	const. att.	Vorübergehende, $P^{D\Diamond w}_{0221}$
(25)	$P^{D\Diamond w}_{0221}$	:=:	P^{Dw}_{02211} weil P^{Dw}_{02212}	
(26a)	P^{Dw}_{02211}	:=:	dFi	tDw02211, lDw02211, p^{Dw}_{02211}
(26b)	p^{Dw}_{02211}	:=:	intend. actum	Vorübergehende, P^{Da}_{02211}
(27)	P^{Da}_{02211}	=	P^{Da}_{021}	
(28a)	P^{Dw}_{02212}	:=:	dFi	tDw02212, lDw02212, p^{Dw}_{02212}
(28b)	p^{Dw}_{02212}	:=:	statum	Vorübergehende, P^{Ds}_{02212}
(29a)	P^{Ds}_{02212}	:=:	dFi	tDs02212, lDs02212/=in gewissen proletarischen Vororten/, p^{Ds}_{02212}
(29b)	p^{Ds}_{02212}	:=:	¬gern haben	die Vorübergehende, den Soldaten

---

(30a) P^{Dw}_{0222} :=: dFi tDw0222, lDw0222, p^{Dw}_{0222}

(30b) p^{Dw}_{0222} :=: sehen ich/=E.M./, P^{Da}_{0222}

(31) P^{Da}_{0222} = P^{Da}_{021}

Symbolliste:

◊	Diamantzeichen. Es markiert komplexe Einheiten,
:=:	Bidirektionales Rückschreibenzeichen,
dFi	Funktor eines propositionsbildenden Prädikatrahmens. Es entspricht dem objektsprachlichen Ausdruck "es ist der Fall, daß",
∘	Kringel. Es markiert ein Hyperonym,
¬	Negationszeichen.

2.3. Die Text-Weltbeschreibung (TwD)

In diesem Abschnitt wird die Beschreibung der Welten des Textes angegeben. Einfachheitshalber habe ich folgenden einfachen Satz ausgewählt: wenige Leute waren auf den straßen. Ferner habe ich angenommen, daß dieser Satz von einem Sprecher (A) zu einem Hörer (B) gesagt und von einem Interpreten (I) gehört wird. Der Interpret glaubt/denkt/betrachtet, daß es der Fall ist bzw., daß es nicht der Fall ist, daß die verschiedenen Welten/Sachverhalte ein Faktum/Fiktum für A sind. Vgl. Fig. 2.

Momentan wird die Verbindung zwischen 'bI' und 'Fk.A' (A in allen acht Fällen) unvollständig gelassen. Eine der folgenden weltkonstitutiven Einheiten könnte dafür gebraucht werden: 'wissen', 'überzeugt sein', 'sich vorstellen', usw..

Was 'eA', 'sA', 'D' angeht, ist es noch zu überlegen, ob sie selbst als Sachverhaltbeschreibung betrachtet werden sollen (d.h. (eA): A erinnert sich daran, daß er gesehen hat, daß wenige Leute auf den Straßen sind; (sA): A sieht, daß wenige Leute auf den Straßen sind; (D): wenige Leute sind auf den Straßen.). Ich gehe davon aus, daß sie als solche betrachtet werden sollen.

Die Beschreibung der Welten des Textes besteht aus der Darstellung der Relationen, die zwischen den einzelnen Sachverhalten und der 'Wissen/Überzeugung/Vorstellung'-Welt von A bestehen/be-

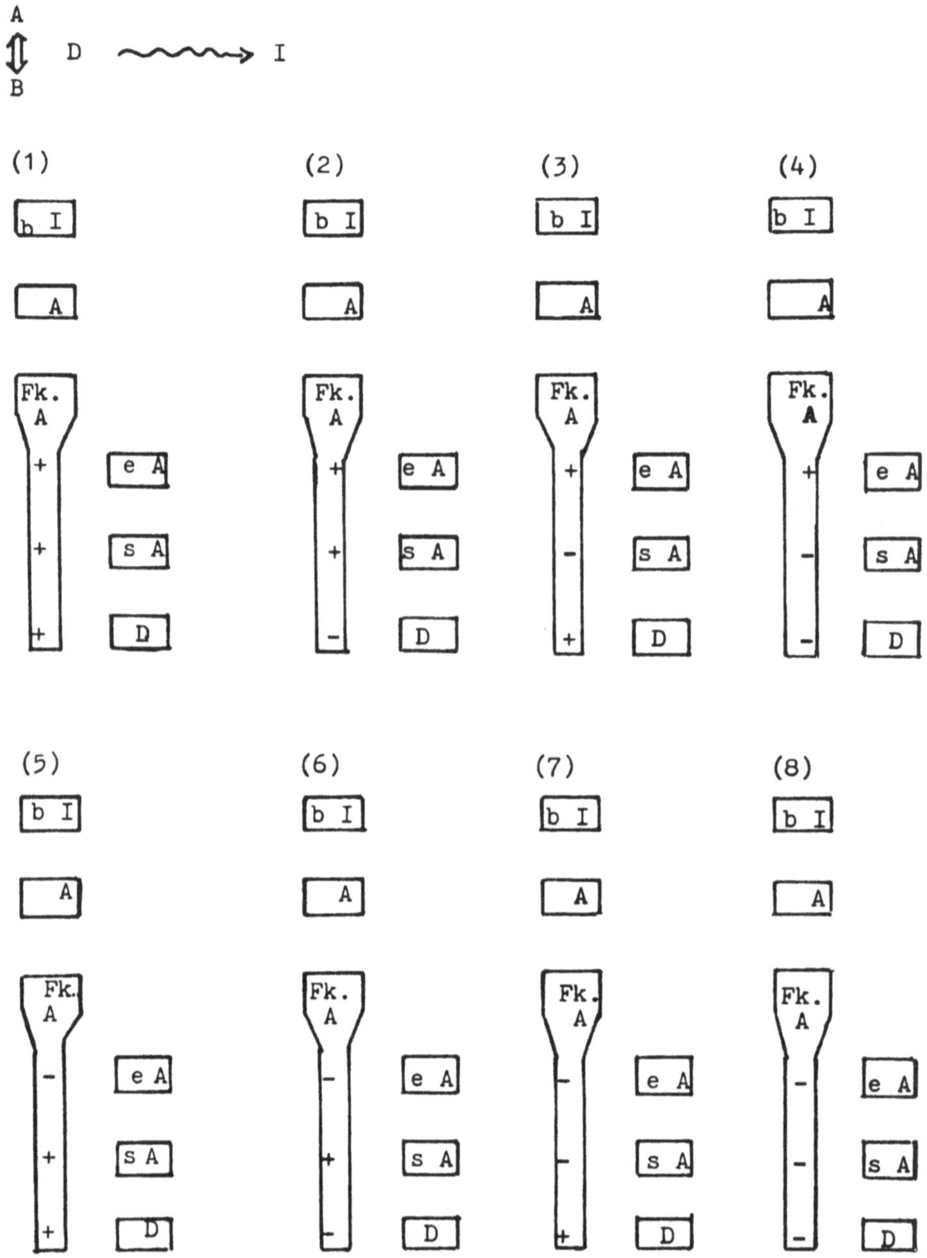

Fig. 2

<u>Anmerkungen</u>

A: Sprecher, B: Hörer, I: Interpret, D: <u>Wenige Leute waren auf den Straßen</u> . Für die Abkürzungen der Welten vgl. die Lesarthinweise .

Lesearthinweise:

(1) ((bI) Betrachtung des Interpreten)

(e A) I betrachtet/glaubt, daß es der Fall ist, daß es ein Faktum/Fiktum (Fk.A) für A ist, daß er sich daran erinnert, daß...

(s A) I betrachtet/glaubt, daß es der Fall ist, daß es ein Faktum/Fiktum für A ist, daß er gesehen hat, daß...

(D) I betrachtet/glaubt, daß es der Fall ist, daß wenige Leute auf den Straßen waren.

(2) (e A) w. o.

(s A) w. o.

(D) I betrachtet/glaubt, daß es nicht der Fall ist, daß wenige Leute auf den Straßen waren.

(3) (e A) w. o.

(s A) I betrachtet/glaubt, daß es nicht der Fall ist, daß es ein Faktum/Fiktum für A ist, daß er gesehen hat, daß...

(D) I betrachtet/glaubt, daß es der Fall ist, daß wenige Leute auf den Straßen waren.

usw.

stehen können.

Es ist zuletzt zu bemerken, daß hier zwei Verfahren möglich sind:

(i) (1) bis (8) als verschiedene TwD zu betrachtetn,

(ii) ausschließlich (1) als TwD zu betrachten, nach dem der Konversation grundlegenden Ehrlichkeitsprinzip, und (2) bis (8) als Möglichkeiten zu benutzen in der Bildung der Modelle, die man in der weltsemantischen Interpretation nimmt.

Anmerkungen

1 Vgl. BIASCI/FRITSCHE 1978.

2 Ich gebe hier die literarische Übersetzung an:

> Es war etwa zwei Uhr, und wie gewöhnlich waren um diese Zeit nur wenige Leute auf der Straße. Keiner der Vorübergehenden beachtet den Soldaten. Die Deutschen waren in diesem Krieg zwar mit den Italienern verbündet, aber in gewissen Arbeitervororten waren sie nicht gern gesehen. MORANTE 1978: 16.

Literatur

BIASCI, Claudia / FRITSCHE, Johannes (Hrsg.) (1978): Texttheorie-Textrepräsentation. Theoretische Grundlagen der kanonischen sinnsemantischen Repräsentation von Texten. Hamburg: Buske.

MORANTE, Elsa (1974): La Storia. Turin: Einaudi.

--- (1978): La Storia. Frankfurt/Main: Fischer Taschenbuch.

PETÖFI, János, S. (1976): Structure and Function of the Grammatical Component of the Text-Structure World-Structure Theory. GUENTHNER, F./ SCHMIDT, S.J. (eds) (in press): Formal Semantics amd Pragmatics fpr Natural Languages. Dordrecht: Reidel.

--- (1978): A Text Theory as a Model for Natural Language Processing. Bergen, C O L I N G 78.

STRUCTURE LINGUISTIQUE ET STRUCTURE POETIQUE

Decreus Freddy

1. Les analyses structurales effectuées par Jakobson témoignent d'une vision particulière de la relation entre la linquistique et la poétique : cette poétique est une partie intégrale de la linguistique et se définit comme "l' étude linguistique de la fonction poétique dans le contexte des messages verbaux en général et dans la poésie en particulier " [1]. N ' acceptant pas que la critique littéraire s'empare des configurations surtout d'ordre phonologique et grammatical , Jakobson voulait à tout prix les insérer dans un modèle linguistique à 6 fonctions, dont une , la fonction poétique , rendrait suffisamment compte des données d'une recherche analytique aux différents niveaux linguistiques. Les mérites de Jakobson dans ce type de recherches sont immenses, mais les dernières études structuralistes ont formulé des réserves sérieuses concernant tant la "valeur poétique" que la notion de "structure" dont Jakobson s'est servi. Le grand problème non résolu est sans doute la pertinence des équivalences trouvées : une analyse linguistique est en effet à même de présenter des quantités de relations d'équivalence [2] et ce suivant la définition bien connue de la fonction poétique:

> "La fonction poétique projette le principe d'équivalence de l'axe de la sélection sur l'axe de la combinaison" (1963: 222).

Une description "complète" de la distribution des diverses classes phonologiques, morphologiques, syntaxiques et lexicales peut en effet offrir des matrices exhaustives que n'importe quel linguiste est capable de rédiger. Du moment pourtant où l'on passe de cette activité analytique à une activité synthétique dans laquelle on détermine la valeur poétique de toutes ces constatations, les résultats de différents linguistes seront loin d'être identiques. A un moment donné en effet, on franchit les limites d'une activité classificatoire et descriptive propre à la linguistique structurale pour arriver à une pratique qui construit selon d'autres principes. La linguistique , dit CULLER (1975:65),"does not provide a method for the discovery of poetic patterns" . Tout compte fait, on peut se poser la question suivante : si ce n'est pas la fonction poétique , qui ou quoi donc projette ce fameux principe d'équivalence de l'axe de la sélection sur l'axe de la combinaison ?

2. Le même poème qui a servi à Jakobson comme pierre de touche pour une de ses investigations structurales, à savoir "Les Chats" de Baudelaire, fut l'objet d'une autre recherche, d'inspiration plutôt behavioriste. L' analyse de RIFFATERRE [3] se basa en effet sur la réception du lecteur ; dans le modèle linguistique de Jakobson ce fut plutôt le destinataire qui était favorisé. R. SCHOLES dit à ce sujet :

> "Having found the Jakobson / Lévi-Strauss reading seriously wanting and resulting in the fraudulent construction of a "superpoem", Riffaterre then presents his own alternative version, answering the "Superpoem" with a "Superreader" (1974: 35).

Les analyses de Jakobson et de Riffaterre datent respectivement des années 1962 et 1966. Leurs travaux se situent à une époque dans laquelle, à la suite des travaux des formalistes russes , on se pose à nouveau la question fameuse des relations entre la linguistique et la critique littéraire. Situées dans une époque où triompha la linguistique structurale américaine et où la linguistique se déclarait parfois incompétente à intégrer les dimensions sémantiques dans son système "objectif et scientifique", ces deux voies de recherche étaient assez incompatibles. D'autre part, dans la critique littéraire même (par ex. le New Criticism) l'attitude fondamentale restait imprégnée de subjectivisme, et ce en dépit du fait que l'oeuvre littéraire - lue par le close reader - fut regardée comme une entité autonome et que les recherches se portèrent sur le texte même [4] .

3. De la même époque datent les analyses de N. RUWET [5] . Il fut un des premiers à reconnaître la grande valeur du livre de LEVIN (1962), Linguistic Structures in Poetry , une étude importante qui propageait la notion de "couplage", une suite logique des

idées de Jakobson. Dans un article paru en 1965 (intitulé "Sur un vers de Charles Baudelaire") , Ruwet mit bon ordre à la pratique de l'analyse structurale. Voici les quatre principes qu'il retint de la linguistique structurale:

(1) la conscience que l'analyse doit porter non sur des termes, mais sur des relations.

(2) la reconnaissance de la notion de niveau linguistique.

(3) la volonté de traiter les problèmes de la substance (au sens hjelmslévien) après avoir résolu les problèmes de forme.

(4) la notion d'équivalence (issue de la définition de la fonction poétique de Jakobson) [6] .

Muni de presque les mêmes concepts méthodologiques que Jakobson, Ruwet ne dépassa guère les résultats obtenus par son éminent collègue américain. Dans un article intitulé " Limites de l'analyse linguistique en poétique", il aboutit à la conclusion suivante:

"La linguistique peut apporter à la poétique un certain nombre de matériaux, mais elle est incapable, à elle seule, de déterminer dans quelle mesure ces matériaux sont pertinents du point de vue poétique ou esthétique" (1972: 211).

D'autre part, dans ses analyses quelques problèmes restaient hors d'une discussion théorique, bien qu'ils fussent employés dans les interprétations concrètes. Les équivalences formelles se combinaient sans beaucoup de rigueur avec les équivalences sémantiques : elles "servaient d'appui", elles "soulignaient " [7] . Les procédures de décision et d'interprétation restaient celles de la linguistique structurale, une linguistique qui, par une application directe de ses méthodes voulait résoudre des problèmes qui se situent hors du domaine qu'elle s'était délimité. La critique fondamentale de RIFFATERRE est toujours valable : "Aucune analyse grammaticale d'un poème ne peut nous donner plus que la grammaire du poème " [8] . Ce type d'analyse structurale se fait souvent par une étude séparée des divers niveaux linguistiques, suivie de conclusions qui ne se basent pas nécessairement sur les matériaux retenus.

4. Dans son article "The linguistic study of literary texts" HALLIDAY (1964:303) proposa les termes "linguistic stylistics" et les définia comme " the description of literary texts, by methods derived from general linguistic theory, using the categories of the description of the language as a whole", mais l'emploi qu'il fit du terme "cohesion" (un regroupement de différents éléments grammaticaux et lexicaux) resta trop exclusivement linguistique. L'activité du même nom (linguistic stylistics) qu'on retrouve chez HENDRICKS (1974: 5-22) dans son article " The relation between linguistics and literary studies" témoigna d'une tout autre conception : il distingua très nettement la stylistique linguistique de la critique littéraire et proposa le terme "stylolinguistics" pour indiquer " the study of the relations between language structure and literary structure ". Mais c'est à FOWLER (1971: 38) que nous devons l'élaboration plus précise de cette relation entre la linguistique et la critique littéraire. Il suggéra trois niveaux d'achèvement : " description, stylistics and criticism " et motiva le passage du premier au second niveau comme suit : " To pass on to stylistics, we must point to patterns which are meaningful ... , ... because they are significant in a comparative context ". Dans le passage au dernier niveau on doit se rendre compte, dit-il, que la signification d'un poème. est plus qu'une simple addition de significations cognitives et formelles et que " perhaps some of the causes of this meaning and value are inaccessible to verbal analysis" . Cette dernière remarque surtout nous paraît significative, puisqu'elle amène à reconsidérer la fonction poétique, telle qu'elle a été formulée par Jakobson. Une des grandes mérites de l'oeuvre de Robert Scholes " Structuralism in Literature" fut sans doute de modifier le modèle linguistique à 6 fonctions de Jakobson, de façon que le contexte acquérait une place plus justifiée dans l'étude de la formation d'une "signification poétique". Il fit ressortir la différence entre communication écrite et communication orale : dans une communication écrite, dit-il, le message est souvent le seul élément à notre disposition. "Poetry may be definable precisely in terms of our having to supply the missing elements in an act of communication ". La

fictivité dans la littérature peut en outre être définie comme un contexte distant. Ainsi, en ce qui concerne la poésie, on aurait affaire à un contexte double ou multiple et on devrait aussi tenir compte d'une duplicité dans les destinateurs et les destinataires. De cette façon, Scholes élargit la fonction poétique de Jakobson, quand il suggéra

> " that the visible signs of literature are to be found at least as frequently in this multiplicity of the nonverbal elements of communication as they are in any special qualities in the grammar and syntax of the message itself. And furthermore, that this duplicity often manifests itself in the message in the form of irony, paradox, and other devices which require for their analysis extended consideration of the full semantic range of the poetic utterance " (1974 : 28).

5. Pour sortir de ce vieux paradoxe de la littérature (j'entends par là les catégories littéraires donnant l'impression d'échapper aux catégories linguistiques dans lesquelles elles sont pourtant communiquées), il suffit d'accepter une dimension littéraire, qui est contenue dans une "hyper-structure" (HENDRICKS 1974: 18) , distincte de la structure linguistique proprement dite. Les trois paragraphes suivantes traiteront des tentatives pour créer une science de la littérature fondée sur <u>le modèle de la linguistique</u> et non plus sur une application directe de ses méthodes.

5. 1. Une des propositions les plus intéressantes à notre avis, mais qui n'a guère été suivie fut l'emploi des 4 catégories hjelmsléviennes par Hans SÖRENSEN [9] . Dans un article qui date déjà de 1958, "Littérature et linguistique", Sörensen proposa l'analogie suivante: du point de vue de la langue littéraire

(1) la langue (proprement dite) figure comme <u>la substance de l'expression</u>;

(2) comme <u>forme de l'expression</u> figurent les éléments stylistiques (= une mise en forme de la substance de l'expression avec un but esthétique) ;

(3) <u>la forme du contenu</u> est représentée par le motif (particularisé à travers les thèmes et résultant dans une composition d'ensemble , comme c'est le cas avec le genre);

(4) la substance du contenu peut être traduite dans l'idée de l'oeuvre d'art, le motif combiné avec un symbole, désignant " ce que le poète a voulu dire".

Cette division, qui a connu peu de succès, rend possible en tout cas une description complète de l'oeuvre littéraire. Un point de vue assez semblable, mais issue d'une toute autre direction de recherche a été fourni par l'esthéticien polonais Roman Ingarden, qui déjà en 1931, discerna dans l'oeuvre littéraire les niveaux suivants (INGARDEN 1972: 26) :

(1) die Schicht der Wortlaute ;
(2) die Schicht der Bedeutungseinheiten ;
(3) die Schicht der Ansichten ;
(4) die Schicht der dargestellten Gegenständlichkeiten.

5. 2. Ces combinaisons d'éléments linguistiques et extra-linguistiques se révèlent des plus effectives lorsqu'elles définissent en plus la littérature comme un système significatif au second degré, c.-à-d. lorsqu'elles comprennent le poème comme un système connotatif et lorsqu'elles définissent le language poétique comme un langage avec un degré d'organisation plus complexe et plus créatif. La conception de la poésie comme " système modélisant secondaire " remonte aux études sémiotiques de Iouri Lotman [10]. Les systèmes modélisants secondaires, dit Lotman, se construisent sur le type du langage et ne reproduisent donc pas tous les aspects des langues naturelles. Dans la poésie, la notion du signe même est différente : en effet, le signe n'a plus de caractère conventionnel (construit sur l'indépendance de l'Expression et du Contenu), mais il est de nature iconique, figurative, c.-à-d. construit sur le principe de la dépendance entre l'Expression et le Contenu. De cette façon, non seulement les signes se limitent différemment, mais la délimitation entre les niveaux de l'Expression et du Contenu devient aussi une chose très complexe. Les éléments de la langue naturelle (phonèmes, morphèmes, ...) peuvent se sémantiser et deviennent à leur tour porteurs de sens. Cette orientation sémiologique permet à Lotman de reformuler la notion d'équivalence : il ne pose pas la question si une équivalence donnée est de nature

poétique ou non, il étudie surtout leur intensité: " die höhere Verknüptheit impliziert nicht nur eine grössere Relevanz, sondern auch eine grössere semantische Sättigung ", dit-il. D'autre part, le texte poétique présente encore autre chose qu'une hiérarchie des niveaux linguistiques : en tant que " langage d'art il contient aussi " une hiérarchie complexe de langues mutuellement corrélés, mais non semblables". Il en suit la pluralité fondamentale de lectures possibles: dans la même "structure" une lecture suivant différents codes sera possible et aboutira à différentes interprétations. C'est à cet instant même qu'on ressent nettement la différence entre une structure linguistique et une structure d'ordre littéraire. Dans la linguistique structurale, on peut en effet aisément reconnaître toutes les unités présentes et calculer leurs combinaisons possibles. Mais, par contre, une structure littéraire, par l'insertion de configurations de nature extra-linguistique, est condamnée à faire éclater les contraintes des structures strictement linguistiques.

5. 3. Une autre façon d'aborder la transition entre la description linguistique d'un texte et la critique littéraire a été rendue possible par la notion de la " compétence littéraire" . En effet, quand on est à la recherche d'éléments littéraires, on se sert de certaines intuitions littéraires, d'intuitions stylistiques, " a rather loosely structured, but often reliable feeling for the quiddity of a writer's linguistic method "[11] . Tout comme la linguistique de Chomsky s'efforce de ramener les intuitions du "native speaker" concernant son langage aux configurations spéciales des éléments linguistiques, cette compétence littéraire se pose comme but de corréler ces intuitions littéraires avec des éléments linguistiques appartenant au modèle systématisant secondaire. Les oeuvres de Barthes et le livre " Structuralist Poetics " de Jonathan Culler représentent une tentative sérieuse pour éclairer ce problème. Culler souligne très fort que l'oeuvre n'a du sens que lorsqu'elle est lue d'une façon spéciale, qui actualise les caractéristiques latentes de l'objet. La tâche qu'il accorde à la "poétique structurale" est de formuler explicitement ce qui est présent d'une façon impli-

cite dans le système sous-jacent de la littérature. Cette poétique structurale ne devient pas une science des contenus, mais une science des conditions de la signification, qui n'interprète pas les symboles, mais en décrit la polyvalence. Cette compétence littéraire se manifeste dans la façon dont on a interiorisé "la grammaire de la littérature" et étudie spécialement la découverte et l'emploi des conventions, par lesquelles l'oeuvre devient intelligible.

Dans cette revue d'études, qui se veulent toutes "structurales", on pourrait distinguer trois acceptions successives de la notion "structure". Tandis que son organisation reste invariable, on a pu constater que sa fonction a changé à trois reprises:

(1) dans une première acception, celle de la linguistique structurale, une structure se définit comme " une entité autonome de dépendances internes" ; elle a pour fonction de rendre compte de l'organisation de la langue.

(2) dans la conception de la poésie comme un modèle modélisant secondaire, le procès de structuration est plus complexe, puisqu'on s'intéresse surtout à la créativité avec laquelle la poiesis joue pour construire des relations sémiotiques "secondaires". Cette deuxième structure est demi-ouverte, puisqu'elle intègre bon nombre d'éléments de nature extra-linguistique.

(3) dans un troisième instant, on tient compte du rôle que joue le langage dans notre vie. La structure, en tant que hiérarchie de codes culturels, devient une structure ouverte et ce dans la mesure où elle nous renvoie à une activité symbolisante de la part de l'auteur. La poétique structurale nous confronte avec les modes de fonctionnement du texte, avec les potentialités que peut réaliser l'homo significans.

Puisque ces trois définitions se rapportent à la pluralité fonctionnelle du langage même, on peut envisager de reconsidérer l'emploi que nous avons fait du mot 'linguistique". Toutes les opérations que nous avons décrites ci-dessus relèvent en effet de l'emploi de notre "faculté de langage"; nous partageons ainsi les idées de JAKOBSON concernant les différentes fonctions linguistiques, mais seulement après avoir élargi sa conception de la fonction poétique.

Notes

1 JAKOBSON (1973: 486) ; dans le "Postscriptum" de ses "Questions de Poétique" il réagit contre certains "préjugés" qu'il a rencontrés dans les critiques de son oeuvre. Il y revient assez souvent sur le rôle même de l'étude linguistique; il dit e.a.: " Ainsi l'idée que l'étude linguistique est enfermée dans les limites étroites de la phrase, et par conséquent rend le linguiste incapable d'examiner la composition des poèmes, se trouve contredite par l'analyse du discours comme l'une des tâches mises de nos jours au premier plan dans la science linguistique" (JAKOBSON 1973: 485-486).

2 DELBOUILLE (1968: 7-22) ; MOUNIN (1968: 155-160) ; ANBEEK (1972: 148-161) ; POSNER (1974: 224-266).

3 RIFFATERRE (1971: 307-364); il justifia son entreprise comme suit:" Ils (sc. Jakobson et Lévi-Strauss) déclarent modestement qu'ils ne se sont intéressés qu'à décrire ce dont le poème est fait. Pourtant, ils n'en tirent pas moins des conclusions sur la signification du poème et tentent de le rattacher à l'esthétique et même à la psyché du poète,... . Une question se pose alors: comment pouvons-nous passer de la description au jugement -- autrement dit de l'étude d'un texte à celle de ses effets sur le lecteur ?" (RIFFATERRE 1971: 309-310).

4 cfr. COHEN (1972: 217-243) ; MAATJE (1974:43-66).

5 RUWET (1972: 151-247).

6 RUWET (1972: 200) ; pour une discussion de ces principes : CULLER (1975: 65-74).

7 POSNER (1974: 235-249).

8 RIFFATERRE (1971: 325); cfr.(1971: 315-325: Non-pertinence de la grammaire).

9 SORENSEN (1958: 182-197) ; HILL (1955: 968-978) proposa d'utiliser la tripartition de la linguistique, faite par TRAGER (prelinguistics ; microlinguistics ; metalinguistics) pour l'analyse littéraire : il discerna " a preliterary level ; a microliterary level ; a metaliterary level (= all those patterned activities outside the poem, but correlated with it").

10 LOTMAN (1972) ; LOTMAN (1973) ; HARDT (1976: 19-33):"Für eine semiotische Konzeption des poetischen Textes".

11 HENDRICKS (1974: 11): citation d'Ohmann.

Littérature

ANBEEK,T. (1972): "De tastbaarheid van de tekens. Over de poëzie-analyse van Roman Jakobson ". Forum der Letteren 13:3: 148-161.

BARTHES, Roland (1964): "L'activité structuraliste". Essais critiques. Paris: Seuil: 213-220.

--- (1968): "L'effet du réel". Communications 11: 84-89.

--- (1970): S/Z . Paris:Seuil.

--- (1972): Le degré zéro de l'écriture, suivi de nouveaux essais critiques. Paris:Seuil.

COHEN, Keith (1972): "Le New Criticism aux Etats-Unis (1935-1950)" . Poétique 10: 217-243.

CULLER, Jonathan (1975): Structuralism, linguistics and the study of literature. London:Routledge.

DECREUS, Freddy (1977): "Strukturele analyse van Catullus' Carmen 76". Handelingen der Koninklijke Zuidnederlandse Maatschappij voor Taal-en Letterkunde en Geschiedenis 31: 51-106.

--- (1978): "Les possibilités d'application de l' analyse structurale dans l'enseignement secondaire". Didactica Classica Gandensia 17-18: 165-190.

DELBOUILLE, Paul (1968): "Analyse structurale et analyse textuelle" . Cahiers d'analyse textuelle 10: 7-22.

FOWLER, Roger (1971): "Linguistics,stylistics ; criticism ?" , in: FOWLER, Roger (1971): The languages of literature. Some linguistic contributions to criticism. London:Routledge.

HALLIDAY, Michael (1964): "The linguistic study of literary texts", in: LUNT, H. (ed.) (1964):Proceedings of the Ninth International Congress of Linguistics. Cambridge 1962. Den Haag:Mouton.

HARDT, Manfred (1976): Poetik und Semiotik. Das Zeichensystem der Dichtung. Tübingen:Niemeyer.

HENDRICKS, William (1974): "The relation between linguistics and literary studies". Poetics 11: 5-22.

HILL, Archibald (1955): "An analysis of the Windhover. An experiment in structural method". Publications of the Modern Language Association 70: 968-978.

INGARDEN,Roman (1972): Das literarische Kunstwerk. Tübingen: Niemeyer.

JAKOBSON, Roman (1963): Essais de linguistique générale. Paris: Minuit.

--- (1973): Questions de poétique. Paris:Seuil.

LEEMAN, A.D. (1971): "Complexiteit en intentie in Vergilius' eerste ecloga". Lampas 4 : 2/4 : 210-224.

LEVIN, Samuel (1962): Linguistic structures in poetry. The Hague:Mouton.

LOTMAN, Iouri (1972): Vorlesungen zu einer strukturalen Poetik. München:Fink.

--- (1973): La structure du texte artistique. Paris:Gallimard.

MAATJE, Frank C. (1974): Literatuurwetenschap. Grondslagen van een theorie van het literaire werk. Utrecht:Oosthoek.

MOUNIN, George (1968): "Baudelaire devant une critique structurale". Annales de la Faculté des Lettres et des Sciences Humaines de Nice 4/5: 155-160.

POSNER, Roland (1974): "Strukturalismus in der Gedichtinterpretation. Textdeskription und Rezeptionsanalyse am Beispiel von Baudelaires "Les Chats", in: IHWE, Jens (ed.) (1974): Literaturwissenschaft und Linguistik,I . Frankfurt : Athenäum.

RIFFATERRE, Michael (1971): Essais de stylistique structurale. Paris: Flammarion.

RUWET, Nicholas (1972): Langage, musique, poésie. Paris : Seuil.

SCHOLES, Robert (1974): Structuralism in literature . An introduction. New Haven: Yale University Press.

SCHRIJVERS, P.H. (1973): "Interpreteren van een Horatiaanse Ode (I 10)". Lampas 6: 2 : 126-152.

SÖRENSEN, Hans (1958) :" Littérature et linguistique". Orbis Litterarum. Suppl. 2 : 182-197.

KOMPONENTEN DES TEXTVERSTEHENS UND IHRE REKONSTRUKTION IN DER TEXTANALYSE

Käthi Dorfmüller-Karpusa

1. Textverständnis im Rahmen einer Texttheorie

1.1. Einen Text verstehen zu können heißt,ihm mindenstens einen eindeutigen Sinn zuordnen zu können.Zur Repräsentation einer eindeutigen Sinnstruktur bedarf es einer wohldefinierten (kanonischen) Sprache.Es hängt von der theoretischen Konzeption ab, wie eine kanonische Sprache aufgebaut ist.

1.2. Im Folgenden werde ich im Rahmen der Text-Struktur-Welt-Struktur-Theorie (TeSWeST) untersuchen,wie viele und was für Sinnstrukturen einem objektsprachlichen Text zugeordnet werden können und wie sie sich zueinander verhalten (PETÖFI 1975 / 1976).Die Repräsentation einer Sinnstruktur nennen wir "intensionalsemantische Repräsentation" (TInR) (BIASCI/FRITSCHE (eds.) 1978,DORFMÜLLER-KARPUSA 1978b).Die intensionalsemantische Repräsentation eines objektsprachlichen Textes besteht aus zwei Komponenten:(i)einer Baumstruktur,die einen atomaren Text (T) oder einen komplexen Text ($T^{\diamond}$) beschreibt und die die elementare Information für die Verarbeitung eines objektsprachlichen Textes gibt - atomare und komplexe Texte sind theoretische Termini -, (ii)einer Beschreibung der Welten,die als im objektsprachlichen Text repräsentiert oder als durch ihn impliziert betrachtet werden können (TWD).Ich beschränke mich hier ausschließlich auf der Baumstruktur.Die elementare Information,die durch das Baumdiagramm gegeben wird,gliedert sich in drei Ebenen:(a)die deskriptive,die einen elementaren oder einen komplexen Sachverhalt beschreibt,(b)die weltkonstitutive,die die Einstellung des Sprechers diesem Sachverhalt gegenüber klarmacht,d.h.ob er ihn vermutet,glaubt,weiß u.a.,(c)die performative-modale,die die performative Modalität,mit der der Sprecher die Sachverhaltsbeschreibung und seine Einstellung ausdrückt,beschreibt,d.h.ob er sie sagt,mitteilt,behauptet u.a. (PETÖFI/KAYSER 1976,DORFMÜLLER-KARPUSA 1978a).

2. Deskriptive Ebene

2.1. Als objektsprachlichen Text nehme ich den folgenden:

(1) Anna hat sich über Peters Geschenk gefreut.

Trotz der großen Vielfalt der möglichen Bedeutungen wird die Mehrheit der "native speakers" eine ganz bestimmte Bedeutung spontan als die allein gültige ansehen.Durch uns unbekannte Mechanismen wird diese herausgegriffen.Der Text (1) wurde zwei Klassen - einer 8. und einer 10. - eines Gymnasiums vorgelegt mit der Frage:"Wie verstehst du die Situation?Erkläre sie bitte kurz!" Ich habe hier anstatt "Sachverhalt" den Ausdruck "Situation" gewählt,weil er für die Schüler einfach zu verstehen ist. Aus den 35 Schülern der 8. Klasse haben 27 die folgende Situation angegeben:"Peter hat Anna ein Geschenk gemacht,und Anna hat sich darüber gefreut." Die selbe Situation wurde von 19 Schülern der 10. Klasse - die Befragten waren 22 - angegeben.Das heißt, daß 80,7% der Befragten eine einzige und die selbe Bedeutung dem angegebenen Text zugeordnet haben.Die Situation,die durch den gegebenen Text repräsentiert ist,ist jedoch keineswegs eindeutig bestimmt.Während der Teil "Anna hat sich gefreut",abgesehen von den Topikalisierungsmöglichkeiten - "Ánna hat sich gefréut" -, explizit eine Situation beschreibt,wird durch den zweiten Teil "über Peters Geschenk" die dem Leser nahegelegte Situation nicht eindeutig beschrieben.Die verschiedenen Möglichkeiten ermitteln zu können,müßen wir den Ausdruck "Peters Geschenk" als Teil des objektsprachlichen Textes analysieren.Die Lexikoneintragung,die wir im Lexikon dem Wort "Geschenk" zuordnen,muß unter anderem die Information enthalten,daß ein "Geschenk" etwas ist,das jemand jemandem gibt.Mit Hilfe dieser Information können wir dem objektsprachlichen Text die folgenden vier möglichen Sinnstrukturen zuordnen:

(1a) Peter gibt Anna ein Geschenk

(1b) Peter gibt jemandem (einem Dritten) ein Geschenk

(1c) Anna gibt Peter ein Geschenk

(1d) Jemand (ein Dritter) gibt Peter ein Geschenk

und

(1k) Anna freut sich

Man könnte sich also vier Situationen vorstellen,in denen Anna einen Grund hätte sich zu freuen:

(2a) Anna hat sich gefreut,weil Peter ihr ein Geschenk gegeben hat.

(2b) Anna hat sich gefreut,weil Peter jemandem (einem Dritten) ein Geschenk gegeben hat.

(2c) Anna hat sich gefreut,weil sie Peter ein Geschenk gegeben hat.

(2d) Anna hat sich gefreut,weil jemand (ein Dritter) Peter ein Geschenk gegeben hat.

Wenn man diese vier Interpretationen mit den oben erwähnten zwei Topikalisierungsmöglichkeiten kombiniert,entstehen mindenstens 32 Lesarten (4x(2x4)).Man könnte sich aber auch Kombinationen von Topikalisierungen vorstellen,die die Anzahl der Lesarten erweitern.Die Unbestimmtheit der Zeitrelationen,die die Zuordnung weiterer intensionalsemantischen Repräsentationen notwendig macht,wird hier nicht behandelt.Ebenso wird eine ironische Lesart hier nicht berücksichtigt.Die Tatsache,daß im allgemeinen mehrere Sinnstrukturen einem Text zugeordnet werden können, werde ich mit dem Terminus "Undeterminiertheit" bezeichnen.Bis jetzt habe ich mich mit der Undeterminiertheit der deskriptiven Ebene beschäftigt.

3. Weltkonstitutive Ebene

3.1. Im Folgenden werde ich mich mit der zweiten Ebene beschäftigen,d.h.mit der weltkonstitutiven.Die sehr oft implizite Existenz von weltkonstitutiven Ausdrücken in objektsprachlichen Texten gibt dem Hörer/Leser wiederum die Möglichkeit diese Undeterminiertheit verschiedenerweise zu interpretieren.Zur vollständigen Analyse muß der objektsprachliche Text (1) in eine kommunikative Situation eingebettet werden:

(1´) Georg sagt Lita,daß
Anna sich über Peters Geschenk gefreut hat.

Im Folgenden wird ein möglicher atomarer Text T_1 konstruiert, der dem objektsprachlichen Text (1) in der kommunikativen Situation (1´) zugeordnet wird.Anschließend werden zur Erläuterung die Symbole sowie die Argumentrollenindikatoren,die in dem atomaren Text T_1 abgekürzt erscheinen,aufgelistet.

(0a) $P^{\mathbb{C}c}$:=: [dFi] {st:tO,sl:L,up:$p^{\mathbb{C}c}$}
(0b) $p^{\mathbb{C}c}$:=: *[KOMMUNIZIEREN]{dp:Georg,go:Lita,eo:T_1:=:$P^{\mathbb{P}c}$}

(1a) $P^{\mathbb{P}c}$:=: [dFi] {st:JETZT/=tO/,sl:HIER/=L/,up:$p^{\mathbb{P}c}$}
(1b) $p^{\mathbb{P}c}$:=: [SAGEN]{ap:ICH/=Georg/,go:DU/=Lita/,eo:P^{Ww}}

(2a) P^{Ww} :=: [dFi] {st:JETZT/=tO/,sl:HIER/=L/,up:p^{Ww}}
(2b) p^{Ww} :=: *[S.ERINNERN]{se:ICH/=Georg/,sp:P^{Dw1}}

(3a) P^{Dw1} :=: [dFi] {st:Dw1t,sl:Dw1l,up:p^{Dw1}}
(3b) p^{Dw1} :=: *[INFORMIERT WERDEN]{se:ICH/=Georg/,cp:X, sp:P^{Dw2}}
(4a) P^{Dw2} :=: [dFi] {st:Dw2t,sl:Dw2l,up:p^{Dw2}}
(4b) p^{Dw2} :=: *[WISSEN] {se:X,sp:P^{Dw3}}

(5a) P^{Dw3} :=: [dFi] {st:Dw3t,sl:Dw3l,up:p^{Dw3}}
(5b) p^{Dw3} :=: *[STATUM] {se:Anna,up:$P^{D\ z}$}

(6) $P^{D\lozenge z}$:=: *[WEIL] {λ_1:P^{Dx},λ_2:P^{Dy}}

(7a) P^{Dx} :=: [dFi] {st:Dxt,sl:Dxl,up:p^{Dx}}
(7b) p^{Dx} :=: [S.FREUEN] {se:Anna}

(8a) P^{Dy} :=: [dFi] { st:Dyt,sl:Dyl,up:p^{Dy} }
(8b) p^{Dy} :=: [GEBEN]{ ap:Peter,go:Anna,ao:Geschenk }

Die sternierten Einheiten kennzeichnen die nicht in expliziter Form im objektsprachlichen Text erscheinenden Einheiten,die auf interpretativem Weg dem Text zugeordnet wurden.

Symbole

dFi : Funktor eines propositionbildenden Prädikatrahmens, entspricht dem objektsprachlichen Ausdruck "es ist der Fall, daß"

$P^{\mathbb{C}c}$: kommunikative Proposition mit kommunikativer Funktion

$p^{\mathbb{C}c}$: kommunikativer Propositionskern mit kommunikativer Funktion

$P^{\mathbb{P}c}$: kommunikative Proposition mit performativ-modaler Funktion

$p^{\mathbb{P}c}$: kommunikativer Propositionskern mit performativ-modaler Funktion

P^{Ww} : weltkonstitutive Proposition mit weltkonstitutiver Funktion

p^{Ww} : weltkonstitutiver Propositionskern mit weltkonstitutiver Funktion

P^{D} : deskriptive Proposition

p^{D} : deskriptiver Propositionskern

P^{Dw} : weltkonstitutive Proposition mit deskriptiver Funktion

p^{Dw} : weltkonstitutiver Propositionskern mit deskriptiver Funktion

Argumentrollenindikatoren (HEYDRICH 1976)

λ_i : Argumentrollenindikator-Variable

ao : affected-object

ap : agent-participant

dp : doer-participant

eo : effected-object

go : goal-object

sl : specification-local

st : specification-temporal

se : state-experiencer

sp : state-participant

3.2. Die Repräsentation des Textes T_1 kann auch in der Form eines Diagramms angegeben werden. Das Diagramm D_1 zeigt schematisch die Relationen zwischen den Welten, die im atomaren Text T_1, sowie von Georg kommuniziert, repräsentiert worden sind. D_1 kann folgendermaßen gelesen werden:

D1

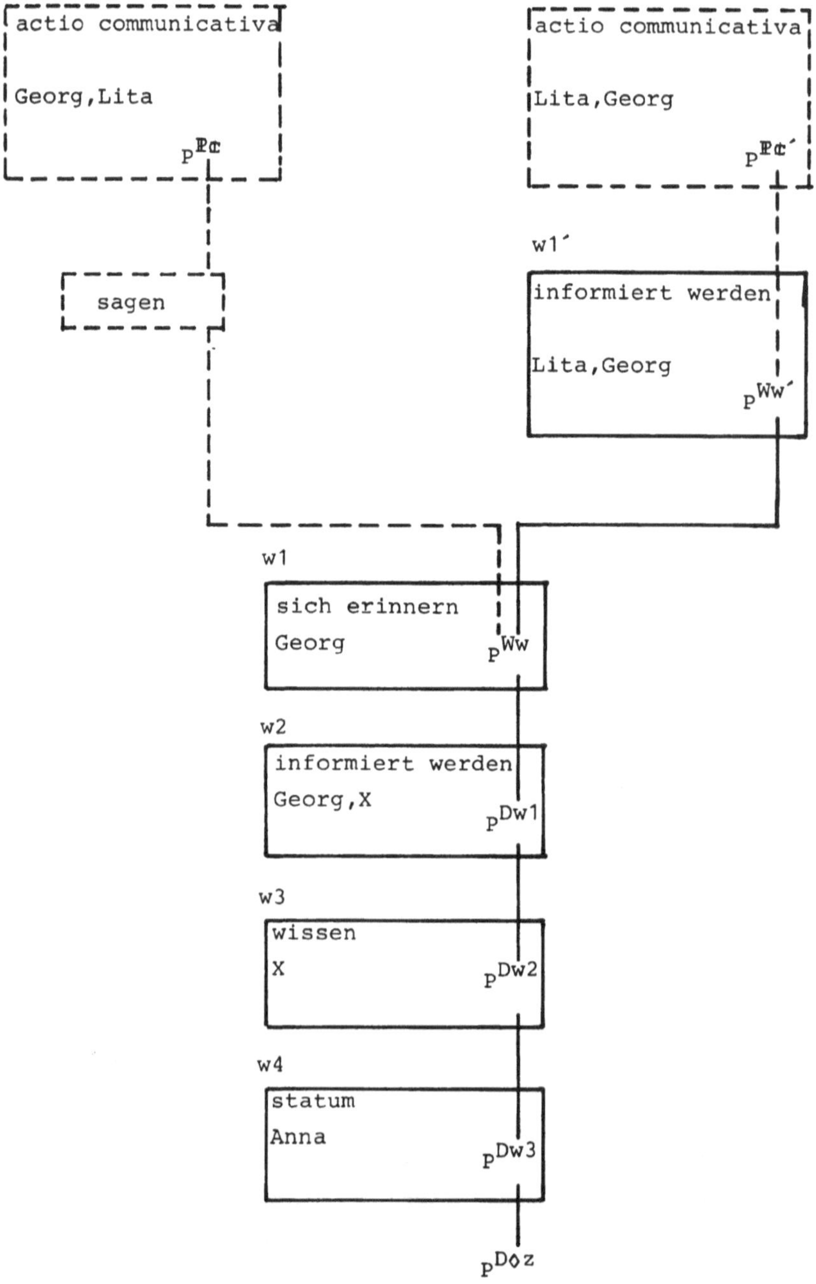
actio communicativa
Georg,Lita
P^{Pc}
actio communicativa
Lita,Georg
$P^{Pc'}$
sagen
w1´
informiert werden
Lita,Georg
$P^{Ww'}$
w1
sich erinnern
Georg
P^{Ww}
w2
informiert werden
Georg,X
P^{Dw1}
w3
wissen
X
P^{Dw2}
w4
statum
Anna
P^{Dw3}
$P^{D\diamond z}$

Jetzt und hier sage ich,Georg,dir,Lita,daß ich mich jetzt und hier erinnere,daß ich von einer Person X informiert worden bin,daß diese Person gewußt hatte,daß Anna im psychischen Zustand gewesen war,daß sie sich freute,weil Peter ihr ein Geschenk gegeben hatte.

3.3. Das Diagramm D_2 entspricht einem anderen atomaren Text T_2 und repräsentiert partiell eine andere mögliche Sinnstruktur des objektsprachlichen Textes (1) im Bezug auf die weltkonstitutive Ebene.D_2 kann folgendermaßen gelesen werden:

Jetzt und hier sage ich,Georg,dir,Lita,daß ich jetzt und hier mich erinnere,daß ich apperzipiert habe,daß Peter Anna ein Geschenk gegeben hat,und ich erinnere mich,daß ich von einer Person X informiert worden bin,daß X gewußt hatte,daß Anna im psychischen Zustand gewesen war,daß sie sich freute,und aus all diesen Gründen erinnere ich mich, daß ich gefolgert habe,daß Anna im psychischen Zustand gewesen war,daß sie sich freute,weil sie im psychischen Zustand (der von Peter Beschenkten) gewesen war,daß Peter intendiert getan hatte,daß er ihr ein Geschenk gegeben hatte.

3.4. Das Diagramm D_3 entspricht wiederum einem anderen atomaren Text T_3,der dem selben objektsprachlichen Text (1) zugeordnet wird.D_3 kann folgendermaßen gelesen werden:

Jetzt und hier sage ich,Georg,dir,Lita,daß ich jetzt und hier mich erinnere,daß ich von einer Person X informiert worden bin,daß X apperzipiert hatte,daß Peter Anna ein Geschenk gegeben hatte,und daß ebenso X gewußt hatte,daß Anna im psychischen Zustand gewesen war,daß sie sich freute, und aus all diesen Gründen folgere ich,Georg,jetzt und hier,daß Anna im psychischen Zustand gewesen ist,daß sie sich freute,weil sie im psychischen Zustand (der von Peter Beschenkten) gewesen ist,daß Peter intendiert getan hatte, daß er ihr ein Geschenk gegeben hatte.

Die Steuerung der Kombinatorik im weltkonstitutiven Bereich wird bestimmt von:(a)den Relationen,die zwischen weltkonstitutiven Einheiten bestehen können,(b)den Relationen,die zwischen welt-

D2

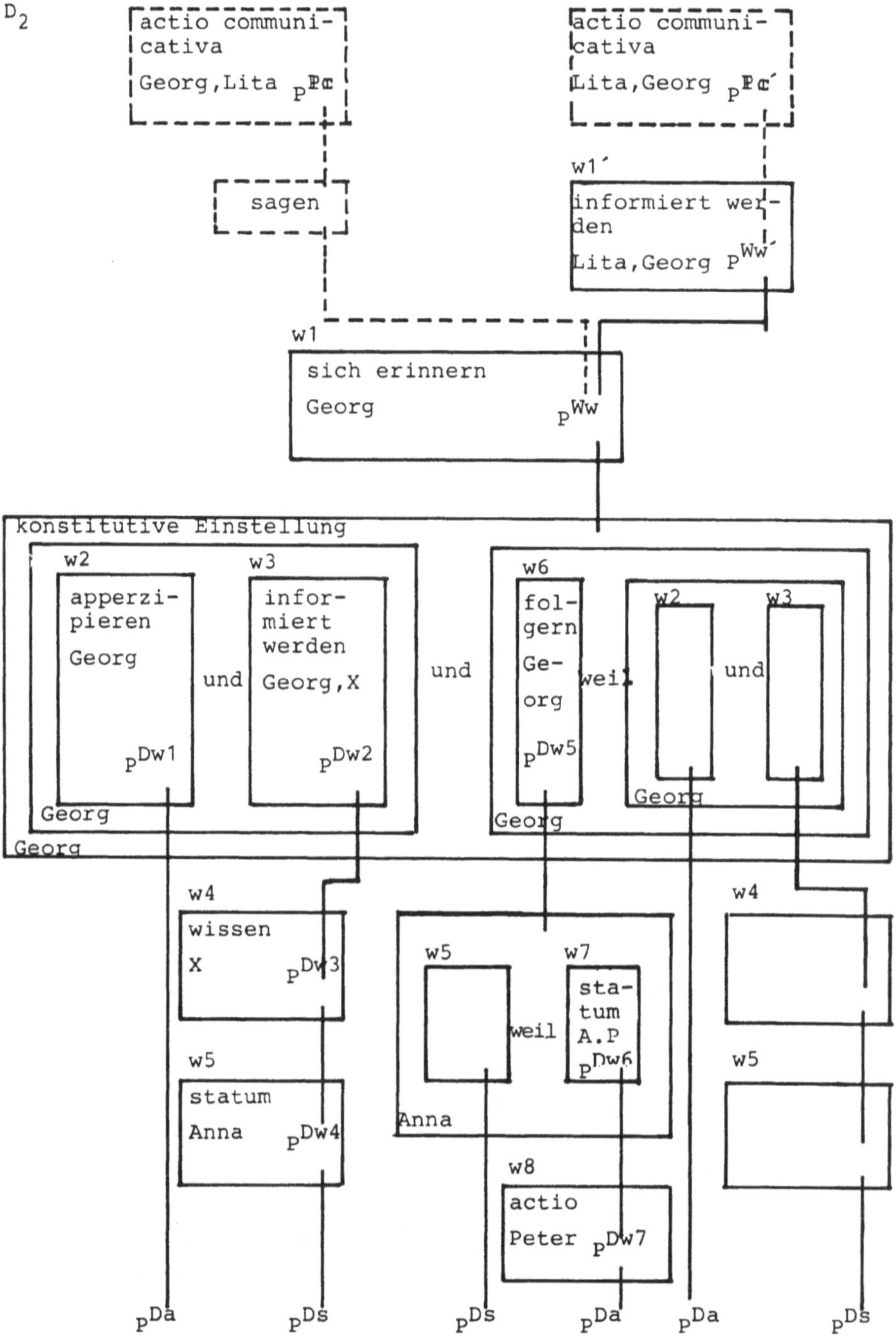

actio communicativa
Georg, Lita pPc
actio communicativa
Lita, Georg pPc´
sagen
w1´
informiert werden
Lita, Georg pWw´
w1
sich erinnern
Georg pWw
konstitutive Einstellung
w2
apperzipieren
Georg
pDw1
und
w3
informiert werden
Georg, X
pDw2
Georg
Georg
und
w6
folgern
Georg
pDw5
weil
w2
und
w3
Georg
Georg
w4
wissen
X pDw3
w5
statum
Anna pDw4
w5
weil
w7
statum
A.P
pDw6
Anna
w8
actio
Peter pDw7
w4
w5
pDa
pDs
pDs
pDa
pDa
pDs

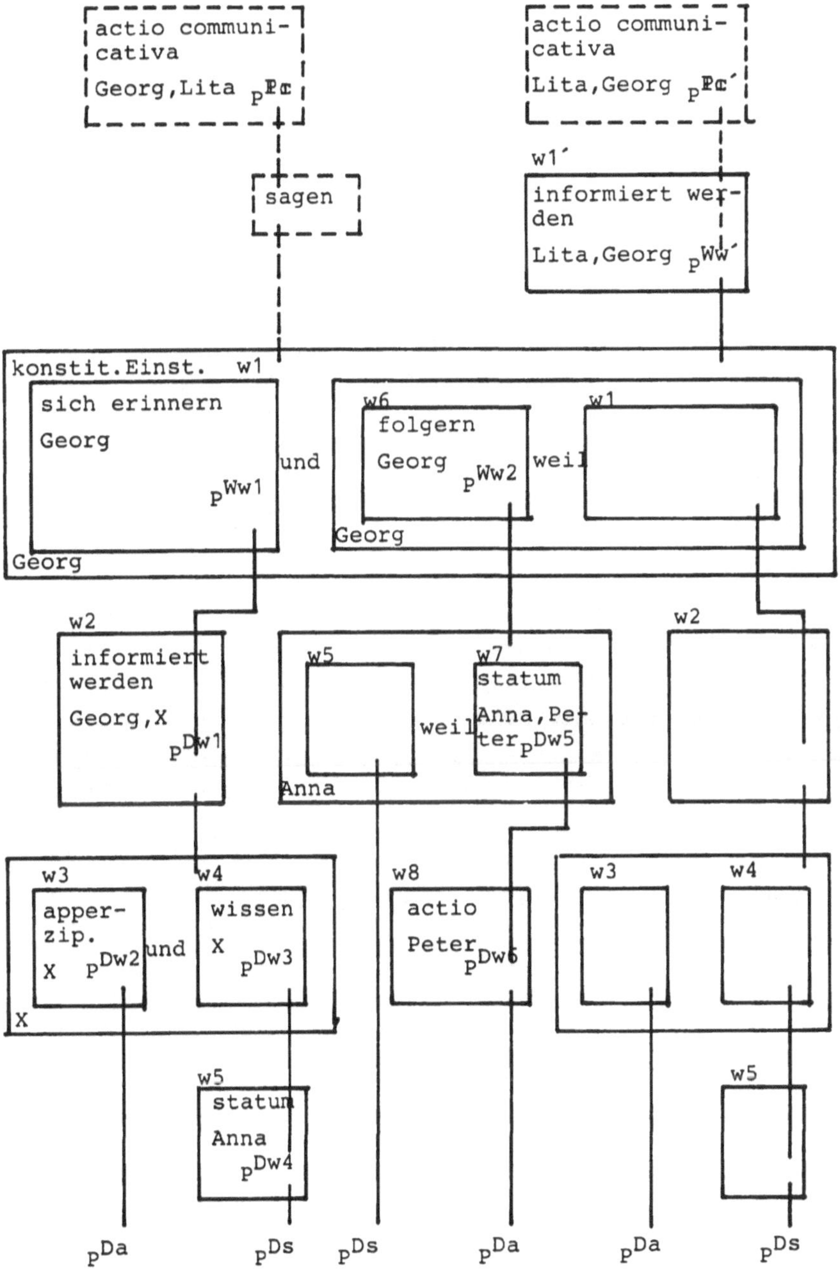
actio communicativa
Georg,Lita $_P$Pc
sagen
actio communicativa
Lita,Georg $_P$Pc´
w1´
informiert werden
Lita,Georg $_P$Ww´
konstit.Einst. w1
sich erinnern
Georg
$_P$Ww1
und
w6
folgern
Georg
$_P$Ww2
weil
w1
Georg
Georg
w2
informiert werden
Georg,X
$_P$Dw1
w5
weil
w7
statum
Anna,Peter $_P$Dw5
Anna
w2
w3
apperzip.
X $_P$Dw2
und
w4
wissen
X $_P$Dw3
X
w8
actio
Peter $_P$Dw6
w3
w4
w5
statum
Anna
$_P$Dw4
w5
$_P$Da
$_P$Ds
$_P$Ds
$_P$Da
$_P$Da
$_P$Ds

konstitutiven und deskriptiven Einheiten bestehen können.

4. Performative-modale Ebene

4.1. Ebenso wird die Kombinatorik im performativen Bereich von den Relationen zwischen den performativen und weltkonstitutiven Einheiten gesteuert.Als indifferente performative-modale Einheit habe ich für den Text (1) "sagen" gewählt,d.h.ein bloßes Sagen ohne weitere Intention.Als weniger indifferente Einheit könnte man "mitteilen" wählen,d.h.ein Sagen mit der Intention des Weitergebens der Information.Man könnte jedoch auch die Einheit "behaupten" wählen,die noch komplexer ist.Diese Undeterminiertheit der performativen Modalität,durch die der Sprecher eine Sachverhaltsbeschreibung und seine Einstellung ihr gegenüber ausdrückt,hat zur Folge,daß der Bereich der möglichen Sinnstrukturen eines Textes sich noch mehr erweitert.

5. Abschließende Bemerkungen

5.1. Ich glaube hiermit gezeigt zu haben,daß die Undeterminiertheit der Intension eines Textes in jeder der drei Ebenen enthalten ist.Das Verstehen eines Textes ist das Produkt von Operationen auf allen drei Ebenen.Die Mechanismen des Funktionierens dieser Operationen sind weitgehend unerforscht und können nur durch Einbeziehung psycholinguistischer und soziolinguistischer Forschung geklärt werden.Es bleibt jedoch bewundernswert,wie trotz all dem wir Menschen uns verständigen können:

> πολλὰ τὰ δεινὰ κοὐδὲν ἀν-
> θρώπου δεινότερον πέλει. (Sophokles,Antigone 332)

Anmerkungen

Ich danke Herrn Studienrat M.Behringer herzlich für die Hilfe bei der Umfrage in seinen Klassen.

Literatur

BIASCI, Claudia / FRITSCHE, Johannes (eds.) (1978): Texttheorie-Textrepräsentation.Theoretische Grundlage der kanonischen Repräsentation von Texten. Hamburg: Buske.

DORFMÜLLER-KARPUSA, Käthi (1978a): "Konnektive Ausdrücke und konnektive Relationen". Erscheint in: Orbis.

--- (1978b): "Kavafis,K.P. 'Μιση ωρα',eine intensionalsemantische Analyse". Arbeitspapier. Universität Bielefeld.

HEYDRICH, Wolfgang (1976): "Zur Einführung von argument-labels". PETÖFI, János S. / BREDEMEIER, Jürgen (eds.) (1977): Das Lexikon in der Grammatik - die Grammatik im Lexikon. Hamburg: Buske.

PETÖFI, János S. (1975): "A Formal Semiotic Text Theory as an Integrated Theory of Natural Language. Methodological Remarks". DRESSLER, Wolfgang U. (ed.) (1977): Current Trends in Textlinguistics. Berlin - New York: W. de Gruyter.

--- (1976): "Structure and Function of the Grammatical Component of the Text-Structure World-Structure Theory". GÜNTHNER, F. / SCHMIDT, Siegfried J. (eds.) (1978): Formal Semantics and Pragmatics for Natural Languages. Dordrecht: Reidel.

PETÖFI, János S. / KAYSER, Hermann (1976): "Sprechhandlungen und semantische Interpretation.Die Rolle der performativ-modalen,weltkonstitutiven und deskriptiven Ausdrücke in der Textinterpretation". MEYER-HERMANN, Reinhard (ed.) (1978): Sprechen - Handeln - Interaktion.Ergebnisse aus Bielefelder Forschungsprojekten zu Texttheorie,Sprechakttheorie und Konversationsanalyse. Tübingen: Niemeyer.

EIN BEITRAG ZUR STRUKTURALEN NARRATIVIK:

SPRACHE – MÄRCHEN – MYTHOS

Hartwig Frankenberg

Für die folgenden Ausführungen[1] bildet das Grimm'sche Märchen 'Hänsel und Gretel' das empirische Ausgangsmaterial. Aus Gründen der Platzersparnis und aus Gründen der breiten Bekanntheit des Märchens, verzichte ich hier auf eine Wiedergabe des Märchentextes. Über eine interpretierende Paraphrase werde ich mich allmählich der Behandlung des Themas nähern.

Das Märchen beginnt mit der Beschreibung des Ausgangszustandes: es herrscht Armut, Mangel an Brot, die Situation ist lebensbedrohend. Dazu werden die vier Hauptfiguren genannt: Vater, Mutter, Hänsel und Gretel.

Die Armut ist nicht mehr zu ertragen. Die Mutter überlegt (aktiv), wie Abhilfe zu schaffen sei, und überredet den Vater (passiv), am nächsten Tag die Kinder im Wald auszusetzen. Das Verhältnis zwischen Vater und Mutter einerseits, und der Mutter und den Kindern andererseits, ist gestört, disharmonisch und sehr lieblos.

Die Kinder sind noch wach und haben das Gespräch der Eltern mitgehört. Während Gretel verzweifelt (passiv) ist, sucht Hänsel durch eine List (aktiv) die mütterliche Entscheidung zu unterlaufen, indem er im Mondlicht Kieselsteine sammelt. Dadurch werden die Mutter und der Sohn zu aktiven Gegenspielern.

Am nächsten Morgen bricht die ganze Familie in den Wald auf. Gretel trägt das mütterliche Brot als Wegzehrung, Hänsel heimlich die Kieselsteine (Batzen/Geld). Um den Rückweg in das elterliche Haus zu ermöglichen, verstreut Hänsel unbemerkt (List) die Kieselsteine. Unter dem Vorwand (List), Holz zu fällen, lassen die Eltern ihre Kinder neben dem Feuer (Ende des elterlichen Bereiches) zurück. Hänsel und Gretel warten bis zum Abend vergeblich auf die Rückkehr der Eltern zu dem Lagerplatz. Hänsel tröstet (aktiv) abermals Gretel. Nach Mondaufgang verfolgen sie den umgekehrten Weg der ausgestreuten Kieselsteine und finden gegen Morgen wieder nach Hause zurück. Während sich der passive Vater über die Rückkehr seiner Kinder freut, ist die überaus aktive Mutter darüber verärgert.

Die Handlung setzt noch einmal neu ein: Zunächst bleiben die Kinder eine Weile zu Hause, bis die Armut (Mangel an Brot) die Mutter erneut dazu treibt, die Kinder im Wald auszusetzen. Zu den ersten Abschnitten entwickelt sich nun ein paralleler Handlungsstrang, der sich allerdings von dem ersten Abschnitt unterscheidet. Hänsel kann nämlich keine Kieselsteine sammeln, da die Mutter die Haustüre verschlossen hat. Eine Interaktion zwischen Mutter und Sohn findet also lediglich statt über das Schema List - verhinderte List - List. Wieder tröstet Hänsel seine Schwester Gretel.

Auch diesmal gibt die Mutter den Kindern je ein Stück Brot als Wegzehrung mit. In Ermangelung von Kieselsteinen verstreut Hänsel sein Brot auf dem Weg in den Wald, in der Hoffnung, den Weg wieder nach Hause zurück zu finden. Wieder wird im Wald eine Feuer für die Kinder gemacht, bei dem die Eltern ihre Kinder zurücklassen. Nach Mondaufgang will Hänsel mit Gretel den umgekehrten Weg der ausgestreuten Brotkrumen nach Hause verfolgen. Hänsel scheitert mit seiner List, so daß beide im Wald umherirren. Nach drei Tagen kommen die Kinder um die Mittagszeit ausgehungert an ein Haus, das sich ihnen zum Verzehr anbietet (Überfluß an Nahrung).

Langsam tritt die Hexe in Erscheinung, die sich zunächst fürsorglich-mütterlich und kinderlieb zeigt. Sie bietet den Kindern an, bei ihr zu übernachten und zu essen. In einem Einschub erfahren wir, daß die Hexe ein bösartiges Wesen ist, das Kinder verschlingt. Am nächsten Morgen zeigt die Hexe ihr wahres Gesicht. Hänsel wird in einen Stall gesperrt (Verlust seiner Aktivität), da ihn die Hexe erst eine Zeit mästen (Überfluß an Nahrung) will, um ihn dann zu verspeisen. Gretel muß Hilfsdienste in der Küche verrichten, Wasser holen, für Hänsel kochen. Gretel muß ale Befehle der Hexe ausführen und befindet hier in einem extremen Zustand der Passivität (Mangel an Nahrung). Trotz des Verlustes seiner Aktivität, weiß sich Hänsel durch eine List der Absicht der Hexe zu entziehen, indem er sich zwar mästen läßt, jedoch durch das 'Knöchelein' den Anschein erweckt, als sei er noch nicht so weit um geschlachtet zu werden.

Vier Wochen später will die Hexe nicht länger warten und beschließt, Hänsel am nächsten Tag zu schlachten. Die letzten Vorbereitungen werden getroffen. Angeblich soll auch gebacken werden. Gretel ist in größter Verzweiflung. Unter dem Vorwand (List), sie hätte zu schwache (passiv) Augen, fordert die Hexe Gretel dazu auf, nach dem Feuer im Backofen zu sehen. Gretel schöpft aber Verdacht, stellt sich dumm (List) und sperrt die Hexe in den Backofen (aktiv). Dann befreit (aktiv) Gretel Hänsel.

Nachdem die Hexe tot ist, entdecken die beiden einen großen Reichtum an Edelsteinen und Perlen (Überfluß an Geld), und sie nehmen mit (beschränkte Menge), was sie tragen können. Sie finden schnell den Weg nach Hause. Der Vater freut sich sehr über ihre Rückkehr. Auch ist er durch seine Kinder reich (begrenzter Reichtum) geworden. Die Mutter war inzwischen gestorben (extremer Gegensatz zu 'aktiv').

Nach dieser Paraphrase des Märchens 'Hänsel und Gretel' sollen nun folgende Fragen beantwortet werden:

1. Welche Merkmale besitzen die Figuren?

2. Unterscheidung zwischen Konstanz und Variabilität der Merkmale?

3. Steht die Konstanz, bzw. Variabilität in irgendeiner Beziehung zur Handlung des Märchens?

4. Ist diese Beziehung strukturbildend?

5. Welche Struktur läßt sich ableiten?

1. Welche Merkmale besitzen die Figuren ?

Vater:	wenig Brot	arm	passiv	kinderlieb
	viel Geld	reich	aktiv (?)	kinderlieb
Mutter:	wenig Brot	arm	aktiv	nicht kinderlieb
	- -	-	tot	- -
Hänsel:	Kieselsteine	?	aktiv	schwesterverbunden
	wenig Brot	?	aktiv	schwesterverbunden

	Überfl.Nahrung	?	passiv	schwesterverbunden
	viel Geld	reich	aktiv (?)	schwesterverbunden
Gretel:	wenig Brot	?	passiv	bruderverbunden
	Hunger	?	passiv	bruderverbunden
	viel Geld	reich	aktiv	bruderverbunden
Hexe:	Überfl. Brot	?	aktiv	kinderlieb
	Überfl. Geld	?	tot	nicht kinderlieb

2. Unterscheidung zwischen Konstanz und Variabilität der Merkmale

Vater:	konstant:	kinderlieb
	variabel:	arm -> reich
		passiv -> aktiv (?)
Mutter:	konstant:	nicht kinderlieb
	variabel:	aktiv -> tot
Hänsel:	konstant:	schwesterverbunden
		aktiv
	variabel:	arm -> reich
Gretel:	konstant:	bruderverbunden
	variabel:	passiv -> aktiv
		arm -> reich
Hexe:	konstant:	Überfluß
	variabel:	Überfluß Brot -> Überfluß Geld
		kinderlieb -> nicht kinderlieb
		aktiv -> tot

3. Steht die Konstanz, bzw. Variabilität in irgendeiner Beziehung zur Handlung des Märchens?

Bezüglich der Beziehung zwischen der Konstanz, bzw. Variabilität der Merkmale und der Handlung ergibt sich eine Dreiphasigkeit des Märchens:

I Mangel an Brot, Armut

II Überfluß an Brot und Geld

III Reichtum an Geld
(begrenzter Reichtum)

Dadurch, daß Gretel im entscheidenden Moment aktiv wird, findet das Märchen seinen Ausgleich in der dritten Phase.

4. Ist diese Beziehung strukturbildend?

Die Beziehung 3. macht deutlich, daß sie fähig ist, eine geschlossene Struktur zu bilden, da in dem Märchen nicht nur eine Entwicklung der Figuren stattfindet, sondern sich diese Entwicklung als ein Funktionsaustausch bestimmter Merkmale (z.B. Brot : Geld, passiv : aktiv) auszeichnet.

5. Welche Struktur läßt sich ableiten?

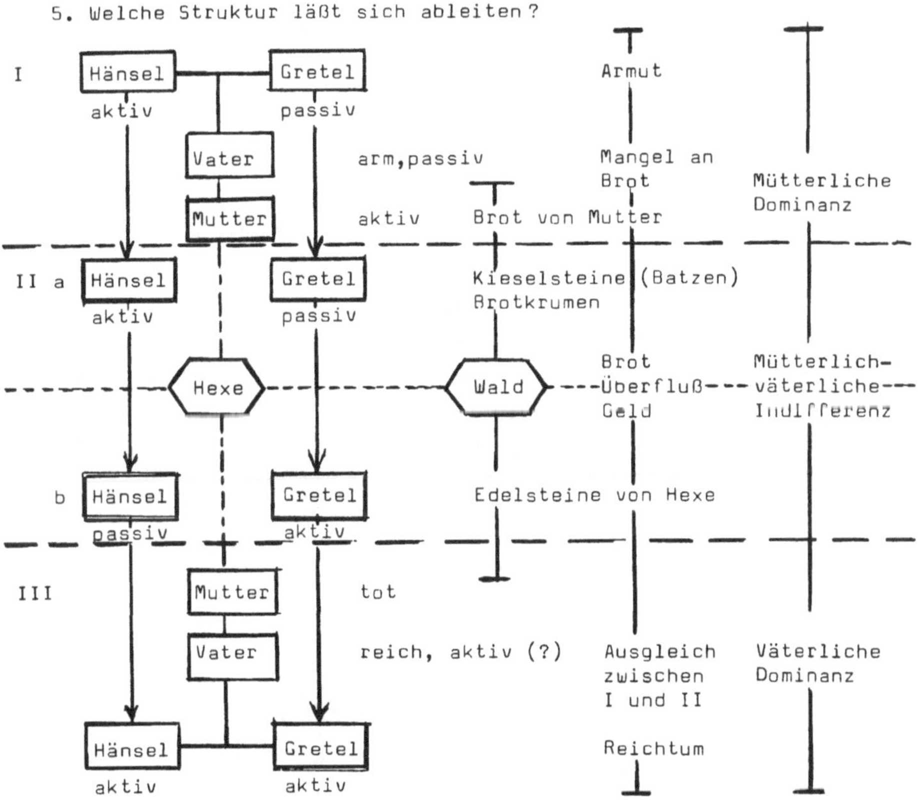

Die oben vorgenommenen Strukturversuche lehnen sich an die strukturale Anthropologie und Mythologie von Claude Lévi-Strauss an. Insbesondere stütze ich mich auf den methodischen Ansatz, wie ihn Strauss in seinem Aufsatz 'Die Geschichte von Asdiwal'[2] anschaulich entwickelt hat. Diese Methode hat Strauss aus der strukturalen Sprachwissenschaft auf die Ethnologie übertragen. Auch die Literaturwissenschaft hat sich um die Entwicklung von strukturalen Methoden[3] bemüht. Diese Methode arbeitet zunächst streng textimmanent, indem sie die signifikanten Elemente eines Textes benennt und deren semantische Merkmale aus ihrer Relation zueinander ableitet.

Strauss wendet die strukturale Methode an, indem er nicht nur die signifikanten Oberflächenmerkmale (parole) eines Textes segmentiert und deren Relationen (bzw. deren Merkmale) zueinander bestimmt, sondern indem er auf abstrakte, einem Text zugrundeliegende Relationen rekurriert. Diese abstrakten Relationssysteme (langue), bzw. ihre Struktur bilden - nach der Auffassung von Strauss - den einem Text zugrundeliegenden Mythos.[4]

Obwohl das Gesamtwerk von Claude Lévi-Strauss durch seine Materialfülle und seine frappierenden Resultate besticht, scheint seine Methode in formaler Hinsicht noch wenig ausgereift zu sein. Die Entwicklung einer effektiven strukturalen Narrativik stellt deshalb ein wissenschaftliches Desiderat dar. Es kann und soll hier allerdings nicht der Anspruch eingelöst werden, eine solche strukturale Narrativik zu entwikkeln. Mir kommt es hier lediglich darauf an, in einem tentativen Ansatz den Zugang zu dem Horizont 'Sprache - Text - Mythos' für eine speziell zu entwickelnde strukturale Narrativik zu eröffnen.

Strauss identifiziert die strukturale Methode mit dem mythischen Denken, indem er behauptet, daß '... das mythische Denken ausgeht von der Bewußtmachung bestimmter Gegensätze und hinführt zu ihrer allmählichen Ausgleichung.'[5] Diese Identifikation soll hier als zunächst ungeprüfte Arbeitshypothese übernommen werden.

Strauss etabliert eine mythische Struktur für solche Texte, die sich in einem geschlossenen System aus Gegensätzen und ihrer Vermittlung aufbaut. Mythos ist also zunächst nur eine bestimmte formale Struktur von Texten, die inhaltlich gesehen eine 'kritische Instanz' darstellt, die '... die Probleme der Gesellschaft auf symbolischer Ebene zu lösen versucht.'[6]

Die Gegensätze und ihre Vermittlung in einem geschlossenen System stellen sich in unserem Beispiel 'Hänsel und Gretel' nun so dar, daß das

Märchen in Phase I mit drei voneinander isolierten Gegensätzen beginnt:

Armut : Wohlstand
Aktivität : Passivität
Herzlosigkeit : Herzlichkeit

Dominierend ist jedoch der Gegensatz Aktivität : Passivität, der in den beiden nächsten Phasen II und III von den Figuren Hänsel und Gretel abgearbeitet wird. Es kommt also in den Phasen II und III zu einer Vermittlung dieser Gegensätze, indem die Figur der Gretel den Hauptanteil dieser Vermittlung trägt. Die Vermittlung findet - auf den Anfang und das Ende des Märchens bezogen - dadurch statt, daß das Merkmal 'aktiv' von der Mutter auf die Tochter übertragen wird. Genauer gesehen, findet sogar eine Spaltung des Merkmals 'aktiv' der überaus energischen Mutter statt, indem es gleichmäßig auf die Tochter und den Vater übertragen wird. Die Übertragung auf den Vater ist hierbei eine vermittelte, da ihn die Umwandlung Brot:Geld reich und damit aktiv macht. Über das Schema Brot:Geld findet zwischen dem Vater und der Mutter eine indirekte Interaktion statt.

Der Ausgleich, bzw. die Vermittlung dieses Hauptgegensatzes wird nun an der zentralen Instanz - der Figur der Hexe - abgearbeitet. Diese zentrale Instanz erscheint uns in allen Märchen (in allen mythischen Texten?) entweder in nur einer Figur verkörpert, wie etwa in 'Frau Holle', oder auf verschiedenen Teilinstanzen verteilt, die der Zentralinstanz vor- und nachgelagert sind, wie etwa in 'Dornröschen' und 'Der gestiefelte Kater'. Je nach struktureller Anlage besitzt die zentrale Instanz entweder insgesamt positive Merkmale, wie z.B. in 'Frau Holle' und 'Dornröschen', oder insgesamt bösartige Merkmale wie in 'Hänsel und Gretel'. Entweder ist diese Instanz überwiegend mit weiblichen Merkmalen versehen wie in 'Frau Holle' oder mit vorwiegend männlichen Merkmalen wie in 'Der gestiefelte Kater' und 'König Blaubart'. Sie kann jedoch auch männlich-weiblich ambivalent sein wie etwa in 'Hänsel und Gretel'. In irgendeiner oder mehrer Hinsichten (Merkmale) bleibt die zentrale Instanz jedoch in jedem Falle ambivalent.

Der Ausgleich der Gegensätze verläuft immer in einer Auseinanderstzung der Hauptfigur(en) mit der zentralen Instanz und ihrer vor- und nachgelagerten Teilinstanzen. Vor Beginn der Auseinandersetzung ist, bzw.

sind die Hauptfigur(en) in Bezug auf ihre wesentlichen Merkmale noch relativ ungefestigt. Erst die Begegnung mit der ambivalenten Zentralinstanz ermöglicht den Erwerb der substantiellen Merkmale der Hauptfigur, bzw. die Unterscheidung von substantiellen und akzidentellen Merkmalen. Dabei bleiben die akzidentellen Merkmale im Märchen meist ungenannt, sie sind auch strukturell nicht angelegt.

In 'Hänsel und Gretel' besitzt die Hexe sowohl explizit weibliche Merkmale: ihr Haus ist eßbar, als auch männliche Merkmale: die Edelsteine, die sich in ihrem Haus befinden. Strukturell vermittelt also die Hexe nicht nur zwischen den Gegensätzen aktiv:passiv, sondern auch zwischen den Gegensätzen weibliche Dominanz:männliche Dominanz. Dieser Gegensatz der weiblich-männlichen Dominanz scheint die dem Märchen zugrundeliegende Relation überhaupt zu sein. Da das Märchen von Entwicklungen erzählt, vermittelt der Mythos dieses Märchens den Gegensatz von Matriarchat:Patriarchat, bzw. den Kampf zwischen dem weiblichen und männlichen Prinzip.

Kehren wir zu unserem Märchen 'Hänsel und Gretel' zurück, so gelingt es uns vielleicht, eine Klasse 'mythischer Text' zu etablieren. Seine Merkmale lassen sich folgendermaßen beschreiben:

1. geschlossene, symmetrische und komplementäre Struktur

2. funktionale Polaritäten bezüglich der Merkmale der Hauptfigur(en)

3. Entwicklung der Figuren durch Merkmalsaustausch: Erwerb der substantiellen Merkmale durch Begegnung mit der Zentralinstanz

Der Anteil der Sprache am Mythos besteht in der strikten Unterscheidung zwischen akzidentellen Merkmalen (dekorativen Attributen) und substantiellen Merkmalen. Die substantiellen Merkmale sind strukturell angelegt, die akzidentellen Merkmale sind es nicht, sie sind frei verfügbar, während die substantiellen Merkmale in der Begegnung der Hauptfigur(en) mit der ambivalent-urmütterlichen Zentralinstanz abgearbeitet und gefestigt werden. Deshalb muß diese Begegnung auch immer als eine Reminiszenz an den Schöpfungsprozeß verstanden werden, der in erster Linie ein Sprachschöpfungsprozeß ist.

ANMERKUNGEN

1 Eine erweiterte Fassung dieses Beitrages mit einem leicht veränderten Titel: 'Sprache - Märchen - Mythos. Ein strukturalistischer Ansatz zur Beschreibung des Menschenbildes im Märchen' habe ich als Vortrag auf der '23. Jahrestagung der Gesellschaft zur Pflege des Märchengutes der europäischen Völker e.V.' in Straßburg (1. - 5. 11. 78) gehalten. Dieser erweiterte Beitrag wird demnächst in einem Sammelband der Märchengesellschaft erscheinen.

2 LÉVI-STRAUSS (1975), S. 169 - 224.

3 Siehe hierzu: GALLAS (1972) und GÜLICH/RAIBLE (1977)

4 Auf die Behandlung anderer Mythos-Begriffe muß hier verzichtet werden.

5 LÉVI-STRAUSS (1972), S. 247.

6 OPPITZ (1975), S. 289.

LITERATUR

GALLAS, H (1972) (Hrsg.): Strukturalismus als interpretatives Verfahren (Darmstadt, Neuwied)

GÜLICH,E./RAIBLE,W.(1977): Linguistische Textmodelle, Grundlagen und Möglichkeiten (München)

OPPITZ,M. (1975): Notwendige Beziehungen - Abriß der strukturalen Anthropologie (Frankfurt)

LÉVI-STRAUSS(1972): Strukturale Anthropologie (I) (Frankfurt)

--- (1975): Strukturale Anthropologie II (Frankfurt)

KAFKAESKE KONVERSATION

Ein Versuch, K.s Mißverstehen zu verstehen[1]

Ernest W.B. Hess-Lüttich

> LICHT (*zum Bedienten*). Es ist dem Herrn Gerichtsrat, will ich hoffen, Nichts Böses auf der Reise zugestoßen? DER BEDIENTE. Je, nun! Wir sind im Hohlweg umgeworfen. ADAM. Pest! Mein geschundner Fuß! Ich krieg' die Stiefeln - LICHT. Ei, du mein Himmel! Umgeworfen, sagt Ihr? Doch keinen Schaden weiter - ? DER BEDIENTE. Nichts von Bedeutung. Der Herr verstauchte sich die Hand ein wenig. Die Deichsel brach. ADAM. Daß er den Hals gebrochen! LICHT. Die Hand verstaucht! Ei, Herr Gott! Kam der Schmied schon? DER BEDIENTE. Ja, für die Deichsel. LICHT. Was! ADAM. Ihr meint, der Doktor. LICHT. Was? DER BEDIENTE. Für die Deichsel? ADAM. Ach, was! Für die Hand. DER BEDIENTE. Adies, ihr Herrn. - Ich glaub', die Kerls sind toll. (*Ab.*) LICHT. Den Schmied meint' ich. ADAM. Ihr gebt Euch bloß, Gevatter. LICHT. Wieso? ADAM. Ihr seid verlegen. LICHT. Was! (*Die erste Magd tritt auf.*) ADAM. He! Liese! Was hast du da? ERSTE MAGD. Braunschweiger Wurst, Herr Richter. ADAM. Das sind Pupillenakten. LICHT. Ich, verlegen! ADAM. Die kommen wieder zur Registratur. ERSTE MAGD. Die Würste? ADAM. Würste! Was! Der Einschlag hier. LICHT. Es war ein Mißverständnis.[2]

"Mißverständnis" ist ein alltagssprachlicher Begriff, der das von jedem erfahrene Phänomen bezeichnet, daß einer in einem Gespräch etwas anderes versteht als sein Dialogpartner gemeint hat. Das Eingangszitat aus Kleists "Der Zerbrochne Krug" enthält mehrere Beispiele für einen Typ von Mißverständnissen, der sich textlinguistisch durch Diskrepanzen der referentiellen Bezüge explizieren läßt. Der Bediente bietet zwei Themen an: Verstauchung der Hand und Bruch der Deichsel. Licht greift beide Themen auf, referiert trotz Wiederholung des ersten Themas auf das zweite, Adam auf das erste. Die den Bedienten verwirrende Verschachtelung der Themenstränge Verstauchung und Doktor, Deichselbruch und Schmied führt hier - anders als beim "Aneinandervorbeireden" (wie in Dialogen bei Pinter oder Kroetz), das nicht zu 'Mißverständnissen' führt, weil aufgrund mangelnder Verstehensmotivation nichts zur Verständnissicherung unternommen wird - zum Abbruch der Kommunikationssequenz ("Adies, ihr Herrn") und bewertender Reaktion ("Die Kerls sind toll"). Lichts nachträglichen Klärungsversuch beantwortet Adam mit metakommunikativem Kommentar, auf den Licht noch referiert, als Adam schon das nächste Mißverständnis (von Liese) provoziert hat: die Fehlinterpretation der pronominalen und deiktischen Verweise auf Würste und Akten. Lichts (dreifach beziehbares) Fazit bezeichnet also einen Typ von Mißverständnissen, der durch Verstoß gegen das Postulat der Herstellung von Referenzidentität konstituiert wird. In der Literaturgeschichte läßt sich dieses Verfahren als ein beliebtes dramaturgisches Mittel zur Erzeugung von Komik verfolgen.

Auch ein zweiter Typ von Mißverständnissen hat in der Literatur meist die Funktion des witzigen Effekts: die Fehlinterpretation einer Aussage aufgrund ihrer kognitiven Nichterfassung. Der gemeinte Sinn wird durch einen unterstellten substituiert, die Diskrepanz führt zum Mißverständnis. In Hauptmanns "Ratten" versucht der Theaterdirektor vergeblich Klarheit über die Identität des Kindes der Frau John zu gewinnen: Zustimmung und Widerspruch seiner Dialogpartnerinnen beruhen gleichermaßen auf dem Nicht-Verstehen des Fremdwortes:

> HASSENREUTER. Und dies hier sei eben das strittige Kind? [...] Das Kind der Frau John kenne ich und bin sicher, daß es mit diesem durchaus nicht identisch ist.
> FRAU KIELBACKE, *respektvoll*. Nee, nee, det muß wahr sind: et is nich identisch.
> DIE PIPERKARCKA. Det Kindeken is janz jenug identisch, wenn ooch bißchen schlecht jenährt und schwächlich is. Det is janz richtig hier mit det Kind! Will Eid schwören, daß richtig identisch is.
> HASSENREUTER. Ich bin sprachlos.[3]

Natürlich kann dieser Typ des Mißverständnisses, der durch den Verstoß gegen das Postulat der prinzipiellen Möglichkeit der Herstellung von Referenzidentität qua kognitiver Kapazität, Vorwissen, Interaktionsgeschichte etc. konstituiert wird, auch sehr viel komplizierter realisiert werden als in den Fehlern der Mrs. Malaprop (die seit Sheridans "Rivals" der Stilfigur den Namen gab) oder der Frau Stöhr (deren Fischsoßenerläuterungen im "Zauberberg" Thomas Mann zu hintergründiger Gesellschaftssatire dienen).

Einen anderen Typ von Mißverständnissen repräsentieren die Konflikte, die dadurch entstehen, daß zwei Dialogpartner ihren gemeinsamen Gesprächsgegenstand unterschiedlich bewerten. In

Kafkas "Prozeß"[4] schätzt K. die Bedeutung seines Prozesses völlig anders ein als seine Dialogpartner (s.u.). Seine Bewertungen - etwa gleich zu Beginn im Dialog mit Frau Grubach sein Kommentar zum Ausbleiben von Fräulein Bürstner: "es kann aber zu weit gehen" (p 23) - kommen anders als beabsichtigt beim Dialogpartner an: Frau Grubach nimmt K.s Kommentar zum Anlaß, beflissen über Fräulein Bürstners Lebensstil zu tratschen, was K. brüsk zurückweist:

> Sie sind auf ganz falschem Weg [...] übrigens haben Sie offenbar meine Bemerkung über das Fräulein mißverstanden, so war es nicht gemeint ... (p 24).

Mißverständnisse stellen also im Unterschied zu kommunikativ ausgetragenen sozialen Konflikten oder Meinungskonflikten eine Kategorie zur Beschreibung von Kommunikationskonflikten dar. Aber nicht alle Kommunikationskonflikte lassen sich so einfach erklären wie die bislang eingeführten und gerade das zum Topos literaturwissenschaftlicher Interpretationen gewordene "Kafkaeske" in dem minutiösen Protokoll des Scheiterns aller Verständigungsversuche K.s vermag die bisherige alltagssprachliche Verwendung des Begriffs 'Mißverständnis' nicht zu fassen. Hierzu sind zunächst präzisere Festlegungen, dann die Erweiterung des Kategorieninventars vonnöten.

Mißverständisse im bis hierher entfalteten Sinne können also aus referentiellen, kognitiven, evaluativen usw. oder (im Status übergreifenden) 'interaktionslogischen' (s.u.) Diskrepanzen in den mutuellen dialogischen Definitionshandlungen (SCHANK 1976) bzw. der auf Objekte gerichteten Bewußtseinsgehalte zweier Dialogpartner (RICHTER 1978) resultieren. Im folgenden verwende ich daher den Begriff 'Mißverständnis' wie folgt:

> B 'mißversteht' die Proposition P einer Äußerung U von A, insofern B eine textlinguistisch mögliche (kontextabstrakt oder oberflächenlogisch zulässige) Proposition P' aus U dekodiert, die von der von A intendierten P aufgrund der Diskrepanz der Objektdefinitionen oder der Verletzung interaktionslogischer Basisregeln signifikant abweicht[5].

Basisregeln betreffen die mutuellen Definitionshandlungen der Dialogpartner, d.h. die Unterstellungen über den Ablauf des kommunikativen Prozesses und die daraus abgeleiteten Handlungsverpflichtungen der Interaktanten hinsichtlich "praktischer Idealisierungen" (SCHÜTZ) zur Aufhebung prinzipieller Unvereinbarkeiten wie

- Inkommensurabilität der beteiligten Selbstidentitäten
- Vagheit der realisierten Zeichensysteme
- Diskrepanz der Zeichendekodate
- Handlungsnotwendigkeit auf der Grundlage unterstellter/antizipierter Handlungsorientierungen des Partners etc.[6]

Die von A intendierte Proposition P in einer Dialogphase i lassen sich in Anlehnung an RICHTER 1978 als Elemente oder Einzelthemen T_j eines geordneten n-tupels P^A darstellen:

$$P_i^A : [T_1, T_2, \ldots, T_j, \ldots, T_n]_i^A$$

Jedes Element des Tupels ist in das von A realisierte, semiotisch manifestierte m und das von B verstandene f zu differenzieren, so daß ein Element (Objektdefinition, Referenzakt, Thema o.ä.) ein je geordnetes Paar bildet:

$$T_{j,i}^A : [m_j, f_j]_i^A$$

Die prinzipielle Divergenz von m_j und f_j wird in alltäglicher Kommunikation "praktisch aufgehoben" oder ihrerseits metakommunikativ thematisiert. As eigene Objektdefinition m_j kann nun wieder als n-tupel aufgefaßt werden, das die Werte von n Abbildungen des von T_j thematisierten Objekts O_j enthält. Wenn wir die bisher genannten herausgreifen, ergeben sich $3 < n$ Möglichkeiten:

$$m_{j,i}^A : [b(O_j), k(O_j), v(O_j)]_i^A$$

Dabei sei b die Beachtung, k die kognitive Erfassung, v die Bewertung des Objekts durch A in der Phase i. Desgleichen kann die unterstellte oder vorausgesetzte Objektdefinition f_j als n-tupel aufgefaßt werden; dann würde entsprechend b' als von A unterstellte Beachtung von O_j durch B, k' als von A unterstellte kognitive Erfassung von O_j durch B, und v' als unterstellte Bewertung von O_j durch B in der Phase i gelten:

$$f_{j,i}^A : = [b'(O_j), k'(O_j), v'(O_j)]_i^A$$

Das Hauptmann-Beispiel könnte also durch die Faktoren $k(O_j)_i^A \neq k(O_j)_i^B$, das Kafka-Beispiel durch die Faktoren $v(O_j)_i^A \neq v(O_j)_i^B$ bzw. $v'(O_j)_i^A \neq v(O_j)_i^B$ bestimmt werden. Statt nun das elaborierte Richtersche Modell des Kommunikationskonflikts, dem wir bis hierher gefolgt sind, zu vervollständigen, sei bereits auf dieser einfachen Ebene seine Kombinierbarkeit mit konversationsanalytischen und literaturwissenschaftlichen Fragestellungen geprüft. Dabei wird der "Versuch, K.s Mißverstehen zu verstehen" vielleicht anschaulicher, wenn ich die bislang postulierten Weisen des Mißverstehens *per exemplum* entfalte.

Die These, daß "das Kafkaeske" wesentlich auf Mißverständnisse provozierende "Interaktionsfehler" zurückzuführen sei, wird durch eine Anwendung der obigen Bestimmungen bestätigt. Die Diskrepanz zwischen K.s Objektdefinitionen und deren Beachtung durch seinen Dialogpartner führt zu zahlreichen Brüchen der semantischen Kohärenz und konstituiert Phasengrenzen durch abrupte Themenwechsel:

K: "Wer sind Sie?" ...
F: "Sie haben geläutet?" (p 7)

Die textlinguistisch erwartbare Anknüpfung an das Thema des Vorsatzes beruht auf dem kommunikationssoziologischen Prinzip der Kooperativität (GRICE 1975). In 1978 habe ich anhand des Dialogs zwischen K. und Titorelli exemplarisch gezeigt, wie solche Konversationspostulate mißachtet werden: Titorelli entspricht K.s Informationserwartungen weder quantitativ noch qualitativ, seine Redebeiträge sind für K.s Handlungsziele weder relevant noch präzise und eindeutig (HESS-LÜTTICH 1978, Ms. pp 15-29). Nach den oben formulierten Postulaten wäre zu prüfen, ob $b(O_j)_i^K = b(O_j)_i^T$, also ob, allgemein formuliert, T.(itorelli) K.s Handlungsschema ratifiziert oder nicht (und *vice versa*: K. ist konversationell meist ebenso unkooperativ wie seine Dialogpartner).

T: "Wollen Sie Bilder kaufen oder sich selbst malen lassen?" ...
K: "Sie arbeiten gerade an einem Bild?" (p 125).

Solche textlinguistisch (durch Lexemrekkurrenz) oder logisch (durch Aufweis von Argumentationsstrategien: HESS-LÜTTICH 1978) "motivierbaren" Abschweifungen, Ablenkungen, Insertionen, Mißachtungen, die in der systematischen Häufung dem Dialogpartner das Gefühl des 'Abprallens' an einer Wand des Mißverstehens (oder unterstellten Mißverstehen-Wollens) geben, kommen immer wieder vor.

Den Divergenzen zwischen $b(O_j)_i^A$ und $b(O_j)_i^B$, die sich in den makrostrukturell aufweisbaren Brüchen der formalen Dialogorganisation und wechselseitigen Verstehensicherung (im Sinne von Sacks, Schegloff usw.) manifestieren, entsprechen die zwischen $v(O_j)_i^A$ und $v(O_j)_i^B$, die sich in den Unterschieden der wechselseitigen pragmatischen oder semantischen Bewertung von Objektdefinitionen spiegeln. So muß K. befremdet zur Kenntnis nehmen, daß alle seine Dialogpartner die Situation des "Angeklagtseins", in der er sich befindet, anders interpretieren als er selbst; die unterschiedlichen Attribute, mit denen die einzelnen Dialogpartner den 'Prozeß' belegen, haben gemeinsam, daß sie seine potentielle Gefährlichkeit erheblich ernster nehmen als K. zu akzeptieren bereit ist; sogar in der Verwendung des fachsprachlichen Registers der juristischen Argumentation bestehen erhebliche Unterschiede zwischen K. einerseits und seinen Dialogpartnern andererseits; das beginnt mit dem Mißverstehen des Terminus 'Verhaftung' und seiner Folgen und endet mit der Identifikation von 'Anklage' und 'Schuld', objektivierbarem Gesetz und mythischer 'Achtung vor der Schrift' im Dom-Dialog.

Die Relation von $k(O_j)_i^A$ und $k(O_j)_i^B$ läßt sich im weiteren Sinne auch auf die unterschiedlichen Gebrauchskonventionen von Ausdrücken anwenden, die sich hier vor allem als Verstoß K.s gegen die essentielle Vagheitsbedingung umgangssprachlicher Kommunikation (cf. z.B. GARFINKEL) ausweist und auf diese Weise Mißverständnisse aufgrund mangelnder Erfassung des vom Partner Gesagten provoziert. Dies gilt z.B. für das Buchstäblich-Nehmen gruppentypisch metaphorisierter Begriffe: K. hält an der umgangssprachlich-wörtlichen Verwendung in ihrem semantischen Gehalt längst verschobener, metaphorisierter Begriffe des Gerichtswesens fest (Anklage, Gesetz, Schuld, Gericht, Recht, Untersuchung etc.).

Solche Divergenzen setzt das Konversationspostulat der beiderseitigen und beim Dialogpartner vorausgesetzten Bereitschaft zur Verständigung auf der Basis geteilter Wissensbestände und Regelinventare außer Geltung, wenn sich z.B. das von K unterstellte Interesse (b') an O_j ("Freispruch") bei T. oder die von K vorausgesetzte Bewertung (v') von O_j ("Prozeß") durch B (Frl. Bürstner) regelmäßig als falsch erweisen. In nicht-pathologischer Alltagskonversation werden signifikante Diskrepanzen zwischen v und v', zwischen B.s Bewertung und von A antizipierter Bewertung B.s durch semantische Instruktionen des Textes ("Schlüsselsymbole", die auch nonverbal sein können!) zu "höhersymbolischer Reinterpretation" der verketteten (oder konkomitanten) Zeichen in alltagssprachlicher Verwendungssituation "motiviert" (SCHÜTZE 1977). Bei Kafka fehlen solche Instruktionen zur praktischen Angleichung der wechselseitigen Relevanzsysteme (SCHÜTZ 1971: Reziprozitätsidealisierungen) entweder völlig, oder die der verschiedenen semiotischen Ebenen widersprechen einander: Frau Grubach signalisiert verbal Sympathie ("bester und liebster Mieter"), ohne dies nonverbal zu bekräftigen ("und vergaß natürlich auch den Handschlag", um den K. zur Besiegelung ihrer Übereinstimmung gebeten hatte, p 23). Ähnliche Interaktionskrisen oder Mißverständnisse ließen sich aus den Dialogen mit Frl. Bürstner, mit Leni, mit der Frau des Gerichtsdieners, mit den Untersuchungsbeamten, mit dem Gefängnisgeistlichen u.a. herauspräparieren.

Am Beginn der meisten Interaktionsbindungen, die K. eingeht, steht ein solches scheinbar unbedeutendes 'Mißverständnis' (dessen jeweiliger Typ sich durch Anlegen des oben entwickelten Rasters genau angeben läßt), das man aber sofort wieder vergißt. Nach einem solchen ersten Mißverständnis normalisiert sich die Verständigung jedoch nicht ohne Rest, sondern bleibt problematisch. Das beginnt mit den Mißverständnissen in den Dialogen mit Frau Grubach und Fräulein Bürstner (Diskrepanzen der Signalinterpretation bzw. der Objektbewertung) und endet bei den Verhören der 'Untersuchung', die nicht zu wechselseitigem Verständnis führen kann, da die Gerichtsbeamten im institutionellen Handlungsrahmen des Gerichtswesens strategisch argumentieren, während K. im alltagsweltlichen Handlungsrahmen seiner persönlichen Situation moralisch argumentiert (cf. ALLEMANNS Differenzierung der Rechtfertigungsmodi bereits 1963, pp 255 seq.). Die völlige Wirkungslosigkeit der Argumentation K.s resultiert aus dieser Divergenz der Objektdefinition, denn *inner*halb eines strategischen Handlungsschemas ist strategische Reaktion Bedingung argumentativen Erfolgs, wenn es nicht selbst zum Gegenstand des 'Diskursus' auf der Metaebene gemacht wird.

Die scheinbar nicht zu durchbrechende Kette der Konsequenzen läßt sich rekonstruieren. Die zunächst vermeintlich unbedeutenden Brüche interaktionslogischer Erwartungen, die in Alltagsdialogen mittels "praktischer Idealisierungen" automatisch aufgehoben werden, führen durch die Nachwirkungen der Diskrepanzen in der Objektdefinition zu einem Mangel an Vertrauen in die eigene Wahrnehmungs- und Typisierungskompetenz. Diese Unsicherheit wiederum führt zu noch geschärfterer und empfindlicherer Umweltperzeption, die, bis in die syntaktische Konstruktion hinein nachweisbar, noch in dem flüchtigsten Gespräch eine Vielzahl von Beobachtungen des Partnersverhaltens sorgfältig registriert (cf. BINDER 1976). Dadurch treten wiederum die mißtrauisch antizipierten Divergenzen der kommunikativen Signalbündel mikroskopisch vergrößert hervor und gewinnen, an der Norm gemessen, dysfunktionales Gewicht. Und schon sind wir mitten im - *in sich* streng logisch ablaufenden! - *circulus vitiosus*, den die Literaturwissenschaft schon immer erahnt und gespürt hat (KOBS 1970), und der m.E. in der Präzision der Darstellung eines progredienten Verlusts alltagsweltlicher Interaktionsbindungen Büchners "Lenz" ebenbürtig ist.

Anmerkungen

1 Die Skizze basiert auf Vorträgen, die ich im Rahmen des 13. Linguistischen Colloquiums am 20.9.78 in Gent und am 22.9. in Paris auf dem 11. Kongreß der Societas Linguistica Europaea gehalten habe. Die Kürzungen betreffen vor allem einen umfangreicheren Teil mit exemplarischen Phasenstrukturanalysen Kafkascher Dialoge. Diesen Teil sowie die Auseinandersetzung mit der literaturwissenschaftlichen Sekundärliteratur konnte ich in meinem Aufsatz "Dialogprozesse im *Prozeß*. Zur Funktion des Kommunikationskonflikts bei Kafka" entfalten, der in HEINDRICHS/RUMP (eds.) 1978 erscheint. Grundlage der folgenden Argumentation ist im wesentlichen eine Verbindung von F. Schützes Versuch, konversationsanalytische Kategorien auf literarische Dialoge anzuwenden (SCHÜTZE 1977), mit Richters Vorschlag eines formalen Modells zur Explikation von Kommunikationskonflikten (RICHTER 1978).

2 Kleist, Werke, p 163

3 Hauptmann, Werke vol. II, p 790

4 Kafka, Werke, vol. II: Zitate nach der 7-bändigen Werkausgabe von Max Brod (Frankfurt 1976) im folgenden durch Seitenangabe im Text

5 Zur Unterscheidung der Basisregeln von der 'Oberflächenlogik'

kommunikativer Interaktion cf. Cicourel 1973, besonders pp 167 seqq. Seine "Definition" des Konzepts ist allerdings noch präzisionsbedürftig ("Die Basisregeln verschaffen ein Gespür von sozialer Ordnung, das für die Existenz oder ... den Aufbau einer normativen Ordnung ... fundamental ist", loc.cit., p 173).

6 Zur genaueren Explikation der Kategorien cf. Kallmeyer/ Schütze 1976; Schütze 1977

Literatur (Auswahl)

ABS = Arbeitsgruppe Bielefelder Soziologen (ed.) 1973: Alltagswissen, Interaktion und gesellschaftliche Wirklichkeit. Reinbek: Rowohlt.

ALLEMANN, B. 1963: "Kafka. Der Prozeß", B. v. Wiese (ed.): Der deutsche Roman. Vom Barock bis zur Gegenwart II. Düsseldorf: Bagel.

BERENS, F.-J. et al. 1976: Projekt Dialogstrukturen. München: Hueber.

BINDER, H. 1976: Kafka in neuer Sicht. Mimik, Gestik und Personengefüge als Darstellungsformen des Autobiographischen. Stuttgart: Metzler.

CICOUREL, A.V. 1973: "Basisregeln und normative Regeln im Prozeß des Aushandeln von Status und Rolle", ABS (ed.) 1973: 147-188.

GARFINKEL, H. 1973: "Das Alltagswissen über soziale und innerhalb sozialer Strukturen", ABS (ed.) 1973: 189-269.

GRICE, H.P. 1975: "Logic and Conversation". P. Cole/J.L. Morgan (eds.): Speech Acts (= Syntax and Semantics vol. III). New York etc.: 41-58.

HAUPTMANN, G. 1962 seqq.: Sämtliche Werke (Centenar-Ausgabe). ed. H.E. Hass/M. Machatzke. Frankfurt/M.: Propyläen.

HEINDRICHS, W./C. Rump (eds.) 1978: Dialoge, Hildesheim: Gerstenberg im Druck .

HESS-LÜTTICH, E.W.B. 1978: "Dialogprozesse im *Prozeß*. Zur Funktion des Kommunikationskonflikts bei Kafka". Heindrichs/ Rump (eds.) 1978 [im Druck]

--- /(ed.) [in Vorb.]: Literatur und Konversation. Sprachsoziologie und Pragmatik in der Literaturwissenschaft. Wiesbaden: Athenaion.

KAFKA, F. 1976: Gesammelte Werke in sieben Bänden. Ed. M. Brod. Frankfurt/M.: M. Fischer.

KALLMEYER, W./F. Schütze 1976: "Konversationsanalyse". Studium Linguistik 1 (1976): 1-28.

KLEIST, H.v. O.J.: Sämtliche Werke. ed. K.F. Reinking. Wiesbaden: Löwit.

KOBS, J. 1970: Kafka. Untersuchungen zu Bewußtsein und Sprache seiner Gestalten. Bad Homburg: Athenäum.

KREIS, R. 1976: Die doppelte Rede des Franz Kafka. Eine textlinguistische Analyse. Paderborn: Schöningh.

NEUMANN, G. 1973: "Umkehrung und Ablehnung: Franz Kafkas 'gleitendes Paradox'". H. Politzer (ed.) 1973: Franz Kafka. Darmstadt : 459-515.

POLITZER, H. 1978: Franz Kafka. Der Künstler. Frankfurt/M.: Fischer.

RICHTER, H. 1978: "Konflikte in Dialogen". Heindrichs/Rump (eds.) 1978 [im Druck].

SACKS, H./E.A. SCHEGLOFF/G. JEFFERSON 1974: "A Simplest Systematics for the Organisation of Turn-Taking for Conversation". Language 50 : 696-735.

SCHANK, G. 1976: "Zur Binnensegmentierung natürlicher Dialoge". Berens et al. 1976 : 35-72.

SCHEGLOFF, E.A. 1968: "Sequencing in Conversational Openings". American Anthropologist 70 : 1075-1095.

SCHÜTZ, A. 1971: Das Problem der Relevanz. Frankfurt/M.: Suhrkamp.

SCHÜTZE, F. 1977: Interaktionspostulate - am Beispiel literarischer Texte. Ms. Bielefeld. (demn. in: Hess-Lüttich [in Vorb.]).

SOME PROPOSALS FOR A CO-TEXTUAL APPROACH TO COHERENCE IN ENGLISH TEXTS[1]

James Monaghan

1. Preamble

Although linguistics has in the last decades been dominated by American transformational grammar, it is in Europe that the study of texts from a linguistic point of view has developed. This is because the transformational-generative approach to language, as an essentially neo-structuralist model (cf. MONAGHAN 1979), is so designed as to make a fruitful approach to texts difficult if not impossible. Moreover, the real achievements of the American tradition have inhibited a clear appreciation of its limitations. I propose to look at two of these limitations which cause text studies the most trouble and then to outline a much less abstract way of showing how readers interpret a non-invented, non-simplified, real-life text.

2. Problems

2.1. The Primacy of the Sentence

The first problem for text linguistics is the wide assumption of the primacy of the sentence. This notion, based ultimately on the idea of language as a system of signs, is accepted uncritically by CHOMSKY (1957:13): "From now on I will consider a language to be a set of sentences". He then defines linguistics as fundamentally the study of sentence structures.

Now, there is no denying the usefulness of the sentence as a linguistic category. The major problem, however, is that it gets harder and harder to isolate sentences as language gets more and more usual. We do not speak in sentences and often do not read in them.[2] Since ordinary conversation makes up the most of our language activity, we find ourselves unable to reduce most of our widely-accepted language activity to fit the definition of language used by Chomsky.

There are two things we can do. We can say the concept sentence is a tool, when it stops being useful we stop using it. Or we can say the sentence is really essential to our theory and try to explain away the difficulties. This latter solution is what has been tried in the transformational-generative tradition and some text linguists have been lured into accepting as essential parts of linguistic theory what are essentially ad hoc attempts to rescue a structuralist approach to language. The main support for the concept of the sentence as the sole valid unit of linguistic description is the theory of competence and performance and the ideal native speaker in a completely homogeneous speech community. The ideal speaker is always deputized for by people like you and me sitting at their desks inventing sentences. A look at some of the 'texts' invented for the purpose of analysis is enough to show us how unlike real texts they are, especially if the real texts are not scientific prose but straightforward pieces like the text reproduced below. This also holds for rewriting of real texts so as to remove most of the significant (but difficult) features to make the analysis work. Since the ideal speaker-hearer (or should it be writer-reader[3]) has only an alibi function and, moreover, is seldom justified, we can move straight to a much more central distinction.

2.2. Competence-Performance

The reason why Chomsky built the distinction of competence and performance into his theory was that he needed a wastebasket to get rid of awkward features of language. Competence is the ability of an ideal native speaker to agree with a particular transformational grammar's output. Performance is that plus every other aspect of your non-idealised linguistic behaviour. People, however, seem mesmerized by the term competence. HYMES talks of "communicative competence" (1971), WUNDERLICH "pragmatische Kompetenz" (1968:9), HARTIG & KURZ elaborate the distinction to "Performanz, Kompetenz und Performanzkompetenz", defined as "die Befähigung zur Performanz" (1971:116). Are all these uses of the distinction the same as Chomsky's? I don't think so. They are puns on competence's everyday meaning. Noone will dispute that speakers of a language are, in the everyday sense, competent to speak that language. But since it is so obvious it is not very useful to say. The truth is the distinction is unnecessary. It was invented to allow the linguist to bring his whole attention to bear on simple matters like structural acceptability. If you want to take a wider view of communicative behaviour it hems you in. The text linguist at least has no use for it because it cuts him off from his texts. Even if you can conceive sentences being produced in an ideal environment, texts are not.

2.3. Texts as Signs

In spite of KATZ & FODOR's attempt (1963) to prove otherwise, sentences do not occur without a context, or if they do they seldom succeed. They occur in texts which are part of situations. While the founders of modern text linguistics were further on than this, they had not got rid of all structuralist lumber. For instance, Hartmann says: "Ausgangspunkt einer Phänomenologie des linguistischen Objekts ist die Texthaftigkeit des originären sprachlichen Zeichens" (quoted in SCHMIDT 1973:12). If texts are to be seen as signs, however, then they are structural units (or objects). This idea encourages a proliferation of constructs filling out that structure, and this process is often derived from a naive belief that language *is* of this or that nature. Theoretical constructs do not, however, have a right to exist. They justify their existence by what the linguist can do with them. Until he does do something with them they cannot be accepted as useful or necessary. So the idea of the text as a structure or the idea of a sentence as basic to linguistics has to be proved in terms of what the linguist does with them.

My alternative to seeing language as an object which exists, is to look at it as a process which happens. It happens in a situation which is defined by a culture. This allows us to make sense of utterances like the following real example (I only make up examples at the last resort.)

Policeman: Haben Sie was gesehen?
Bricklayer: Es gibt Leute die arbeiten müssen.

Policeman: Schade, ich habe Maurer normalerweise sehr gern.

Bricklayer: Du mich auch!

Here we have a text where inner-linguistic coherence is scarce but we have no difficulty following what is meant.

3. Texts as Events

When we learn language we learn it as texts in situations. We use what we have heard in one clause to predict what is coming and thus we can understand what we don't hear. When listening to a conversation you often do not hear individual words, but this doesn't trouble you too much, because you can guess what is coming next. This is a difficulty with foreign languages. When you don't speak the language very well, you can't guess as well as you can in your own language. In understanding language, we make certain important assumptions. Among these the notion that the speaker or writer is going to tell us something we don't know. The speaker or writer complements this by telling us what he is doing. Even if he doesn't, we try to form connections. We do this so automatically and imaginatively that we often make connections which are not intended. Another collected example shows this. It is the case of the two signs photographed in England beside a narrow alley: GENTLEMEN and CYCLISTS PLEASE DISMOUNT. Although the signs are not meant to be taken together, the point of the joke is our attempt to do so. What makes us all agree on how we would relate these two signs if they were a unified text?

4. Signalling Clause Relations

In the mid-sixties E.O. Winter while working on a large corpus of scientific English developed a method of classifying the fact that contiguous sentences are understood in terms of each other. He found that the easiest way of doing this was to ask which question connected each pair of sentences. (Applying this to our examples above, the answer is a question of the form: "What is the consequence of this alley being (or containing) a toilet".[4] The more he did this the more he found that the same questions kept cropping up (WINTER 1974). Moreover, the questions were found to contain a set of words which characterized the clause relations concerned.

Even more important, these words were often used in the texts themselves to signpost the very relationship which the (metalinguistic) questions were exposing. This material was published rather fitfully until 1977. In his book of that year, Winter ties in his work to some basic notions in the Neo-Firthian tradition.

He points out that traditionally there is a distinction made between "open class" and "closed system" words. Closed system words form small, finite groups whose main function is to organize the open-class words (WINTER 1977:2). He draws particular attention to those closed-system words which connect sentences or clauses. There are what he calls Class 1 words or subordinators and Class 2 words or sentence adjuncts. What struck Winter about the recurring vocabulary of his questions was that there seemed to be a definite number of them and that they stood in paraphrase relation to words in Class 1 and Class 2. He concluded from this, that these words represent a third way of signposting the relationships expressible by subordination and sentence adjunction. He further suggests that the possible relations between sentences (or clauses) "are finite and few in number and can be named by this special vocabulary" (WINTER 1977:2). These words are semantically a closed systems but lexically an open set. Let's look at examples. There are, says Winter, "two rule-governed ways in which we interpret one sentence in the light of another" (WINTER 1977:6). Either we compare them or we relate them in terms of change in time or space. These two relationships he calls Matching and Logical Sequence. If two sentences are related by a Matching relation they presuppose a linking question which has a meaning we can express by words like <u>same</u>, <u>different</u>, <u>compare</u>, <u>true</u>, <u>contrast</u>. If they are Logical Sequence they have the semantic relation expressible in words like <u>consequence</u>, <u>cause</u>, <u>reason</u>. You will notice that, as in prepositional phrases, these words function as clause adjuncts i.e. Class 2 items e.g. <u>in comparison</u>, <u>in conclusion</u>. Matching and Logical Sequence relations can be combined in one sentence pair and since the relationship is semantic it is not affected (at this level) by whether, say, the relation is signalled by <u>difference</u>, <u>differentiate</u>, <u>different</u> or <u>differ</u>.

All clauses in a coherent text (i.e. one which doesn't strike you immediately as badly written or disjointed) are therefore semantically connected to other clauses. This connection can be left implicit. In this case, we supply the link. It can and usually is made explicit especially by the use of subordination, sentence connection or outright labelling with Class 3 words. When the last method is used we get a situation which Winter calls anticipation. When a clause anticipates, it makes explicit in advance what the next clause relation will be (WINTER 1977:8). Class 3 words are used to signal relationships of this type.

Winter has collected just over 100 words which are used as Vocabulary 3 items. He sub-classifies them into signposts for four types of clause relation. The vast majority of the words signal the Matching or Logical Sequence Relations. Most of these items can paraphrase items of Vocabularies 1 and 2. There are three other small groups which express special connective roles in discourse structure (WINTER 1977:18). The first of these are words like _situation_, _problem_, _solution_ and _evaluation_ (I would add here, and their synonyms). These signal the larger clause relation of the Metastructure of an argument, as we will see below. Secondly, there are the words which express what Winter calls the Attitudinal Function. These are words expressing meanings like _surprise_, _expect_, etc. and paraphrase sentence connectors like _fortunately_, but also _but_, _or_, _and_ etc. Then there are what WINTER (1977:19) calls the anaphoric connectors of the clause (the same as HALLIDAY & HASAN's "General Nouns") such as _action_, _event_, _do_, _happen_, _move_. Winter points out that these are merely ways of talking about the fundamental information content of a preceding clause or sentence (WINTER 1977:19), they do not connect facts as such.

What Vocabulary 3 words have in common is not only that they represent the notions underlying clause relation "made explicit by Vocabularies 1 and 2 but also that they themselves may be used to make the relation explicit by saying what the relation is" (WINTER 1977:22). Note that the three vocabularies do not represent three different systems. They are three different grammatical environments for the same semantic system, the clause relations of English. A corollary of the fact that these items label relations is that in addition to normal collocational restraints, Vocabulary 3 words must have somewhere in their context a realization which fulfils the expectations which they awake.

5. Sample Analysis

Let us look at our text from this point of view. It is from "The Observer" of 5th March 1978. It was chosen at random and at first glance it is not 'difficult'. It illustrates, however, several of my points. (You will be horrified by the illogicality of my sentence numbering. I find Miss Whitehorn's punctuation almost as arbitrary. I have only numbered sentences and paragraphs so that I can refer to paragraph 2 sentence 1 etc. as 2.1.)

Sentence 1 in the first paragraph (1.1.) is connected internally by the logical sequence relation expressed by subordination. "What would be the result if I said...?" and "For what reason wouldn't we know?". Sentence 1.2 is matched to 1.1 by the Vocabulary 2 item otherwise. This relation of comparative denial is underlined by the evaluation of the situation think it absurd and sentence 1.2 is matched to sentence 1.3 by Vocabulary 2 so.

Parallel to the, as it were, internal sentence connection is the author's attitudinal markers. In the second sentence (1.2.) And of course and in the third (1.3.) but tell us that the predication of sentence 1.2. is not surprising in the light of sentence 1.1., whereas 1.3. is a surprising addendum to sentence 1.2.

If we go on to look at the text as a whole, we get the argumentative structure clearly labelled. The facts of the first paragraph are brought into relation with the problems they throw up. The word problems occurs in 2.1. and in the key sentence 3.1. It is paraphrased by (no) trouble in 2.2. and by pest in 4.2. The anticipation set up by listing something as a problem is satisfied by 7.2. where the answer is labelled as such.

6. Conclusion

This sketch of the signposting in the text looks both pedantic and incomplete. Pedantic because it reflects how we automatically understand a text. Incomplete because every clause is related to every other clause with or without overt signals. Clause relations may be across or within sentence boundaries. As with information structure in spoken texts, the sentence is not relevant as a primary unit of textual analysis. The text is not a structural unit, it is a speech act. If we stop the ongoing development of meaning in such a speech event by isolating one clause or sentence, we find a finely woven network of meanings relating forwards and backwards. A text is best seen as a semantic unit. The successes of modern syntactic theories do not mean all other branches of linguistics have to ape syntax. (Remember 'sememe', 'prosodeme', 'texteme' etc. as spin-offs from the success of 'phoneme'!) Rather we need methods, like the above, which allow us to look at real texts and analyse them as they stand.

NAMING THE GAME KATHARINE WHITEHORN

Paragraph 1. (1) IF I SAID Henry Maginnis was flying in, accompanied by Pierre Sinclair and met by Peter Prout, you wouldn't know who I meant; since they've gone on working under their maiden names: Kissinger, Trudeau, Walker. (2) And of course they'd think it absurd to do otherwise. (3) But so, it seems, do quite a few women, who choose to hang on to their own names when they marry.

Paragraph 2. (1) Jean Coussins of the National Council for Civil Liberties (herself married to A.N. Other) is getting out a pamphlet in May on the rights and problems you have if you do this. (2) The law, it seems, is no trouble: you are what you are 'commonly' known as - even if it's Tootsy-Baby or Red Rum. (3) And you can have several: Mrs. Goodbody at home, Miss Fixit in the office; even Mrs. Antonio Peccadillo in Brighton.

Paragraph 3. (1) There are certainly problems. (2) Actresses, writers and such riff-raff have long kept their own names, if only because Miss Ellen Terry sounds so much *younger* than settled Mrs. Watts (not that she settled on him for long). (3) Many Fleet Street women are stuck with the name, not of their father or current husband, but the man they were married to when they first got a decent job; Messrs. Conran and Blandford are so much water under the bridge now, but Linda Blandford and Shirley Conran, several husbands on, are still known as such (and Linda, who deplores Ms, insists in *Miss* Blandford, at that).

Paragraph 4. (1) I write under my maiden name myself: the original idea was that as two journalists with separate names we could quarrel with different editors. (2) It didn't work, of course, since everyone in Fleet Street knows who is married and partly married to who; and in many ways it's been a pest. (3) It's funny to be made to write for the school magazine because you're the notorious Miss Whitehorn - and then have the by-line Lyall because that 's the pupil's name; but it's not funny when charming classmates call your kid bastard, which has happened. (4) People get muddled and call me Mrs. Whitehorn, which is my mother, or call Gavin Mr. Whitehorn, which enrages me - all right, so I admit that I'm unreconstructed too. (5) It reached its most chaotic when I used to work from a hotel in Paris as Whitehorn (Veetahorron, rather). (6) When Gavin Lyall turned up for the weekend they thought I'd found a lover at last, but if he didn't and friends rang asking for Mrs. Lyall, they supposed I was a pair of Lesbians sharing a room on the cheap.

Paragraph 5. (1) If I were starting again, I frankly wouldn't think it worth the trouble. (2) But then I haven't had to suffer - no working woman has - this awful blanking-out of identity which so many women feel, when they feel the world sees them only as an adjunct to the man and the children. (3) Nor, thank heaven, the misery of being chucked back into the open market - with the SOLD label indelibly attached.

Paragraph 6. (1) But what, anyway, is this Whitehorn handle? (2) My Father's name. (3) Wnat about my mother's? (4) (I had thought my mother's name, Gray, would make a great name for a daughter. (5) Gray Lyall, I said, lovely. (6) Gray lisle stockings, said a friend.) (7) Our grandmothers and great-grandmothers gave us as much as our fathers, but their line is lost in any case - unless one of them happened to be Queen Victoria, handing on haemophilia, to half the royalty in Europe. (8) Or, of course, in the trade which (according to social historian Peter Laslett) was handed down, with the name, from mother unto daughter; it was even, I'm afraid, spelt HOARE.

Paragraph 7. (1) I suspect the most sensible answer, clumsy though it is, is the pre-feminist American one. (2) A woman simply adds on her husband's surname at the end of her own - Anne Morrow Lindbergh, Clare Booth Luce - and can subtract it again if necessary. (3) And when a child comes of age, there's no reason, even now, why it shouldn't choose its mother's name not its father's - if it can face filling in enough forms.

Paragraph 8. (1) When the designer Chanel was asked why she kept on saying no to the Duke of Westminster, her reply was succinct. (2) 'Duchesses of Westminster!' she said, 'there are three of them. (3) But there's only one Coco Chanel'.

Notes

1 I wish to thank L. Lipka and W. Sedlak for detailed comments on earlier versions of this paper.

2 The reader might like to try some introspection here. What do we normally write? It is probably fair to say that it is unusual for the average language-user to write anything. People who write for a living vary in their use of sentences. Scholarly papers do tend to consist of sentences. Private letters less so. Journalists, novelists and poets are even less tied to the sentence, as can be seen from our text.

3 The assumption that spoken and written texts are produced and understood in the same way is extremely misleading. One result is that many linguists talk all the time about speakers but analyse only written language (cf. note 2 above).

4 The answer, if the two sentences were a text, would be that cyclists must dismount. Paraphrasing, we would then have a text specifying that cyclists should not attempt to enter or use a toilet while mounted.

References

CHOMSKY, Noam (1957): Syntactic Structures. The Hague etc.: Mouton.

HARTIG, M. & KURZ, V. (1971): Sprache als soziale Kontrolle. Neue Ansätze zur Soziolinguistik. Frankfurt/Main:Suhrkamp.

HALLIDAY, Michal A.K. & HASAN, Ruquaiya (1976): Cohesion in English. London: Longman.

HYMES, Dell H. (1971): On Communicative Competence. Philadelphia: University of Pennsylvania Press.

KATZ, Jerald, J. & FODOR, Jerry A. (1973): The Structure of a semantic theory. Language 39, 170-210.

MONAGHAN, James (1979): The Neo-Firthian Tradition and its Contribution to General Linguistics. Tübingen: Niemeyer.

SCHMIDT, Siegfried J. (1973): Texttheorie. München: W.Fink.

WINTER, Eugene O. (1974): Replacement as a Function of Repetition: a Study of Some of its Principal Features in the Clause Relations of Contemporary English. Ph.D. Dissertation for the University of London: Senate House.

--- (1977): A Clause-Relational Approach to English Texts. (Special Issue of Instructional Science Volume 6 No. 1).

WUNDERLICH, Dieter (1968): Pragmatik, Sprechsituation, Deixis. Universität Stuttgart. Lehrstuhl Linguistik: Papier 9.

POUR UNE METHODE INDUCTIVE EN SYNTAXE

Application au moyen français

Rika Van Deyck

1. Le sujet traité concerne la mise au point d'une recherche méthodologique. L'élaboration d'une technique d'analyse syntaxique en forme le but. Dans une perspective plus large, la problématique soulevée est celle de l'établissement d'une syntaxe de l'ancienne langue, plus précisément de celle du moyen français. L'objectif qui s'impose de façon prioritaire est la détermination d'une procédure qui permette l'induction du système idiolectal d'un auteur particulier des manifestations de sa parole. Une première application de la technique envisagée à une oeuvre concrète, celle de François Villon - selon le ms. Coislin -, forme l'aspect pragmatique de l'investigation[1]. L'accent sera mis ici sur l'aspect méthodologique de même que sur la justification théorique du travail.

2. Le fait que le système idiolectal peut être considéré comme un jalon intermédiaire entre la parole et la langue forme la motivation même de la recherche menée. Le terme est pris ici dans l'acceptation d'"Idioglosse" que Theodor BERCHEM explicite comme suit :

> das jedem Sprecher eigene, inhärente Strukturmuster und Laut-, Flexions-, Syntax-, Wortschatzarsenal der Muttersprache bzw. eines erlernten fremden Idioms, das sich im Idealfall mit der <u>langue</u> vollkomen decken würde, praktisch aber immer wohl nur einen mehr oder minder grossen Prozentsatz von ihr ausmachen wird[2].

Dans cette perspective, le système idiolectal forme un témoin partiel de la langue et fournit, de ce fait, une contribution importante dans l'approche de celle-ci. Il apparaît plus précisément comme une composante de la langue; dès lors cette dernière peut résulter, par recoupements successifs, de l'étude cohérente du système d'une classe d'idiolectes. Selon la conception de H.-H. LIEB[3] le système idiolectal occupe le rang inférieur dans une hiérarchie d'états de langue("Sprachstadien"). Il est lui-même un élément d'un système d'idiolectes, qui est à son tour un élément du système de la langue d'une période donnée; enfin, cette dernière constitue elle-même un élément de la langue globale. Ces divers états de langue reflètent donc une hiérarchie de systèmes qui se recoupent et qui progressent du partiel au complet.

D'autre part, le système idiolectal se laisse dégager des manifestations de la parole d'un individu puisqu'elles reflètent l'usage que fait celui-ci de l'état de langue qui lui est propre. Il convient donc d'établir une technique qui puisse pratiquer le passage de la parole au système idiolectal avant d'aborder l'étude de la langue de la période envisagée.

3. L'entreprise proposée exige, au préalable, une justification théorique de l'existence d'une linguistique de la parole.

Ici, la reconnaissance par Eric BUYSSENS, par Klaus HEGER et par Hans-Heinrich LIEB (et auparavant déjà par Karl BUHLER)[4] de niveaux d'abstraction intermédiaires entre la parole et la langue s'avère fondamentale.

La parole en tant qu'acte de langage, lié de façon indissociable à une situation spécifique, ne se prête pas à l'analyse et par suite, elle ne peut former l'objet d'une étude scientifique. En effet, toute actualisation de la parole se trouve nécessairement associée à un contexte, déterminé par les conditions mêmes de l'enchaînement discursif. Le caractère éphémère de ces conditions fait que, même si elle comporte des données récurrentes, toute séquence de l'énoncé reste spécifique et unique eu égard à la situation discursive, toujours différente. C'est ce déroulement "sui generis" qui caractérise la parole progressive et dynamique en voie de réalisation.

Par contre, dissocier les faits de parole de leur situation concrète c'est le rendre, par la même occasion, accessibles à l'analyse. Par cette seule opération d'abstraction ils deviennent objets d'analyse, de classification, de quantification. L'on accède donc à un niveau d'abstraction qui reflète la parole non pas dans son déroulement progressif, mais en tant que produit achevé, c'est-à-dire en tant que somme d'actualisations. C'est ce que traduit la notion de Σparole introduite par Klaus HEGER qui oppose l'expression momentanée ("Vorkommen") à une actualisation-type ("Typus-Vorkommen")[5]. C'est à ce niveau que le phénomène de la récurrence devient opérant. En effet, la somme des manifestations de la parole d'un auteur constitue une oeuvre accomplie à l'intérieur de laquelle il est possible de dégager des types d'actualisations, soit des entités linguistiques quantifiables.

L'opposition entre "Sprechhandlung" et "Sprachwerk" de Karl BUHLER apparaît donc comme essentielle pour l'élaboration d'une linguistique de la parole.

Cette justification théorique étant admise, il est permis de prendre la "somme-parole", et a priori l'un de ses aspects - sa syntaxe - comme objet d'analyse en vue d'y découvrir le système organisé correspondant. C'est admettre qu'il est possible d'opérer le passage de la parole d'un auteur donné à son système idiolectal.

Le choix d'une analyse formelle basée sur un corpus d'auteur, le choix d'une codification des faits de parole au plan d'analyse syntaxique retenu et d'un classement distributionnel des termes codifiés, enfin le choix d'une étude des fonctions syntaxiques forment les principales options méthodologiques.

4. Avant de préciser chacunes de ces options, il y a lieu de retracer brièvement la démarche suivie afin de pouvoir raccrocher chaque élément de la méthode à sa finalité pragmatique. La première phase de la recherche comprend une description des faits de parole relevés. Vu leur diversité et leur complexité, une telle description exige le recours à un certain niveau d'abstraction : c'est dans ce but qu'est conçue la codification des données, en alliance avec un classement distributionnel des matériaux. Cette technique descriptive aboutit à une typologie exhaustive des schémas de la parole, au plan d'analyse choisi (ici le groupe verbal). Dans cette première étape il est fait abstraction de la situation concrète du discours puisque l'on considère le corpus en tant que produit dissocié du contexte "hic et nunc" et que l'on y dégage le seul aspect syntaxique. C'est donc partir de la

parole d'un individu (P_1) pour aboutir à un aspect partiel de la Σparole correspondante. La seconde phase de la recherche nous conduit alors de la Σparole au système idiolectal de l'auteur envisagé (S_1) grâce à l'application d'une procédure de réductions. Les propriétés des fonctions syntaxiques fournissent ici les facteurs de réduction puisqu'elles aboutissent à l'établissement d'équivalences fonctionnelles. C'est le degré d'équivalence des schémas qui détermine leur degré de réductibilité. Schématiquement, la démarche se présente donc comme suit :

P_1	⟶	Σ_{nl}parole	⟶	S_1
PAROLE		PAROLE		SYSTEME
contextuelle		somme		IDIOLECTAL
VORKOMMEN		TYPUS-VORKOMMEN		IDIOGLOSSE
(Kl. Heger)		(Kl. Heger)		(Th. Berchem)

l = François Villon
n = auteur particulier[6]

5. Une justification des options méthodologiques peut se faire à présent en fonction du but visé.

Une approche formelle a reçu le préférence étant donné que l'auteur ne peut qu'avoir, face à l'ancienne langue, une compétence limitée. L'aspect sémantique, solidaire de l'aspect formel, est réduit à fournir un cadre nécessaire à la compréhension et à l'articulation du texte sous examen.

5.1. Le corpus est formé par tous les groupes verbaux du ms. de référence dont le support est une forme verbale conjuguée.

Le fait de préférer, à défaut d'autographe, le texte d'un manuscrit témoin à celui d'une édition critique se justifie dans la perspective du processus de la communication : en effet une copie produit un message qui, même s'il ne traduit pas exactement l'encodage voulu par l'auteur (émetteur du message), sera décodé en tant que tel par le lecteur contemporain (récepteur). L'établissement même de ce circuit d'information entre le copiste, source d'information et le destinataire, lecteur compétent, confère à la copie sa cohérence linguistique. C'est ce message qui prend vie dans l'esprit des lecteurs successifs et qui reçoit d'eux son (ses) interprétation(s).

Le choix d'une oeuvre poétique présente l'avantage de favoriser une exploitation maximale du code grammatical. Tout en se conformant au système de la langue, afin de garantir l'efficacité du message, le langage poétique transgresse à l'occasion la norme sociale[7] responsable de restrictions par rapport au système. Il en est donc moins esclave, si bien que l'on accède dans cette zone privilégiée qui sépare la norme sociale de la langue et où les possibilités théoriques sont, non pas maximales, mais moins limitées.

Le choix du groupe verbal lors de la délimitation d'un plan d'analyse syntaxique particulier reflète une exigence purement pratique. Si la notion de syntaxe est prise dans sa valeur large de discipline dont l'objet est l'analyse des rapports qui relient les unités fonctionnelles d'une structure déterminée, elle peut s'appliquer tant aux unités minimales (phonèmes) qu'à l'unité maximale représentée par le discours intégral. Le groupe verbal a été retenu ici en raison du fait qu'il forme un plan d'analyse intermédiaire où la création individuelle, responsable de la variété des chémas syntaxiques garde une place importante en regard des plans inférieurs tout en évitant l'écueil de déboucher sur la complexité de l'analyse transphrastique.

Opter pour une étude fondée sur un corpus - ce qui ne signifie pas limité à lui - implique une démarche inductive. C'est aussi reconnaître l'existence d'une structure immanente. Il reste que l'analyse garde inévitablement un caractère partiel dans la mesure où tous les phénomènes possibles ne sont pas attestés.

5.2. Nous avons adopté une technique descriptive qui comporte une codification des faits de parole en alliance avec un classement distributionnel des matériaux afin de rendre compte, le plus exactement possible, de leur multiplicité. Elle prévoit la substitution de codes appropriés aux termes du groupe verbal, ceci avec respect de leur ordre d'apparition dans l'énoncé. Ces unités codifiées font alors l'objet d'un classement distributionnel centré sur le support verbal.

5.3. Le distributionalisme s'applique donc dans ce cas aux fonctifs entre lesquels s'établissent les rapports syntaxiques, soit aux termes d'un réseau verbal. Ceux-ci adoptent une forme particulière disposée dans une suite linéaire déterminée. Le principe de la co-occurrence d'unités formelles discrètes, propre au distributionalisme est donc d'application de la même façon que pour les phonèmes, pour les morphèmes ou pour les classes de mots. Effectivement, ce ne sont pas les fonctions abstraites du réseau syntaxique qui font l'objet d'une distribution, mais les unités formelles entre lesquelles elles s'établissent[9]. Ces fonctifs ne coïncident pas nécessairement avec les membres de classes de mots : dans la séquence Lancellot, le roy de Behaygne, ou est il ? (Testament, vv. 378-9), le groupe nominal constitue un seul terme du groupe verbal envisagé, mais il renferme cinq membres de classes de mots; par ailleurs, certaines parties du discours sont inaptes à fonctionner comme termes d'un groupe verbal (p. ex. les prépositions, les introducteurs nominaux). Ce sont des syntagmes de grandeur variable. La disposition linéaire des fonctifs permet une étude de leurs possibilités combinatoires. Le nombre de termes reflète la latitude combinatoire des faits de parole.

La technique descriptive ainsi conçue aboutit à une typologie exhaustive des groupes verbaux, avec un classement concentrique des fonctifs autour du centre distributionnel. Elle explicite de cette manière la teneur exacte de la somme-parole syntaxique.

5.4. Enfin, l'option d'une analyse des fonctions syntaxiques est destinée à fournir les facteurs de réduction susceptibles de dégager le système idiolectal à partir de la typologie distributionnelle. Elle repose sur le principe de l'équivalence fonctionnelle de structures différentes.

Il convient non seulement de déterminer la hiérarchie des fonctifs dans leur dépendance structurale, mais encore d'examiner la nature même des rapports syntaxiques; l'une révèle l'ordre des réductions, l'autre en détermine les modalités en fonction du degré d'équivalence reconnu.

Ici aussi, il n'est fait usage que de critères formels tels

que la faculté de pronominaliser un rapport syntaxique donné du réseau verbal, la corrélativité de certains emplois adverbiaux et nominaux, l'action syntaxique d'opérateurs sur un (des / tous les) rapport(s) préexistant(s) dans le réseau (cf. la négation, la coordination).

6. A partir de chacun des jalons du mouvement d'abstraction qui conduit de la situation concrète au système organisé (P_1, Σ_{n1} parole, S_1), il est possible de s'engager dans une dimension différente qui prévoit le passage de l'individuel au collectif, ceci par une multiplication d'investigations analogues à partir d'un nombre élevé de corpus d'auteurs. L'expression individuelle qu'est la parole s'avère dès lors réductible à la langue à condition de procéder dans les deux dimensions d'abstraction mentionnées à l'analyse d'une classe d'idiolectes et de confronter les systématisations obtenues.

Notes

1 Il s'agit plus précisément de la syntaxe des groupes verbaux du manuscrit Coislin, étude qui paraîtra prochainement. On trouvera une brève démonstration de la technique proposée dans Van Deyck (1977): "Une méthode syntaxique en diachronie". Communication & Cognition 10,1: 97-112.

2 Cf. BERCHEM 1968: 25-26.

3 Cf. LIEB 1970: 12 svv. et 159 svv.

4 Cf. BUHLER 1934: § 4., BUYSSENS 1942-1943 : 17-23, LIEB 1970: 11 svv. et HEGER 1976: 24 svv.

5 Cf. HEGER 1976: 24-28. Précisons que dans ses travaux récents H.-H. Lieb désigne cette somme d'actualisations par le terme "idiolecte" par opposition à celui de "système idiolectal" qui marque le début de la hiérarchie systématique.

6 Cf. HEGER 1976: 26.

7 Le terme "norme" reçoit ici la valeur que lui donne COSERIU 1962: 62 svv.

8 Le groupe verbal étant défini comme le réseau syntaxique qui relie toutes les unités fonctionnelles qui entrent en relation directe ou indirecte avec un support verbal.

9 Dans cet ordre d'idées il nous semble dangereux d'adopter des codes qui désignent des fonctions syntaxiques (tels que <u>S</u> pour sujet, <u>O</u> pour objet, <u>C</u> pour complément, etc.), vu qu'une telle

codification risque de confondre les fonctifs et les fonctions et d'attribuer aux uns les propriétés des autres.

Littérature

BERCHEM, Theodor (1968) : "Sprachstruktur und Sprachwandel". Romanistisches Jahrbuch 19: 21-33.

BUHLER, Karl (1934) : "Sprachtheorie". Jena: Fischer.

BUYSSENS, Eric (1942-1943): "De l'abstrait et du concret dans les faits linguistiques : La parole - le discours - la langue". Acta Linguistica 3: 17-23.

COSERIU, Eugenio (1962): "Teoria del lenguaje y linguistica general". Madrid: Gredos.

HEGER, Klaus (1976): "Monem, Wort, Satz und Text". Tübingen: Niemeyer.

HJELMSLEV, Louis (1971): "Prolégomènes à une théorie du langage". "La structure fondamentale du langage". - trad. fr. Paris: Minuit.

LIEB, Hans-Heinrich (1970): "Sprachstadium und Sprachsystem". Stuttgart: Kohlhammer.

GENERATIVE STYLISTICS

An algorithmic approach to stylistic and source data retrieval problems based on generative syntax

Reinhard Wonneberger

1. State of the art

When computers were introduced in the field of literature, both modern and classic - the latter of which will be our concern here -, two major paths of research were followed that even today still prevail, as may be seen from the periodical information about actual projects given in CALCULI and COMPUTERS AND THE HUMANITIES.

One of them is to make concordances, a weary task if done by hand, and thus constrained before to a very limited number of works like Shakespeare and the Bible. As it is much effort to bring texts into machine readable form, one should prefer to perform more operations on them than just making concordances. And as it is a basic ability of computers to count things, the second path has been, what is called "sylistic analysis".

Computerized analysis of style, however, has so far been almost synonymous with statistic analysis. While statistical craftsmanship is impressive, its results have been rather poor, and non-computational scolars often find it difficult to take any enlightenment out of them (cf. BUJARD (1973) 231f).

First, stylistic indicators are chosen according to whether they can easily be distinguished by a program. Though there are a good lot of them, almost none of them can say anything substantial about style. Second, this is due to the fact that, from a linguistic point of view, the choice of these indicators is arbitrary, because they are not embedded in a linguistic theory of style.

From a linguistic point of view, the concept underlying this kind of research might be called the "fingerprint approach", because its main goal is to identify an author through his text itself. Here the scolar can choose his criteria at will indeed, as long as they only pass the test of significance. Moreover, this arbitrary choice seems to exclude the influence of subjective interpretation, coming close to the ideals of science. This idea of exactness may explain to a certain extent, why a large amount of these projects are done in North America (cf.WEIZENBAUM (1976); WONNEBERGER 1978b). Unfortunately this concept will not be applicable to texts that are suspected to have undergone redaction, as is the case for most books of the Bible (cf. the sophisticated approach of DEIST (1977)).

On the other hand, the word "style" is used to describe literary peculiarities of individual writers, to accumulate them in a canon of rhetorical figures (cf. LAUSBERG (1960)) or judge the value of a certain literary behaviour. Though this kind of style description has had a long tradition rooted in Greek philosophy, it only covers a special sector of texts, contributing little if anything at all to the explanation of institutional texts, i.e. texts that are not written with literary ambitions, but reflect the needs of every day institutions or are, as is the case in many books of the Bible, the result of a complicated process of tradition and redaction. It was left to the School of Formgeschichte to coin the term of Gattungsstil. Stylistic features are no longer regarded as isolated phenomena, but investigated in their connection to the Gattung of the text (cf. KOCH 1964/3.A.1974:18-20).

In the field of modern linguistics, an analogous concept of style was proposed by SANDIG (1978); she tries to describe what she calls "conventional style"; one of her examples is the language of horoscopes.

2. Consequences

In this situation, two ways of research should be followed. First, the ideas of style should be incorporated into the theory of speech acts. As this theory is not so much a specific linguistic algorithm like generative syntax, but much more a philosophical concept about what is done in speaking, it should not only confine itself on what is said (illocution) and what is aimed at (perlocution), but also include, h o w it is said.

In recent times, several scolars have proposed to introduce another aspect into the theory of speech acts, called "Allokution" (SCHECKER (1977); cf. RAIBLE (1977) 143f.), "Kollokution" (KELLER (1977)) and "Relokution" (WONNEBERGER (1977a) and (1978a)). Though these ideas were not introduced out of an interest in style, they may provide a basis to incorporate the dimension of style into speech act theory.

Second, practical methods should be developped to analyse conventional style. The qualitative kind of analysis implied by the idea of conventional style can be reduced to the problem of source data retrieval, which can well be solved on the computer, if only we are able to specify the search criteria in an operational form. If we can find a linguistic theory that is able both to specify the stylistic figures we are interested in, and to do that in an operational way, then our problem is solved, at least in principle.

It has been seen before long, that these requirements are met by the concept of generative grammar (see KLEIN (1971); OHMANN (1971)), but so far, generative grammars for the classical languages have not been available.

For Latin, a transformational theory has been developped by BOEDER (1972), and a dependence oriented theory of the verbal complex has been published in HAPP (1976).

For the language of the Old Testament, Hebrew, a PSG fragment has been given in DEIST 1977. Surprisingly enough the rules are not intended so much to describe syntactic rules as such, but to provide a base for transformations serving as stylistic indicators.

For the language of the New Testament and the Septuagint, the so-called κοινή -Greek, a generative sytax is provided in WONNEBERGER (1979a), which is complete in so far as it covers all types of constructions. This grammar allows to describe the syntax of Greek sentences by formal structures of syntactic positions. Then both the positions and their relations in a structure can be interpreted as indicators of style (for further details see WONNEBERGER (1979a) 289-294); on the other hand, these descriptions are well suited for the computer because their graph structures can easily be stored and handled as nested lists.

3. Outlines of a system

To perform style analysis according to this idea, a computer system will have to combine the following components:

a. a parsing module, which will help the scolar to analyse the sentences; automatic analysis is not intended, because too much work would have to be wasted on technical problems of morphology and ambiguity, and because only doing the analysis himself will bring the scolar into a contact with the text close enough to foster heuristic ideas.

b. a data base accumulating the text as well as the results of analysis. Because sentence descriptions are neither scalar nor vector data but graphs, i.e. two dimensional networks, the data base will have to be built up on the concept of nested lists.

c. a retrieval language to define the items to be searched for in the data base. We shall specify some possible properties of such a language in our next chapter.

Further details are discussed in WONNEBERGER 1977b. A system of this kind will enable the scolar to find text material for any stylistic indicator that can be described completely or even only in part in the terms of generative grammar. As an additional benefit the complexity of sentences can be measured (see STRASSNER (1973)). This measure as well as the occurrence of certain constructions are considered as an important factor in sociolinguistic research (cf. AMMON (1973/1977) 68-70).

4. Retrieval Language

The retrieval language mentioned before could have the following features:

a. Operators like <u>any</u> and <u>all</u> to specify the range of elements to be found.

b. The well-known logical operators like <u>and</u>, <u>or</u>, <u>not</u> to combine different search categories.

c. A set of descriptors. Classical descriptors will be words, i.e. entries to the lexicon, while descriptors typical of the present system are the symbols of the syntax.

d. Operators denoting relations between descriptors, saying e.g. that position b is a dependent of position a. These operators will allow to specify parts of a stemma (tree patterns) in the search.

e. A structuring mechanism like parentheses to allow nested expressions.

f. Some special elements, e.g. to find out the distance between two hits measured by the number of sentences or words.

To show what questions are to be solved by this retrieval system and how it may be used to find the solutions, we shall discuss some examples now.

5. Examples

The following examples have been taken from different areas of normal, i.e. non-computational research. For some of them solutions could easily be provided by hand, but this is due only to the fact that there are a lot of extraordinary working tools available for the Bible. These examples are meant as analogies, because one of the aims of the proposed system is to help researchers deal with texts not so well covered as the canonical ones and so to make use of the still growing depositories of texts in machine readable form.

5.1 Genitivus absolutus

The construction called genitivus absolutus is found in the following example:

τῶν ἐχθρῶν νικηθέντων οἱ φίλοι ἀναπαύονται

the enimies having been defeated, the friends take a rest

According to classical rules of style, the subject of the genitivus absolutus should not at all occur in the main sentence (see BLASS/DEBRUNNER §423). In later koine-Greek, this restriction is relaxed. It may be interesting to see how a certain author or text behaves with respect to this restriction. Now how could an investigation of this problem be done? The first step would be to find out sentences with a genetivus absolutus. This cannot be achieved by grammar either, because grammars

only give examples and not complete lists for a whole text.
In our system, the following steps would be necessary to determine the subset of sentences:

1. We look into the grammar to find out, how a genetivus absolutus is described there. We find that it is represented by the symbol pk (participle construction).
2. From the directory of symbols we may see, that the symbol pk might occur in other positions, too. To make sure it is a genitivus absolutus, it should be specified as dependent on the symbol a (adverbial phrase), and the congruence index x should have a value of 4 for genitive.
3. The search question could then be posed by the following subset definition:

   all S (a r pk:x=4)

   This would have to be read: find all sentences with an adverbial phrase realized as a participle construction in genitive.

From this formula the search compiler could derive the idea that only such sentences will qualify that have an a - type construction at all. The second component in the formula is dependent on the first; this leads to the idea, that it will do to look it up in only those sentences that were hit by the first criterion. This will be a more restrictive search question than if we had formulated:

all S (a and pk:x=4).

To solve this question, the compiler would only have to perform a and-operation on the search tables of both positions.

Having thus found out all the genitivi absoluti in the text, we are supposed to look ourselves whether there is a common subject in them or not. It is quite obvious that this problem cannot be solved automatically because reference identity may depend on more factors than merely syntax or even semantics. Yet the system can be used to sort the subset of genitivus absolutus-sentences into different groups. We might look if there are pronouns in both the genitivus absolutus and the main sentences, and if they coincide in person and number:

ἠνῖν, μὴ σκοποῦντων ἡμῶν (2.Cor 4,18)

Or we might look if the subject of the main clause is elliptic:

μνησθευθείσης τῆς μητρὸς ... εὑρέθη (Mt 1,18)

As these questions may be asked and answered without delay, we are very quick to check our hypotheses.

5.2 Kataphoric pronouns

Pronouns cannot only represent subjects and things in reality or fiction, but linguistic units (see BLASS/DEBRUNNER §290):

εἰδῶς τοῦτο ὅτι (1.T 1,9)

knowing this, that...

The only way to find such constructions would be to thumb all entries of the pronom in the concordance and verify them in the text, because context lines in normal concordances are often too short to show whether there is a subordinate clause or not.

The first part of our definition will be that the position <u>pf</u> should be present. But this criterion results in about the same number of sentences as that which is given in the concordance. Now if such a clause is present, it will be subset to the position <u>apa</u> (appositional attribute). If we just add this criterion, it will still give us the examples with any appositional attribute at all. So we should further specify that <u>apa</u> should be of type <u>sf</u>, which allows only for member-clauses.

When I analysed 2.Cor 5,2f, it turned out, that it forms a new sub-type of this construction, combining the pronouns with a whether-clause (See WONNEBERGER Beitrag (1975). This connection had not been seen before. Let us assume e.g. that some scolar is suspicious of the analyses in the date base, and wants to see, if there are not any connections that have been overlooked. The only thing he has got to do is to relax the search criterion; the formula might be:

S(PF <u>and</u> SF:h=4,5)

now specifying the set of sentences having both pronoun and a member sentence. By this criterion he will get a good many useless sentences, but will also be sure that no possible hit is missed.

5.3 Translational Greek

Special linguistic problems are posed by translational Greek. It may be important to show that a construction is uncommon in ordinary Greek, because it may strengthen the evidence that it is influenced by a semitic original and hence is to be considered as a translation. Let us study an example in the story of Bel and the dragon, an extension to Daniel in the LXX (cf. KOCH (1979); WONNEBERGER (1979b)). the version reads:

οὐ δύνασαι εἰπεῖν ὅτι οὐκ ἔστιν οὗτος θεὸς ζῶν
καὶ προσκύνησον αὐτῷ (Dan 14,24)
you cannot say that this one is not a living god,
and worship him.

This is as deviant in Greek as it sounds in English.

Linguistic analysis of the deviation shows that it is not a question on the syntax but on the text level. The illocutionary aspect of the two speech acts is different in a way that does not allow for conjunction with and.

Although the system is not designed to handle speech act Problems, it may help to find analogous cases. An appropriate search process might be:

all s (nek(καί) + (sf:h=a r va:k=imp.))

The plus sign denotes that the nective and occurs in conjunction with sf. The imperative type is specified by va:k=imp (see WONNEBERGER (1979a) for detail). Nothing is said about the previous sentence, but that is not worth while, because that sentence will probably be affirmative. If it was specified the search process would have to be expanded on more than one sentence at once.

5.4 Hymnic style

Romans 3,24 - 26 has often been said to be "hymnic" (see WONNEBERGER (1979a)). In fact there is a significant bulk of prepositional expressions. One way to find additional material would be to look for phrases with more than one preposition. But syntactic analysis has shown that the stylistic elements might be due to another phenomenon: the nominal parts of the prepositional phrases are realized as Translationsketten (tk), i.e. nominal phrases derived from verbal phrases by the word formation morpheme -σις. So it might even be more interesting to look for the tk - position instead of or even in addition to prepositions.

In the same text, Christ is characterized by a relative sentence that again might evoke the idea of hymnic style. When reading through Romans, one gets the impression that the occurrence of God or Christ often triggers additional characterization in the form of a relative clause. This problem might be solved by the concordance on the assumption that the relative clause will immediately follow the proper noun. What cannot be seen from the concordance, however, is whether there are other expressions besides these proper names that are qualified by such relative clauses. But only if we knew this background, we could fairly judge the stylistic significance of the construction.

5.5 An example of pragmatics and fuzzy set description

Modern research in the field of pragmatics has shown certain elements of a text to be signals indicating sort of hierarchy in the text (see GUELICH (1976)). Among them, special interest is paid to performative sentences. A traditional example is

therefore, my brothers, I implore you by God's mercy (R 12,1)

In performative sentences, what is said and what is done both have to be identical, thus they have to be in 1.person present. This would also be the search criterion. Whether the hits are performative in fact would be left to the scolar to decide.

If such phrases are to indicate a new part of the text, they often contain additional phrases like

by God's mercy

These additional parts may vary to some extent, which makes it impossible to give a precise formula definition. An appropriate description could be given using fuzzy set algebra. The formula is then described as a fuzzy set. Its elements are taken from the syntactic analysis of the basic pattern of the corresponding sentences. The grades of membership for each element would be calculated from the results of data base search. Because of its algorithmic structure, fuzzy set algebra can be easily used in connection with the system and may become an important additional tool in the description of style.

Bibliography

AMMON, Ulrich (1973): Probleme der Soziolinguistik. Germanistische Arbeitshefte 15. Tübingen. 2.A. 1977.

BLASS/DEBRUNNER (1896): Grammatik des neutestamentlichen Griechisch. Göttingen/Zürich: Vandenhoeck 1896 14.A. 1976.

BOEDER, Winfried (1972): Untersuchungen zur lateinischen Satzeinbettung. Hamburg (unver.)

BUJARD, Walter (1973): Stilanalytische Untersuchungen zum Kolosserbrief als Beitrag zur Methodik von Sprachvergleichen. Göttingen: Vandenhoeck & Rupprecht.

CALCULI: Bimonthly Newsletter, supported by the American Philological Association. Stephen V.F.Waite (ed.). Hanover, New Hampshire. 1967ff.

CHum, Computers and the Humanities (Zeitschrift). Flushing, New York 1965ff.

DEIST, Ferdinand (1977): "Stilvergleich als literarkritisches Verfahren". Zeitschrift für die Alttestamentliche Wissenschaft 89: 325-357.

GUELICH, Elisabeth (1976): "Ansätze zu einer kommunikationsorientierten Erzähltextanalyse (am Beispiel mündlicher und schriftlicher Erzähltexte)". HAUBRICHS, W. (ed.). Erzählforschung 1. Zeitschrift für Literatur und Linguistik, Beiheft 4: 224-256.

HAPP, Heinz (1976): Grundfragen einer Dependenzgrammatik des Lateinischen. Göttingen: Vandenhoeck & Rupprecht

IHWE, Jens (ed.) (1971): Literaturwissenschaft und Linguistik. Ergebnisse und Perspektiven. Frankfurt. 3 Bde.

KELLER, Rudi (1977): "Kollokutionäre Akte". Germanistische Linguistik 1-2 (1977) 3-50.

KLEIN, S.(1971): "Stilkontrolle mit einer generativen Grammatik". IHWE (ed.) I 234-252.

KOCH, Klaus (1964): Was ist Formgeschichte? Methoden der Bibelexegese. Neukirchen: Neukirchener V., 3.A. mit einem Nachwort: Linguistik und Formgeschichte 1974.

KOCH, Klaus/ NIEWISCH, Till/ TUBACH, Jürgen (1979): Das Danielbuch. Erträge der Forschung. Darmstadt: Wissenschaftliche Buchgesellschaft.

LAUSBERG, H. (1960): Handbuch der literarischen Rhetorik. Eine Grundlegung der Literaturwissenschaft. 2 Bde. München: Hueber.

OHMANN, R. (1971): "Generative Grammatik(en) und der Begriff literarischer Stil", IHWE (ed.) I 213-233.

RAIBLE, Wolfgang (1977): "Argumentationen als allokutionäre Sprechakte? Bemerkungen zur These von Michael Schecker". Schecker (ed.) 139-148.

SANDIG, Barbara (1978): Stilistik. Sprachpragmatische Grundlegung der Stilbeschreibung. Berlin/New York: de Gruyter.

SCHECKER, Michael (ed.) (1977): Theorie der Argumentation. Tübingen:

- - - (1977b): "Argumentationen als allokutionäre Sprechakte", Schecker (ed.) 75-138.

STRASSNER (1973): "Textverständlichkeit und Textvergleich". Deutsche Sprache 1: 42-52.

WEIZENBAUM, Joseph (1976): Computer Power and Human Reason. From Judgement to Calculation. - Transl.: Die Macht der Computer und die Ohnmacht der Vernunft. Frankfurt: Suhrkamp, 1977.

WONNEBERGER, Reinhard (1975): "Der Beitrag der generativen Syntax zur Bibelexegese". Ein Beispiel (2.Kor 5,2f.) und neun Thesen. Bijdragen 36: 312-317.

--- (1977a): "Relokution, negativer Sprechakt und wo der Römerbrief anfängt". Linguistische Arbeiten 50: 287-295.

--- (1977b): Computer Analysis of Greek and advanced Stylistics. Paper read at 3rd ICCH Conference, Waterloo, Ontario 1977 (unpublished)

--- (1978a): "Relokution und Partikel am Beispiel 1.Korinther 6,7. Linguistische Arbeiten 62: 177-185.

--- (1978b): "Wie Herr K. vor Kafkas Schloß". Gefährden Datensysteme die Grundrechte des Bürgers? Lutherische Monatshefte 17: 609-613.

--- (1979a): Syntax und Exegese. Eine generative Theorie der griechischen Syntax und ihr Beitrag zur Auslegung des Neuen Testamentes, dargestellt am 2.Korinther 5,2f. und Römer 3,21-26. Frankfurt: Peter Lang.

--- (1979b): "Zur Syntax und Pragmatik der Partikel in der griechischen Bibel" WEYDT, Harald (ed.): Die Partikeln der deutschen Sprache. Berlin: de Gruyter (to appear).

VERZEICHNIS DER AUTOREN

APELDOORN, Nicolaas W.M. : Institut de Phonétique, Univ. Libre de Bruxelles, Franklin Rooseveltlaan 50, B-1050 Brussel. Geb. 1944 - Studium: niederländische, allgemeine und angewandte Linguistik an der Universiteit van Amsterdam - wiss. Assistent an der ULBruxelles - Hauptarbeitsgebiete : niederländische und allg. Sprachwiss.

APELTAUER, Ernst : Germanistisches Institut der Westfälischen Wilhelms-Universität, Johanisstrasse 1- 4, D-4400 Münster. Studium: Germanistik, Pädagogik, Philosophie, Geschichte - Wiss. Assistent - Hauptarbeitsgebiete: Semantik, Pragmatik, Gesprächsanalyse.

BAGHDIKIAN, Sonia : 132, avenue Général-Médécin Derache (Bte. 14), B-1050 Bruxelles. Geb. 1947 - Studium: germanische Philologie, Anglistik und allg. Sprachwissenschaft in Brussel (V.U.) und Newcastle upon Tyne - Wiss. Assistentin an der V.U.Brussel - Hauptarbeitsgebiete : englische Syntax; Sprachwandel; Übersetzung.

BIASCI, Claudia : Fakultät für Linguistik und Literaturwissenschaft, Univ. Bielefeld, Universitätsstrasse 1, D-4800 Bielefeld 1. Geb. 1950 - Studium: "Lingua e Letterature stranieri", Literaturtheorie, Linguistik in Pisa, Paris und Bielefeld - Doktorandin in Bielefeld - Hauptarbeitsgebiete: Texttheorie, Semantik.

BIERE, Bernd Ulrich : Germanistisches Institut der RWTH Aachen, Lehrstuhl für deutsche Philologie, D-5100 Aachen. Geb. 1947 - Studium: Germanistik und Romanistik in Tübingen und Heidelberg - Wiss. Assistent in Aachen - Hauptarbeitsgebiete : Kommunikationsanalyse; Spracherwerb, Kindersprache; Semantik und Pragmatik.

BRUNNER, Gisela : Seminar für Allgemeine Sprachwissenschaft Universität Düsseldorf, Universitätsstrasse 1, D-4000 Düsseldorf 1. Geb. 1951 - Studium: klassische Philologie, Germanistik, Sprachwiss. in Hamburg und Osnabrück - Wiss. Mitarbeiterin im DFG-Projekt "Modalitäten" - Hauptarbeitsgebiete : linguistische Pragmatik; Kommunikationssoziologie; Wissenschaftstheorie.

COULMAS, Florian : Seminar für Allgemeine Sprachwissenschaft Universität Düsseldorf, Universitätsstrasse 1, D-4000 Düsseldorf 1. Geb. 1949 - Studium in Berlin(FU), Paris, Oberlin(Ohio), Bielefeld - Mitarbeiter im Forschungsprojekt über "Routineformeln und standardisierte Kommunikationssituationen" Düsseldorf - Hauptarbeitsgebiete : kontrastive Pragmatik; anthropol. Linguistik; Zeichentheorie.

DECREUS, Freddy : Vlaanderenstraat 102, B-9000 Gent. Geb. 1949 - Studium: klassische Philologie, osteuropäische Literatur und Geschichte in Gent - Wiss. Assistent an der R.U.Gent - Hauptarbeitsgebiete : Didaktik der klassischen Sprachen; Textanalyse.

DORFMÜLLER-KARPUSA, Käthi : Fakultät für Literaturwissenschaft und Linguistik Universität Bielefeld, Postfach 8640, D-4800 Bielefeld. Geb. 1935 - Studium: klassische und neugriechische Philologie, Philosophie, Linguistik - Lehrbeauftragte für neugriechische Sprache an dem Sprachenzentrum der Univ. Bielefeld - Hauptarbeitsgebiete : Semantik; Textlinguistik.

FIEHLER, Reinhard : Ebbendorferstrasse 13, D-4517 Hilter 2.
Geb. 1949 - Studium: Sprachwissenschaft, Soziologie und Literaturwissenschaft in Hamburg, Freiburg und Osnabrück - Doktorand im Fach Sprachwissenschaft an der Univ. Osnabrück - Hauptarbeitsgebiete : linguistische Pragmatik, Kommunikationssoziologie, Wissenschaftstheorie.

FRANKENBERG, Hartwig : Cheruskerstrasse 80, D-4000 Düsseldorf 11.
Geb. 1944 - Studium: Germanistik, Anglistik, allgem. Sprachwiss. - Wiss. Assistent an der PH Rheinland, Abt. Neuss - Hauptarbeitsgebiete : ling.Pragmatik, Erzähltextforschung, Strukturale Mythologie.

GAHN, Annette : Witzlebenstrasse 2, D-1000 Berlin 19.
Geb. 1950 - Studium: Mathematik und Informatik - Studentin - Hauptarbeitsgebiete : theoretische Informatik, Künstl.Intelligenz.

GOOSSENS, Louis : Dept. Germaanse Universitaire Instelling Antwerpen, Universiteitsplein 1, B-2610 Wilrijk. Geb. 1937 - Studium: germanische Philologie an der R.U.Gent - Professor für englische Sprache und Linguistik in Antwerpen - Hauptarbeitsgebiete : Studium der englischen Sprache; kontrastive Linguistik Engl.-Ndl.

HABEL, Christopher : Wilhelmstrasse 155, D-1000 Berlin 20.
Geb. 1950 - Studium: Mathematik, Linguistik in Göttingen und Osnabrück - Wiss. Angestellter am Projekt "Automatische Erstellung semantischer Netze" in Berlin (TU) - Hauptarbeitsgebiete : theoretische Linguistik, maschinelle Sprachverarbeitung, Künstl.Intelligenz.

HAMANS, Camiel : Gabriël Metsustraat 24^{bv}, NL-Amsterdam.
Geb. 1948 - Studium: niederländische Sprache und Literatur, allgem. Sprachwissenschaft an der Universiteit van Amsterdam - Wiss. Assistent an der Univ. Leiden - Hauptarbeitsgebiete : Phonologie und Semantik des Niederländischen.

HESS-LÜTTICH, Ernest W.B. : Fachbereich 16 Germanistik, Freie Universität Berlin, Habelschwerdter Allee 45, D-1000 Berlin 33.
Geb. 1949 - Studium: Germanistik, Anglistik, allgem. und angew. Sprachwiss., Philosophie, Pädagogik, Soziologie in Bonn, Köln, London - Wiss. Assistent für deutsche Philologie/Linguistik an der FU Berlin - Hauptarbeitsgebiete : deutsche Philologie, Intersektionslinguistiken ; Textlinguistik, ling. Pragmatik usw.

HINDELANG, Götz : Germanistisches Institut, Johannisstrasse 1-4, D-4400 Münster. Geb. 1948 - Wiss. Assistent - Hauptarbeitsgebiet : Pragmatik.

HOFFMANN, Ludger : Philippistrasse 12, D-4400 Münster.
Geb. 1951 - Studium: Germanistik und Theologie in Münster - Wiss. Assistent an der PH Münster - Hauptarbeitsgebiete : Pragmatik, Soziolinguistik, Sprachdidaktik.

KLEIN, Klaus-Peter : Kemnader Strasse 249, Postfach 250305, D-4630 Bochum 1. Geb. 1948 - Studium in Würzburg und Bochum - Wiss. Assistent für deutsche Sprache und Literatur an der PH Münster - Hauptarbeitsgebiete : Pragmatik/Handlungstheorie; Texttheorie; Rezeptionsästhetik; Sprachdidaktik.

KOHRT, Manfred : Meinertzstrasse 21, D-4400 Münster.
Geb. 1947 - Studium: Germanistik und Geschichtswiss. in Göttingen und Marburg - Wiss. Assistent am German. Institut der Westf. Wilhelms-Univ. Münster - Hauptarbeitsgebiete : Theorie der Schrift-Sprache, Syntax, Theorie und Geschichte der Linguistik.

MONAGHAN, James : Institut für Englische Philologie Universität München, Schellingstrasse 3, D-8000 München 40.
Geb. 1944 - Studium: Anglistik,Mediävistik,Linguistik in Glasgow, Keele und Frankfurt/Main - Wiss. Assistent in München - Hauptarbeitsgebiete: zeitgenössische Texte;neo-Firthian linguistics.

NIKOLAUS, Kurt : Löhleinstrasse 17, D-1000 Berlin 33.
Geb. 1954 - Studium: Germanistik, Soziologie und Psychologie in Berlin (FU) - Doktorand - Hauptarbeitsgebiete : Textsemantik und kognitive Verarbeitung von Texten.

PAVLIDOU, Theodossia : Roermonderstrasse 363, D-5100 Aachen.
Geb. 1947 - Studium: Physik in Carleton Coll.,Northfield Minn., USA und Kommunikationsforschung,Informatik und Logik in Bonn - Freie Mitarbeiterin beim Rundfunk und Verlag - Hauptarbeitsgebiete : Kommunikations-,Sprechhandlungs-,Argumentationstheorie.

PLATTEAU, Frank : Lanteernhofstraat 52, B-2200 Borgerhout.
Geb. 1956 - Studium: germanische Philologie und Neurolinguistik - Wiss. Assistent an der Universitaire Instelling Antwerpen - Hauptarbeitsgebiete : Neurolinguistik,Pragmatik,Montague-Grammatik, Komputerlinguistik.

REDDER, Angelika : Seminar für Allgemeine Sprachwissenschaft Universität Düsseldorf, Inst.Gruppe III, Universitätsstrasse 1, D-4000 Düsseldorf. Geb. 1951 - Studium: allgem. Sprachwiss., Germanistik, Geographie - Wiss. Mitarbeiterin im DFG-Projekt "Modalitäten" in Düsseldorf - Hauptarbeitsgebiete : linguistische Pragmatik,Sprachpsychologie,Wissenschaftstheorie.

REITER, Norbert : Institut für Balkanologie, Hertastrasse 7,D-1000 Berlin 37. Geb. 1928 - Studium: Slavistik,Indogermanistik,Psychologie - Professor für Balkanologie und slavische Sprachwissenschaft an der FU Berlin und Direktor des Instituts für Balkanologie - Hauptarbeitsgebiete : vergleichende Grammatik der Balkansprachen; theoretische und geschichtliche Untersuchungen; Semantik; Soziolinguistik; Nationalismusforschung; Didaktik.

ROEGIEST, Eugeen : Goed ten Pauw 17, B-9130 Lochristi.
Geb. 1946 - Studium: romanische Philologie in Gent - Wiss. Assistent an der R.U.Gent und Dozent am HIVET,Gent - Hauptarbeitsgebiete : Romanistik, Hispanistik : Syntax und Semantik; kontrastive Linguistik Spanisch-Französisch-Niederländisch.

ROLLINGER, Claus-Rainer : Fachbereich 20 Institut für angewandte Informatik Technische Universität Berlin, Otto-Suhr-Allee 18-20, D-1000 Berlin 10. Geb. 1950 - Studium: Informatik,Mathematik, Jura - Wiss. Angestellter im Projekt "Automatische Erstellung semantischer Netze" - Hauptarbeitsgebiete : Künstliche Intelligenz,maschinelle Sprachverarbeitung,Rechtsinformatik.

ROMBOUTS, Jos : Moerstraat 98, B-2320 Hoogstraten.
Geb. 1956 - Studium: germanische Philologie an der UIAntwerpen - Hauptarbeitsgebiete : Semantik und Pragmatik; Soziolinguistik.

SCHMIDT, Arno : Institut für angewandte Informatik Technische Universität Berlin, Otto-Suhr-Allee 18-20, D-1000 Berlin 10.
Geb. 1950 - Studium: Informatik - Wiss. Angestellter im Projekt "Automatische Erstellung semantischer Netze" - Hauptarbeitsgebiete : maschinelle Verarbeitung natürlicher Sprache, Frage-Antwort-Systeme, Repräsentation von Wissen.

SÖKELAND, Werner : Pferdeweide 3a, D-2000 Hamburg 55.
Geb. 1951 - Studium : Germanistik, Soziologie, Anglistik, Pädagogik in Erlangen, Edinburgh, Hamburg - Doktorand - Hauptarbeitsgebiete : Argumentations- und Sprechakttheorie; strukturale Semantik; Kasusgrammatik.

SØRENSEN, Viggo : Karensvej 13, DK-8220 Brabrand.
Geb. 1941 - Studium: nordische Sprachen und Literatur; neuere dänische Sprache - Lektor am Institut für nordische Sprachen und Literatur an der Universität Århus - Hauptarbeitsgebiete : Textlinguistik; Text- und Argumentationsanalyse; Pragmatik; Dialektologie.

THYS, Guido : Grote Steenweg 566, B-2600 Berchem.
Geb. 1956 - Studium: germanische Philologie, Philosophie, Neuro- und Psycholinguistik in Antwerpen (UIA,UFSIA),Leuven (KUL) und Brussel (VUB) - Wiss. Assistent an der Universitaire Instelling Antwerpen - Hauptarbeitsgebiete : Sprachphilosophie und -psychologie: Sprache als Kommunikationsmittel.

VAN DER AUWERA, Johan : Dept. Germaanse Universitaire Instelling Antwerpen, Universiteitsplein 1, B-2610 Wilrijk. Geb. 1953 - Studium: Linguistik und Philosophie - Stipendiat des NFWO - Hauptarbeitsgebiete : Semantik; Pragmatik; Logik und Sprachphilosophie; Sprachuniversalien.

VANDEWEGHE, Willy : siehe das Herausgeberverzeichnis.

VAN DEYCK-BAUWENS, Rika : Fakulteit Letteren en Wijsbegeerte R.U.G., Blandijnberg 2, B-9000 Gent. Geb. 1944 - Studium: romanische Philologie - Wiss. Assistentin an der R.U.Gent - Hauptarbeitsgebiete : französische Syntax; Textlinguistik.

VAN NOPPEN, Jean-Pierre : Reikemstraat 21, B-3009 Winksele-Delle.
Geb. 1946 - Studium: germanische Philologie - Wiss. Assistent an der Université Libre de Bruxelles und Vrije Universiteit Brussel - Hauptarbeitsgebiete : religiöse Sprache; Lexikologie und -graphie; Theorie der Metapher.

VERSCHUEREN, Jef : Dept. Germaanse Universitaire Instelling Antwerpen, Universiteitsplein 1, B-2610 Wilrijk.
Geb. 1952 - Studium: germanische Philologie und Linguistik in Antwerpen und Berkeley,Cal.,USA - Stipendiat des NFWO - Hauptarbeitsgebiete : linguistische Pragmatik, lexikalische Semantik.

WEIGAND, Edda : Leithenstrasse 9, D-8521 Marloffstein/Erlangen.
Studium: klassische Philologie und Germanistik in Tübingen - Wiss. Assistentin am Germanistischen Institut der Ruhr-Universität Bochum - Hauptarbeitsgebiete : Morphologie, Syntax und Semantik des Deutschen; Textgrammatik; Sprachdidaktik.

WILLEMS, Dominique : Fakulteit Letteren en Wijsbegeerte R.U.G., Blandijnberg 2, B-9000 Gent. Geb. 1948 - Studium: romanische Philologie in Gent - Wiss. Assistentin an der Rijksuniversiteit Gent - Hauptarbeitsgebiete : Syntax und Semantik des Französischen; Komputerlinguistik; Pragmatik und Konversationsanalyse.

WONNEBERGER, Reinhard : Sodentwiete 2, D-2000 Hamburg 63.
Geb. 1946 - Studium: Theologie in Erlangen, Basel und Heidelberg - Wiss. Assistent für Altes Testament in Hamburg - Hauptarbeitsgebiete : Exegese; Textlinguistik,Pragmatik; Informatik.

VANDEWEGHE, Willy : St.-Pietersnieuwstraat 88, B-9000 Gent. Geb. 1948 - Studium : germanische Philologie an der R.U.Gent - Wiss. Assistent beim "Seminarie voor Nederlandse Taalkunde en Vlaamse Dialektologie" der Rijksuniversiteit Gent - Hauptarbeitsgebiete : Semantik und Pragmatik; Syntax des Ndl.

VAN DE VELDE, Marc : Prof. P.Fredericqstraat 42, B-9000 Gent. Geb. 1944 - Studium : germanische Philologie an der R.U.Gent - Dozent am Hoger Instituut voor Vertalers en Tolken HIVET (Dolmetscherinstitut) in Gent - Hauptarbeitsgebiete : Syntax Deutsch-Niederländisch; Soziolinguistik.

FSC
www.fsc.org
MIX
Papier | Fördert gute Waldnutzung
FSC® C083411

Kolibri 360 GmbH
Ferdinand-Jühlke-Straße 7
99095 Erfurt, Deutschland
produktsicherheit@kolibri360.de